Texts and Monographs in Physics

Texts and Monographs in Physics

Series Editors:
R. Balian W. Beiglböck H. Grosse E. H. Lieb W. Thirring

From Microphysics to Macrophysics I + II Methods and Applications of Statistical Physics By R. Balian

Variational Methods in Mathematical Physics
A Unified Approach
By P. Blanchard and E. Brüning

Quantum Mechanics: Foundations and Applications
3rd enlarged edition By A. Böhm

The Early Universe
Facts and Fiction 3rd corrected and enlarged edition By G. Börner

Geometry of the Standard Model of Elementary Particles
By A. Derdzinski

Random Walks, Critical Phenomena, and Triviality in Quantum Field Theory By R. Fernández, J. Fröhlich and A. D. Sokal

Quantum Relativity
A Synthesis of the Ideas of Einstein and Heisenberg
By D. Finkelstein

Quantum Mechanics I + II
By A. Galindo and P. Pascual

The Elements of Mechanics
By G. Gallavotti

Local Quantum Physics Fields, Particles, Algebras
Corrected 2nd printing
By R. Haag

Elementary Particle Physics
Concepts and Phenomena
By O. Nachtmann

Inverse Schrödinger Scattering in Three Dimensions
By R. G. Newton

Scattering Theory of Waves and Particles 2nd edition
By R. G. Newton

Quantum Entropy and Its Use
By M. Ohya and D. Petz

Generalized Coherent States and Their Applications
By A. Perelomov

Essential Relativity Special, General, and Cosmological Revised 2nd edition
By W. Rindler

Path Integral Approach to Quantum Physics
An Introduction By G. Roepstorff

Advanced Quantum Theory and Its Applications Through Feynman Diagrams 2nd edition
By M. D. Scadron

Finite Quantum Electrodynamics
By G. Scharf

From Electrostatics to Optics
A Concise Electrodynamics Course
By G. Scharf

Large Scale Dynamics of Interacting Particles By H. Spohn

General Relativity and Relativistic Astrophysics By N. Straumann

The Dirac Equation By B. Thaller

The Theory of Quark and Gluon Interactions 2nd completely revised and enlarged edition By F. J. Ynduráin

Günter Scharf

From Electrostatics to Optics

A Concise Electrodynamics Course

With 26 Figures

Springer-Verlag
Berlin Heidelberg New York
London Paris Tokyo
Hong Kong Barcelona
Budapest

Professor Dr. Günter Scharf

Institut für Theoretische Physik
Universität Zürich-Irchel
Winterthurer Strasse 190
CH-8057 Zürich, Switzerland

Editors

ISBN 3-540-57682-7 Springer-Verlag Berlin Heidelberg New York
ISBN 0-387-57682-7 Springer-Verlag New York Berlin Heidelberg

Library of Congress Cataloging-in-Publication Data. Scharf, Günter, 1938- . From electrostatics to optics : a concise electrodynamics course / Günter Scharf. p. cm. – (Texts and monographs in physics) Includes bibliographical references and index.
ISBN 3-540-57682-7 (Berlin: acid free paper). – ISBN 0-387-57682-7 (New York: acid free paper)
1. Electrodynamics. I. Title. II. Series. QC631.S33 1994 537.6–dc20 94-1192

Typesetting: Data conversion by K. Mattes, Heidelberg

SPIN: 10127723 55/3140-5 4 3 2 1 0 - Printed on acid-free paper

Preface

We live in the electronic age. Millions of people stare at television sets; computers penetrate our working life; and the heartbeat of many older people is regulated by a pace-maker. Even music is becoming electric. The good old mechanical Swiss watch has meanwhile been replaced by an electronic watch made in Hong Kong, which is ten times as accurate and costs a tenth of the price. (The cheap Swiss watches are today also electric and the parts are probably made in Hong Kong.)

In physics, the retirement of mechanics as the basis of science already took place in the 19th century, when it was found that light and electromagnetism must be described by a field theory and not by a mechanical theory. That was a radically new idea: The basic objects in Newton's mechanics are point particles with coordinates $q_i(t)$ and momenta $p_i(t)$ depending on time. If the index i runs from 1 to f, the system has finitely many (f or $2f$) degrees of freedom. The basic quantities of field theory, on the other hand, are a few vector fields $\boldsymbol{E}(t, \boldsymbol{x})$, $\boldsymbol{B}(t, \boldsymbol{x})$, $\boldsymbol{v}(t, \boldsymbol{x})$ which depend on time and space. The velocity field $\boldsymbol{v}(t, \boldsymbol{x})$ is fundamental in continuum mechanics, which is certainly a field theory like electrodynamics, not a mechanical particle theory. Since at every place $\boldsymbol{x}$ in $\mathbb{R}^3$ the fields at a given time are free to take on values, the system has infinitely many degrees of freedom. Consequently the fundamental equations are partial differential equations in $(t, \boldsymbol{x})$, whereas in point mechanics one deals with ordinary differential equations in time. More important, the forces in point mechanics act at a distance, and they depend on the coordinates and momenta of the particles at a fixed time. In field theory only local interactions appear and the mechanism by which the action propagates through the space is described in detail.

For many branches of physics both a particle and a field theory exist: The field theory corresponding to classical mechanics is fluid mechanics and elasticity theory. Optics can be geometrical optics, which is a particle theory, or wave optics, or electrodynamics as the corresponding field theory. Newton's theory of gravitation is a mechanical action-at-a-distance theory; the corresponding field theory is Einstein's general relativity theory. Finally, the fundamental theories of weak, electromagnetic and strong interactions are quantized versions of gauge field theories, for which electrodynamics was the prototype. Although not yet completely understood, these theories are the basis of particle physics, and are indeed being confirmed to increasing degree by experiments. Thus, even in particle physics the field concept is more fundamental than particles. For this reason electrodynamics is a central part of any physics course.

The book is divided into five chapters. It starts with electrostatics in vacuum. It should be stressed that the vacuum equations are much more fundamental than the phenomenological electrodynamics in matter. They are valid also in the microscopic world (after refinement by quantum theory), whereas the phenomenological equations are no more then an approximate description, which is valid only for simple materials under favorable smooth conditions. Therefore we avoid any mixing of the two. In Chap.2, Maxwell's equations are derived from electrostatics and the relativity principles. This is really the simplest way, conceptually and mathematically, to obtain these equations. In the third chapter, applications of Maxwell's equations in vacuum are discussed, in particular radiation phenomena. We treat these applications in a manner that elucidates the connection between the mathematical structure (given by Maxwell's equations) and the real world. We thus concentrate on the vital points; for further details of applications there exist many special monographs. In Chap.4 the phenomenological Maxwell's equations just criticized are derived by spatial averaging, assuming a simple classical model for the atomic structure of matter. In this derivation there are certain ambiguities, so that the equations are only meaningful if so-called constitutive relations are given in addition. The latter must, in principle, be furnished by a microscopic many-body theory. This shows once more the totally different character of the phenomenological equations. Optics in Chap.5 is not an independent field, but rather an application of Chap.4. Optics then appears "back-to-front": we start with wave optics and light scattering and end up with geometrical optics and scalar diffraction theory, which are treated as approximations to wave optics. A short discussion of the laser is also included. This lies a bit beyond the scope of the book, but because of its great importance it must appear somewhere in a main course. The selection of topics was motivated by our desire to travel from electrostatics to optics on a single ticket and by the limits posed by a one-semester course with four hours a week. The book closes with an epilogue on quantum electrodynamics to stress the fact that the story has a second part and to give the interested reader an idea of it.

Acknowledgements. I wish to thank S.Parrott and above all R.Balian for a careful reading of the manuscript from start to finish, for their criticism, and for many valuable suggestions and stimulating remarks. Many fruitful ideas emerged from discussions with my colleague A.Thellung, who also helped me in correcting the manuscript. I am grateful to W.Beiglböck for important comments and to the staff at Springer-Verlag for the excellent collaboration.

Zürich, 1993 G. Scharf

Contents

0. Preliminaries

In this short preparatory chapter we first give a historical introduction which leads to the plan of the course in a natural way. We then collect some important results of vector analysis in $\mathbb{R}^3$. We will present this rather elementary material in a form best suited for later use in electrodynamics. Then we need not interrupt the physical discussions in the body of the book by mathematical considerations. If the reader is well trained in analysis, he may directly jump from the historical introduction to Chap.1. But otherwise we strongly recommend the study of this basic chapter. The reader may also skip the parts printed with small letters in the text for the first reading. We have decided to number preliminary material by zero. To be consistent we therefore start as follows.

0.0 Historical Introduction

History of science can be concerned with the life and work of scientists and with the history of ideas and concepts. For physicists the latter is more interesting. Electrodynamics is indeed a field where the development of new fundamental physical concepts took place: the advent of field theory in the 19th century and the discovery of special relativity at the beginning of the 20th century. According to T.Kuhn such moments are called scientific revolutions (*T.Kuhn, The Structure of Scientific Revolutions, 1962*). This sounds dramatic and attracts the interest of the public. On the other hand, Einstein and Infeld were more modest, when in 1938 they published their popular book with the title "The Evolution of Physics". We therefore want to look at the history of electrodynamics, taking these two different possibilities of interpretation (revolution versus evolution) into consideration. We follow one particular line of development, which, in retrospective view, was the most successful one. There are many other aspects concerning the reception of Faraday's and Maxwell's electrodynamical field theories. For further historical studies we recommend the book by *Jed Z.Buchwald, From Maxwell to Microphysics, University of Chicago Press 1985.*

An important witness from the end of the 19th century is A.Sommerfeld. He writes at the beginning of his "Elektrodynamik" (1948): "To illustrate the change of concepts which Faraday's and Maxwell's theories have brought about, I best tell about my university years in 1887-91.... Those were the years of Hertz' experiments. However, first electrodynamics was presented in the old style: Besides Coulomb's and Biot-Savart's laws we learnt Ampère's law for the force between two current elements and its rivals, the laws of Grassmann, Gauss, Riemann, Clausius, and as a culmination the law of Wilhelm Weber. They all were action-at-a-distance laws of Newtonian type. This presentation of electrodynamics was clumsy, incoherent and not at all self-contained. Professors and students tried hard to assimilate the new results of Hertz' experiments, and to understand them on the basis of the difficult original theory in Maxwell's Treatise. The scales fell from my eyes when I read the most important paper by Hertz "Über die Grundgleichungen der Elektrodynamik für ruhende Körper" (*Göttinger Nachrichten, March 1890, and Ann. der Physik Bd. 40, 577*). Here Maxwell's equations as purified by Heaviside and Hertz are the axiomatic basis. The totality of electromagnetic phenomena is deduced from them in a systematic way."

Sommerfeld here describes the change in understanding from the mechanical action-at-a-distance theory to field theory which, apparently, was far from straightforward. To see this in detail, let us look into the paper by Hertz (cited above). Hertz writes: "Maxwell starts from the assumption of direct forces at a distance. He investigates the laws about how these forces influence the hypothetical polarizations of the dielectric aether. He finishes with the proposition that the polarizations really change according to these laws, but the action at a distance forces are not the true reason for it. This causes the unsatisfactory feeling that either the final result or the line of reasoning must be wrong." Hertz finally arrives at the conclusion that the second alternative is true: Maxwell's equations are correct but there is no satisfactory derivation. The latter can only be given by using relativity theory. Before we come to this subject, let us see how Hertz has "purified the equations" as Sommerfeld said.

Hertz continues: "... the derivation of the formulae leaves various superfluous rudimentary quantities in the equations, which had a meaning only in the old action-at-a-distance theory. As such rudimentary quantities of physical nature I mention the dielectric displacement of the aether, in distinction from the generating electric force, and the ratio of the two which is the dielectric constant of the aether.... As a rudimentary quantity of mathematical nature I mention the appearance of the vector potential in the basic equations... I have made efforts for some time to sort Maxwell's formulae and to separate their essential meaning from the accidental form in which they first appeared.... Mr. Oliver Heaviside has worked in the same direction since 1885. The quantities which he eliminates from Maxwell's equations, are the same that I also eliminate.... In this respect Mr. Heaviside has the priority."

Was that a revolution ? No. It was hard working of physicists like Faraday, Maxwell, Heaviside and Hertz over many years. The crucial test of the theory had to be given by experiments which Hertz carried out in 1886-87. But the experimental confirmation of Maxwell's theory was not the end of the story. The aether as the carrier medium for light and electricity, first introduced by Newton in his "Opticks" (1704) and the source of imagination for Maxwell, was long-lived. H.A. Lorentz in his memoir of 1904 (*Proc. Acad. Sci. of Amsterdam 6, 1904, p.11*) still speaks of the "dielectric displacement in the ether $\boldsymbol{D}$" instead of the electric field strength. Even after relativity in 1913 he said in a lecture (*H.A. Lorentz, Das Relativitätsprinzip, Teubner, Leipzig 1920, p.23*): "As far as this lecturer is concerned, he finds a certain satisfaction in the older interpretation, according to which the aether possesses at least some substantiality, space and time can be sharply separated, and simultaneity without further specification can be spoken out." Independent of this tenacity of the aether, the final purification of electrodynamics was achieved by Einstein in 1905.

Einstein was bothered by an asymmetry in electrodynamics. He starts his famous paper of June 1905 as follows: "It is known that Maxwell's electrodynamics – as usually understood at the present time – when applied to moving bodies, leads to asymmetries which do not appear to be inherent in the phenomena. Take, for example, the reciprocal electrodynamic action of a magnet and a conductor. The observable phenomenon here depends only on the relative motion of the conductor and the magnet, whereas the customary view draws a sharp distinction between the two cases in which either the one or the other of these bodies is in motion. For if the magnet is in motion and the conductor at rest, there arises in the neighbourhood of the magnet an electric field with a certain definite energy, producing a current in the conductor. But if the magnet is stationary and the conductor in motion, no electric field arises in the neighbourhood of the magnet. In the conductor, however, we find an electromotive force, to which in itself there is no corresponding energy, but which gives rise – assuming equality of relative motion in the two cases discussed – to the same electric current as produced by the electric force in the former case." A detailed discussion of this example is given in Chap.2 (2.5.13). The notion "electrodynamics – as usually understood at the present time" indicates that Einstein was not sure that Maxwell's theory is entirely correct. He had good reasons for this uncertainty: Shortly before, in March 1905, he had published his paper on the light-quantum hypothesis, where he had to leave classical electrodynamics to understand the photoelectric effect. This work and not the relativity theory gained him the Nobel prize in 1922.

In the June paper he continues as follows: "Examples of this sort, together with the unsuccessful attempts to discover any motion of the earth relatively to the 'light medium', suggest that the phenomena of electrodynamics as well as of mechanics possess no properties corresponding to the idea of absolute rest. They suggest rather that ... the same laws of electrodynamics and optics will be valid for all frames of reference for which the equations of mechanics hold good. We will raise this conjecture (hereafter called the 'Principle of Relativity') to the status of a postulate, and also introduce another postulate ... that light is always propagated in empty space with a definite velocity c which is independent of the status of motion of the emitting body." These two principles are the basis of relativity theory (Sect.2.1).

In contrast to Lorentz and Poincaré, Einstein does not use electrodynamics as the starting point for the reasons mentioned above. He goes the way in the other direction when he continues: "These two postulates suffice for the attainment of a simple and consistent theory of electrodynamics of moving bodies based on Maxwell's theory for stationary bodies. The introduction of a 'luminiferous ether' will prove to be superfluous..." Here the long life of the aether ends. At the same time a new approach to electrodynamics is opened. Einstein refers here to macroscopic electrodynamics of moving matter. We shall follow a somewhat different route: We will obtain microscopic electrodynamics in vacuum from electrostatics by means of the relativity theory. There is no doubt that this is conceptually the most satisfactory way to electrodynamics.

Was Einstein's special relativity a revolution? Many people say yes, because the basic concepts of time and space have been changed. But this is a misunderstanding. Time and space today are the same as before 1905, namely measured by clocks and metres. What did change are the properties of clocks and metres under transformation between moving reference systems. A.Pais in his book "Subtle is the Lord ..." (1982) calls the birth of special relativity "an edge of history". Einstein himself refers to it as "den Schritt", the step (*Pais, p.163*). "Revolutionary" he has only called his paper of March 1905 on the light-quantum hypothesis (*Pais, p.376*). Indeed, if there actually exists a revolution in physics, it is in quantum theory.

0.1 Some Topics of Analysis in $\mathbb{R}^3$

With boldface letters $\boldsymbol{A}$, $\boldsymbol{B}$ etc. we denote vectors in three-dimensional Euclidean space, $\boldsymbol{A} = (A_1, A_2, A_3)$. $A_j \in \mathbb{R}$ are the Cartesian components. The vectors form a real linear space with scalar product

$$\boldsymbol{A} \cdot \boldsymbol{B} = \sum_j A_j B_j, \tag{0.1.1}$$

which is positive definite: $\boldsymbol{A} \cdot \boldsymbol{A} > 0$ for $\boldsymbol{A} \neq \boldsymbol{0}$. We shall often omit the summation symbol if the summation goes over double indices (Einstein's summation convention). The two vectors are orthogonal, if $\boldsymbol{A} \cdot \boldsymbol{B} = 0$.

In $\mathbb{R}^3$ there exists a second product of two vectors, the vector product

$$\boldsymbol{A} \wedge \boldsymbol{B} = \boldsymbol{C}, \tag{0.1.2}$$

defined by

$$C_1 = A_2B_3 - B_2A_3, \quad C_2 = A_3B_1 - B_3A_1, \quad C_3 = A_1B_2 - B_1A_2. \tag{0.1.3}$$

This can be written in a single equation by means of the 3-dimensional ε-tensor

$$C_j = \varepsilon_{jkl}A_kB_l, \tag{0.1.4}$$

which is equal to

$$\varepsilon_{jkl} = \begin{cases} 1 & \text{if } j,k,l \text{ is an even permutation of 1,2,3} \\ -1 & \text{if } j,k,l \text{ is an odd permutation} \\ 0 & \text{otherwise.} \end{cases} \tag{0.1.5}$$

The product of two ε-tensors can be expressed by Kronecker δ's:

$$\varepsilon_{jkl}\varepsilon_{mnl} = \delta_{jm}\delta_{kn} - \delta_{jn}\delta_{km}. \tag{0.1.6}$$

The vector product is linear in both factors and antisymmetric

$$\boldsymbol{A} \wedge \boldsymbol{B} = -\boldsymbol{B} \wedge \boldsymbol{A} \tag{0.1.7}$$

which implies $\boldsymbol{A} \wedge \boldsymbol{A} = \boldsymbol{0}$. Furthermore, we note the following properties

$$\boldsymbol{A} \wedge (\boldsymbol{B} \wedge \boldsymbol{C}) = \boldsymbol{B}(\boldsymbol{A} \cdot \boldsymbol{C}) - \boldsymbol{C}(\boldsymbol{A} \cdot \boldsymbol{B}) \tag{0.1.8}$$

$$(\boldsymbol{A} \wedge \boldsymbol{B}) \cdot (\boldsymbol{C} \wedge \boldsymbol{D}) = (\boldsymbol{A} \cdot \boldsymbol{C})(\boldsymbol{B} \cdot \boldsymbol{D}) - (\boldsymbol{A} \cdot \boldsymbol{D})(\boldsymbol{B} \cdot \boldsymbol{C}). \tag{0.1.9}$$

The triple product

$$\boldsymbol{A} \cdot (\boldsymbol{B} \wedge \boldsymbol{C}) = \boldsymbol{C} \cdot (\boldsymbol{A} \wedge \boldsymbol{B}) = \boldsymbol{B} \cdot (\boldsymbol{C} \wedge \boldsymbol{A}) \tag{0.1.10}$$

is equal to the volume of the parallelepiped of the three vectors, consequently it vanishes if two vectors are parallel. An affine tensor T of second rank is a linear transformation in $\mathbb{R}^3$

$$T\boldsymbol{A} = \boldsymbol{B}: \quad \sum_k T_{jk}A_k = B_j, \tag{0.1.11}$$

for all $\boldsymbol{A} \in \mathbb{R}^3$. It is represented by a 3×3 matrix.

In field theory most quantities depend on space and time. We therefore consider scalar fields $\Phi(\boldsymbol{x})$, vector fields $\boldsymbol{A}(\boldsymbol{x})$ and tensor fields $T(\boldsymbol{x}) = (T_{jk}(\boldsymbol{x}))$, which associate to every $\boldsymbol{x} \in \mathbb{R}^3$ a real number, a vector or a tensor, respectively. This $\boldsymbol{x}$-dependence is assumed to be partially differentiable, sufficiently many times. Then the following important differential operators can be introduced. The gradient

$$\operatorname{grad} \Phi(\boldsymbol{x}) = \left(\frac{\partial \Phi}{\partial x_1}, \frac{\partial \Phi}{\partial x_2}, \frac{\partial \Phi}{\partial x_3}\right) \tag{0.1.12}$$

changes a scalar field into a vector field. The divergence

$$\operatorname{div} \boldsymbol{A}(\boldsymbol{x}) = \frac{\partial A_1}{\partial x_1} + \frac{\partial A_2}{\partial x_2} + \frac{\partial A_3}{\partial x_3} \tag{0.1.13}$$

generates a scalar field from a vector field and the curl

$$\operatorname{curl} \boldsymbol{A}(\boldsymbol{x}) = (\partial_2 A_3 - \partial_3 A_2, \partial_3 A_1 - \partial_1 A_3, \partial_1 A_2 - \partial_2 A_1) \tag{0.1.14}$$

transforms a vector field into another vector field. We will not use the nabla notation because it sometimes gives rise to misunderstandings. A further differential operator is

$$(\boldsymbol{A} \cdot \operatorname{grad})\boldsymbol{B} = \boldsymbol{C} \tag{0.1.15}$$

which is defined by

$$C_j = (A_1\partial_1 + A_2\partial_2 + A_3\partial_3)B_j, \quad j = 1, 2, 3. \tag{0.1.16}$$

Finally the second order operator

$$\operatorname{div} \operatorname{grad} \Phi = \frac{\partial^2 \Phi}{\partial x_1^2} + \frac{\partial^2 \Phi}{\partial x_2^2} + \frac{\partial^2 \Phi}{\partial x_3^2} = \triangle \Phi \tag{0.1.17}$$

is the usual Laplace operator which will also be used for vector fields

$$\triangle \boldsymbol{A} = (\triangle A_1, \triangle A_2, \triangle A_3). \tag{0.1.18}$$

These differential operators obey many identities which we list here for later reference:

$$\operatorname{curl} \operatorname{grad} \Phi = 0, \quad \operatorname{div} \operatorname{curl} \boldsymbol{A} = 0 \tag{0.1.19}$$

$$\operatorname{div}(\Phi \boldsymbol{A}) = \Phi \operatorname{div} \boldsymbol{A} + (\operatorname{grad} \Phi) \cdot \boldsymbol{A} \tag{0.1.20}$$

$$\operatorname{div}(\boldsymbol{A} \wedge \boldsymbol{B}) = \boldsymbol{B} \cdot \operatorname{curl} \boldsymbol{A} - \boldsymbol{A} \cdot \operatorname{curl} \boldsymbol{B} \tag{0.1.21}$$

$$\operatorname{curl}(\Phi \boldsymbol{A}) = \operatorname{grad} \Phi \wedge \boldsymbol{A} + \Phi \operatorname{curl} \boldsymbol{A} \tag{0.1.22}$$

$$\operatorname{curl}(\boldsymbol{A} \wedge \boldsymbol{B}) = (\boldsymbol{B} \cdot \operatorname{grad})\boldsymbol{A} - (\boldsymbol{A} \cdot \operatorname{grad})\boldsymbol{B} + \boldsymbol{A} \operatorname{div} \boldsymbol{B} - \boldsymbol{B} \operatorname{div} \boldsymbol{A} \tag{0.1.23}$$

$$\operatorname{grad}(\Phi_1 \Phi_2) = \Phi_1 \operatorname{grad} \Phi_2 + \Phi_2 \operatorname{grad} \Phi_1 \tag{0.1.24}$$

$$\operatorname{grad}(\boldsymbol{A} \cdot \boldsymbol{B}) = (\boldsymbol{A} \cdot \operatorname{grad})\boldsymbol{B} + (\boldsymbol{B} \cdot \operatorname{grad})\boldsymbol{A} + \boldsymbol{A} \wedge \operatorname{curl} \boldsymbol{B} + \boldsymbol{B} \wedge \operatorname{curl} \boldsymbol{A} \tag{0.1.25}$$

$$\operatorname{curl}(\operatorname{curl} \boldsymbol{A}) = \operatorname{grad} \operatorname{div} \boldsymbol{A} - \triangle \boldsymbol{A}. \tag{0.1.26}$$

The last identity (0.1.26) holds in cartesian coordinates only (we have not defined the differential operators in curvilinear coordinates here). It is straightforward to prove these identities in coordinates, writing the vector products with help of (0.1.4).

A vector field $\boldsymbol{B}(\boldsymbol{x})$ is called source-free if

$$\operatorname{div} \boldsymbol{B} = 0, \tag{0.1.27}$$

whereas if

$$\operatorname{curl} \boldsymbol{B} = 0, \tag{0.1.28}$$

the vector field is called vortex-free. The reason for this terminology which is mainly used by physicists, will become clear below. Very often we will have to compute derivatives of the euclidean distance

$$r = |\boldsymbol{x}| = \sqrt{x_1^2 + x_2^2 + x_3^2}. \tag{0.1.29}$$

The gradient

$$\operatorname{grad} r = \frac{\boldsymbol{x}}{r} \tag{0.1.30}$$

is the radial unit vector,

$$\operatorname{grad} \frac{1}{r} = -\frac{\boldsymbol{x}}{x^3} \tag{0.1.31}$$

and we have

$$\operatorname{grad} f(r) = f'(r)\frac{\boldsymbol{x}}{r} \tag{0.1.32}$$

for an arbitrary differentiable scalar function f.

We shall often apply the integral theorems of the calculus.

Theorem of Gauss. Let K_3 be a compact region in $\mathbb{R}^3$ with smooth boundary ∂K_3 and $\boldsymbol{A}(\boldsymbol{x})$ a continuously differentiable vector field ($\in C^1(K_3)$ for short). Then

$$\int\limits_{K_3} \operatorname{div} \boldsymbol{A}(\boldsymbol{x})\, d^3x = \int\limits_{\partial K_3} \boldsymbol{A}(\boldsymbol{x}) \cdot d\boldsymbol{\sigma}. \tag{0.1.33}$$

The surface integral

$$\int\limits_{\partial K_3} \boldsymbol{A}(\boldsymbol{x}) \cdot d\boldsymbol{\sigma} = \int\limits_{\partial K_3} A_n(\boldsymbol{x})\, d\sigma \tag{0.1.34}$$

is the flux of the vector field through the surface ∂K_3, A_n is the component of $\boldsymbol{A}$ normal to the surface and $d\sigma$ the scalar surface element. If we let K_3 shrink to a point $\boldsymbol{x}$, we get

$$\operatorname{div} \boldsymbol{A}(\boldsymbol{x}) = \lim_{K_3 \to \boldsymbol{x}} \frac{1}{V(K_3)} \int\limits_{\partial K_3} \boldsymbol{A} \cdot d\boldsymbol{\sigma}, \tag{0.1.35}$$

where $V(K_3)$ is the volume of K_3. This is the infinitesimal flux. If $\operatorname{div} \boldsymbol{A} = 0$ in some region, the flux through any closed surface in this region vanishes. Then the region is free of sources, $\operatorname{div} \boldsymbol{A}$ is a measure of the strength of the sources. If

$$\boldsymbol{A}(\boldsymbol{x}) = \operatorname{curl} \boldsymbol{a}(\boldsymbol{x}), \tag{0.1.36}$$

then $\boldsymbol{A}(\boldsymbol{x})$ is source-free, $\operatorname{div} \boldsymbol{A} = 0$, as a consequence of (0.1.19). Conversely, we will see below that every source-free vector field can be represented in this way.

We note some corollaries of Gauss' theorem.

Corollary 1. Let $\Phi(\boldsymbol{x}) \in C^1(K_3)$ be a scalar field, then

$$\int_{K_3} \operatorname{grad} \Phi(\boldsymbol{x})\, d^3x = \int_{\partial K_3} \Phi(\boldsymbol{x})\, d\boldsymbol{\sigma}. \tag{0.1.37}$$

Proof. Take $\boldsymbol{A}(\boldsymbol{x}) = \Phi(\boldsymbol{x})\boldsymbol{a}$ with an arbitrary constant vector $\boldsymbol{a}$, then

$$\operatorname{div} \boldsymbol{A} = \operatorname{grad} \Phi \cdot \boldsymbol{a}$$

and (0.1.33) implies

$$\int_{K_3} \operatorname{grad} \Phi(\boldsymbol{x}) \cdot \boldsymbol{a}\, d^3x = \int_{\partial K_3} \Phi(\boldsymbol{x})\, d\boldsymbol{\sigma} \cdot \boldsymbol{a}.$$

Since $\boldsymbol{a}$ is arbitrary, this gives Gauss' theorem for scalar fields (0.1.37). □

Corollary 2. Let $\boldsymbol{A}(\boldsymbol{x}) \in C^1(K_3)$ be a vector field, then

$$\int_{K_3} \operatorname{curl} \boldsymbol{A}(\boldsymbol{x})\, d^3x = \int_{\partial K_3} d\boldsymbol{\sigma} \wedge \boldsymbol{A}(\boldsymbol{x}). \tag{0.1.38}$$

Proof. For the 1-component we have

$$\int_{K_3} (\partial_2 A_3 - \partial_3 A_2) d^3x = \int_{K_3} \operatorname{div} \boldsymbol{a}\, d^3x,$$

with $\boldsymbol{a} = (0, A_3, -A_2)$. By Gauss' theorem this is equal to

$$= \int_{\partial K_3} \boldsymbol{a} \cdot d\boldsymbol{\sigma} = \int_{\partial K_3} (A_3 d\sigma_2 - A_2 d\sigma_3).$$

□

Corollary 3. Let $T_{jk}(\boldsymbol{x}) \in C^1(K_3)$ be a tensor field, then

$$\int_{K_3} \partial_j T_{jk}(\boldsymbol{x})\, d^3x = \int_{\partial K_3} T_{jk}(\boldsymbol{x})\, d\sigma_j. \tag{0.1.39}$$

Proof. We apply Gauss' theorem to $\boldsymbol{A}(\boldsymbol{x}) = T(\boldsymbol{x})\boldsymbol{a}$, where $\boldsymbol{a}$ is an arbitrary constant vector. Since

$$\operatorname{div} \boldsymbol{A} = \partial_j T_{jk}(\boldsymbol{x}) a_k,$$

we get

$$\int_{K_3} \partial_j T_{jk}(\boldsymbol{x})\, d^3x\, a_k = \int_{\partial K_3} T_{jk}(\boldsymbol{x})\, d\sigma_j a_k.$$

This proves (0.1.39) because $\boldsymbol{a}$ is arbitrary. Using $\boldsymbol{A}(\boldsymbol{x}) = \boldsymbol{a}T(\boldsymbol{x})$, we obtain in the same way

$$\int_{K_3} \partial_k T_{jk}(\boldsymbol{x})\, d^3x = \int_{\partial K_3} T_{jk}(\boldsymbol{x}) d\sigma_k. \tag{0.1.40}$$

□

Corollary 4. Green's theorem. Let $f, g \in C^2(K_3)$, then

$$\int_{K_3} (f \triangle g - g \triangle f) d^3x = \int_{\partial K_3} (f \operatorname{grad} g - g \operatorname{grad} f) \cdot d\boldsymbol{\sigma}. \tag{0.1.41}$$

Proof. Apply Gausss' theorem to

$$\boldsymbol{A}(\boldsymbol{x}) = f(\boldsymbol{x}) \operatorname{grad} g(\boldsymbol{x}) - g(\boldsymbol{x}) \operatorname{grad} f(\boldsymbol{x}).$$

Calculate $\operatorname{div} \boldsymbol{A}$ by means of (0.1.20)and use (0.1.17). □

This theorem holds also in the case of non-compact regions if f or g tends rapidly to 0 at infinity.

The last integral theorem is the

Theorem of Stokes. Let K_2 be a compact, 2-dimensional, oriented smooth surface in $\mathbb{R}^3$ with boundary ∂K_2, and $\boldsymbol{A}(\boldsymbol{x}) \in C^1(K_2)$. Then

$$\int_{K_2} \operatorname{curl} \boldsymbol{A} \cdot d\boldsymbol{\sigma} = \int_{\partial K_2} \boldsymbol{A} \cdot d\boldsymbol{s}. \tag{0.1.42}$$

The line integral along the boundary

$$\int_{\partial K_2} \boldsymbol{A} \cdot d\boldsymbol{s} = \int_{\partial K_2} A_t ds$$

is the circulation of the vector field along this closed path, A_t is the tangential component. If K_2 shrinks to a point $\boldsymbol{x}$, we get

$$\operatorname{curl} \boldsymbol{A}(\boldsymbol{x}) = \lim_{K_2 \to \boldsymbol{x}} \frac{1}{F(K_2)} \int_{\partial K_2} \boldsymbol{A} \cdot d\boldsymbol{s}, \tag{0.1.43}$$

where $F(K_2)$ is the area of K_2. This is the infinitesimal circulation. If $\operatorname{curl} \boldsymbol{A} = \boldsymbol{0}$ in some region, the circulation along every closed curve there vanishes. Then $\boldsymbol{A}(\boldsymbol{x})$ is called vortex-free in this region, $\operatorname{curl} \boldsymbol{A}$ is a measure of the vortex strength.

Corollary 5. If $\operatorname{curl} \boldsymbol{A} = \boldsymbol{0}$ in a region containing $\boldsymbol{0}$, then the line integral

$$\int_{\boldsymbol{0}}^{\boldsymbol{x}} \boldsymbol{A} \cdot d\boldsymbol{s} \stackrel{\text{def}}{=} \Phi(\boldsymbol{x}) \tag{0.1.44}$$

is independent of the path. It defines a potential $\Phi(\boldsymbol{x})$ for $\boldsymbol{A}(\boldsymbol{x})$:

$$\boldsymbol{A}(\boldsymbol{x}) = \operatorname{grad} \Phi(\boldsymbol{x}). \tag{0.1.45}$$

The potential is unique up to a constant which depends on the fixed initial point $\boldsymbol{0}$ in (0.1.44). This corollary means that a vortex-free vector field is a potential field (or conservative).

0.2 The Laplace Equation

The integration of the inhomogeneous Laplace, Poisson or potential equation

$$\triangle \Phi(\boldsymbol{x}) = h(\boldsymbol{x}) \tag{0.2.1}$$

in $\mathbb{R}^3$ is an application of Green's theorem.

Theorem 1. Let $\Phi(\boldsymbol{x})$ be a solution of (0.2.1) with

$$\Phi(\boldsymbol{x}) \to 0, \quad |\boldsymbol{x}| \operatorname{grad} \Phi \to \boldsymbol{0}, \quad \text{for} \quad |\boldsymbol{x}| \to \infty. \tag{0.2.2}$$

Then it follows

$$\Phi(\boldsymbol{x}) = -\frac{1}{4\pi} \int \frac{h(\boldsymbol{y})}{|\boldsymbol{x} - \boldsymbol{y}|} d^3 y, \tag{0.2.3}$$

if the integral converges.

Proof. We want to prove the following equation

$$\Phi(\boldsymbol{x}) = -\frac{1}{4\pi} \int \frac{\triangle_y \Phi(\boldsymbol{y})}{|\boldsymbol{x} - \boldsymbol{y}|} d^3 y, \tag{0.2.4}$$

which obviously allows to conclude (0.2.3) from (0.2.1).

In Green's theorem (0.1.41) we take

$$f(\boldsymbol{y}) = \frac{1}{|\boldsymbol{x} - \boldsymbol{y}|}, \quad g(\boldsymbol{y}) = \Phi(\boldsymbol{y}). \tag{0.2.5}$$

We compute for $\boldsymbol{y} \neq \boldsymbol{x}$

$$\triangle_y \frac{1}{|\boldsymbol{x}-\boldsymbol{y}|} = \sum_j \frac{\partial^2}{\partial y_j^2} \frac{1}{\sqrt{(x_1-y_1)^2+(x_2-y_2)^2+(x_3-y_3)^2}}$$

$$= \sum_j \frac{\partial}{\partial y_j} \frac{x_j - y_j}{r^3} = -\frac{3}{r^3} + 3\sum_j \frac{(x_j-y_j)^2}{r^5} = 0. \tag{0.2.6}$$

As compact region K_3 in Green's theorem, we choose

$$K_3 = \{\boldsymbol{y} \in \mathbb{R}^3 \mid |\boldsymbol{y}-\boldsymbol{x}| \geq \varepsilon,\ |\boldsymbol{y}| \leq R\}, \tag{0.2.7}$$

in order to exclude $r = 0$. Then

$$\int_{K_3} \frac{1}{|\boldsymbol{x}-\boldsymbol{y}|} \triangle \Phi(\boldsymbol{y})\, d^3y = \int_{\partial K_3} \Big(\frac{1}{|\boldsymbol{x}-\boldsymbol{y}|} \operatorname{grad} \Phi(\boldsymbol{y}) - \Phi(\boldsymbol{y}) \operatorname{grad} \frac{1}{|\boldsymbol{x}-\boldsymbol{y}|} \Big) \cdot d\boldsymbol{\sigma}. \tag{0.2.8}$$

The boundary ∂K_3 consists of the surfaces of a big and a small sphere. Let us first consider the integral over the big sphere $|\boldsymbol{y}| = R$. Choosing $R > 2|\boldsymbol{x}|$, we find

$$|\Phi(\boldsymbol{y})| \leq \varepsilon(R), \quad |\operatorname{grad} \Phi(\boldsymbol{y})| \leq \frac{\varepsilon(R)}{R}, \quad \text{where } \lim_{R\to\infty} \varepsilon(R) = 0,$$

on the sphere $|\boldsymbol{y}| = R$, according to (0.2.2). Hence,

$$\Big| \int_{|\boldsymbol{y}|=R} \frac{1}{|\boldsymbol{x}-\boldsymbol{y}|} \operatorname{grad} \Phi(\boldsymbol{y}) \cdot d\boldsymbol{\sigma} \Big| \leq \frac{1}{R/2} \frac{\varepsilon(R)}{R} 4\pi R^2 = 8\pi\varepsilon(R) \to 0$$

and

$$\Big| \int_{|\boldsymbol{y}|=R} \Phi(\boldsymbol{y}) \frac{\boldsymbol{x}-\boldsymbol{y}}{|\boldsymbol{x}-\boldsymbol{y}|^3} \cdot d\boldsymbol{\sigma} \Big| \leq \varepsilon(R) \frac{1}{(R/2)^2} 4\pi R^2 = 16\pi\varepsilon(R) \to 0$$

for $R \to \infty$, so that this integral gives no contribution. Next we consider the integral over the small sphere $|\boldsymbol{y}-\boldsymbol{x}| = \varepsilon$. Let $\boldsymbol{n}$ be the inward radial unit vector on the surface of this sphere, which is the outer normal vector for ∂K_3, then

$$\Big(\operatorname{grad}_y \frac{1}{|\boldsymbol{x}-\boldsymbol{y}|} \Big) \cdot \boldsymbol{n} = \frac{\boldsymbol{x}-\boldsymbol{y}}{|\boldsymbol{x}-\boldsymbol{y}|^3} \frac{\boldsymbol{x}-\boldsymbol{y}}{|\boldsymbol{x}-\boldsymbol{y}|} = \frac{1}{|\boldsymbol{x}-\boldsymbol{y}|^2} = \frac{1}{\varepsilon^2}.$$

Hence, we get

$$\int_{|\boldsymbol{y}-\boldsymbol{x}|=\varepsilon} \ldots = \frac{1}{\varepsilon} \int \frac{\partial \Phi}{\partial n}\, d\sigma - \frac{1}{\varepsilon^2} \int \Phi(\boldsymbol{y}) d\sigma.$$

For $\varepsilon \to 0$ the first term goes to 0, but the second one has a finite limit $-4\pi\Phi(\boldsymbol{x})$. Substituting this into (0.2.8) and taking the limits $R \to \infty$ and $\varepsilon \to 0$, we arrive at (0.2.4). □

The equation (0.2.4) can be interpreted in the sense of distributions as follows. We assume Φ to be an infinitely differentiable test function (with compact support, $\in C_0^\infty(\mathbb{R}^3)$ for short). Then

$$\int \frac{1}{|\boldsymbol{x}-\boldsymbol{y}|} \triangle_y \Phi(\boldsymbol{y})\, d^3y = \left\langle \frac{1}{|\boldsymbol{x}-\boldsymbol{y}|}, \triangle\Phi \right\rangle \tag{0.2.9}$$

is a continuous linear functional on C_0^∞, a distribution (see *L.Hörmander, The Analysis of Linear Partial Differential Operators I, Springer Verlag 1990*, for example). Using the notion of distributional derivative (formal partial integration), this is equivalent to

$$= \left\langle \triangle_y \frac{1}{|\boldsymbol{x}-\boldsymbol{y}|}, \Phi \right\rangle.$$

Now (0.2.4) can be written in the following way

$$-4\pi\Phi(\boldsymbol{x}) = \left\langle \triangle_y \frac{1}{|\boldsymbol{x}-\boldsymbol{y}|}, \Phi \right\rangle, \tag{0.2.10}$$

or

$$\triangle_y \frac{1}{|\boldsymbol{x}-\boldsymbol{y}|} = -4\pi\delta(\boldsymbol{x}-\boldsymbol{y}), \tag{0.2.11}$$

where δ is the three-dimensional δ-distribution. That this vanishes for $\boldsymbol{x} \neq \boldsymbol{y}$ has been calculated above (0.2.6), however, at $\boldsymbol{x} = \boldsymbol{y}$ we have obtained a point measure, the Dirac delta-distribution. According to (0.2.11) one calls $-1/4\pi|\boldsymbol{x}-\boldsymbol{y}|$ the fundamental solution or Green's function of the Laplace equation.

Now we will show that the solution (0.2.3) of the inhomogeneous Laplace equation (0.2.1) is unique. If Φ' is a second solution, then

$$\triangle(\Phi - \Phi') = 0. \tag{0.2.12}$$

To conclude that $\Phi - \Phi' = 0$ we need the mean value theorem.

Theorem 2. Let $\Phi(\boldsymbol{y})$ be a harmonic function (i.e. $\triangle\Phi = 0$) in a sphere

$$K_R = \{\boldsymbol{y} \in \mathbb{R}^3 \mid |\boldsymbol{y}-\boldsymbol{x}| < R\}, \quad \boldsymbol{x} \quad \text{fixed}, \tag{0.2.13}$$

which is continuous in the closed sphere $\overline{K_R}$. Then $\Phi(\boldsymbol{x})$ is equal to the mean value over the surface of the sphere:

$$\Phi(\boldsymbol{x}) = \frac{1}{4\pi R^2} \int\limits_{\partial K_R} \Phi(\boldsymbol{y})\, d\sigma. \tag{0.2.14}$$

Proof. Again we use Green's theorem

$$\int_{K_R} (f \triangle g - g \triangle f) d^3y = \int_{\partial K_R} \Big(f \frac{\partial g}{\partial n} - g \frac{\partial f}{\partial n} \Big) d\sigma,$$

where

$$\partial/\partial n \overset{\text{def}}{=} \boldsymbol{n} \cdot \text{grad} \tag{0.2.15}$$

is the normal derivative. For $f = 1$ and $g = \Phi$ we have

$$0 = \int_{\partial K_R} \frac{\partial \Phi}{\partial n} \, d\sigma. \tag{0.2.16}$$

For $f = 1/|\boldsymbol{x} - \boldsymbol{y}|$, $g = \Phi$ we get in the same way as in Theorem 1

$$4\pi \Phi(\boldsymbol{x}) = \int_{\partial K_R} \Big(\frac{1}{|\boldsymbol{x} - \boldsymbol{y}|} \frac{\partial \Phi}{\partial n_y} - \Phi(\boldsymbol{y}) \frac{\partial}{\partial n_y} \frac{1}{|\boldsymbol{x} - \boldsymbol{y}|} \Big) d\sigma. \tag{0.2.17}$$

Since

$$\boldsymbol{n} \cdot \text{grad} \frac{1}{r} = -\frac{1}{r^2},$$

we finally obtain

$$\Phi(\boldsymbol{x}) = \frac{1}{4\pi} \int_{\partial K_R} \Big(\frac{1}{R} \frac{\partial \Phi}{\partial n} + \frac{\Phi(\boldsymbol{y})}{R^2} \Big) d\sigma = \frac{1}{4\pi R^2} \int_{\partial K_R} \Phi(\boldsymbol{y}) d\sigma, \tag{0.2.18}$$

by means of (0.2.16). □

The uniqueness (0.2.12) is now a consequence of the following

Corollary 6. If Φ is harmonic in $\mathbb{R}^3$ and $\Phi \to 0$ for $|\boldsymbol{x}| \to \infty$, then it follows $\Phi = 0$. This is obtained by taking $R \to \infty$ in (0.2.14).

A further consequence of the mean value theorem 2 is the maximum principle:

Theorem 3. If $\Phi(\boldsymbol{x}) \not\equiv$ const. is harmonic in the compact open region K and continuous in the closure $\overline{K}$, then Φ assumes its maximal and minimal values on the boundary ∂K:

$$\min_{y \in \partial K} \Phi(\boldsymbol{y}) < \Phi(\boldsymbol{x}) < \max_{y \in \partial K} \Phi(\boldsymbol{y}) \tag{0.2.19}$$

for all $\boldsymbol{x} \in K$.

Proof. Let us assume that Φ were maximal at an inner point $\boldsymbol{x}_0 \in K$,

$$\Phi(\boldsymbol{x}_0) = M = \max_{x \in \overline{K}} \Phi(\boldsymbol{x}). \tag{0.2.20}$$

Then it follows that $\Phi(\boldsymbol{x}) = M$, in any sphere K_δ around $\boldsymbol{x}_0$ in K of radius δ, because otherwise Theorem 2 would imply

$$\Phi(\boldsymbol{x}_0) = \frac{1}{4\pi\delta^2} \int\limits_{\partial K_\delta} \Phi(\boldsymbol{y}) d\sigma < \frac{M}{4\pi\delta^2} \int\limits_{\partial K_\delta} d\sigma = M,$$

which is a contradiction. If we pack K with spheres K_δ, it follows that $\Phi(\boldsymbol{x}) = M = \text{const.}$ in K. This is again a contradiction. For the minimum we consider $-\Phi(\boldsymbol{x})$. □

Corollary 7. If Φ is harmonic in K and $= 0$ on ∂K, then $\Phi = 0$ in the whole of K.

Finally we want to decompose a vector field into a source-free and a vortex-free part.

Theorem 4. Let $\boldsymbol{A}(\boldsymbol{x}) \in C^2(\mathbb{R}^3)$ be a vector field with

$$|\boldsymbol{A}(\boldsymbol{x})| \leq \frac{C}{|\boldsymbol{x}|^{1+\varepsilon}}, \quad \varepsilon > 0 \tag{0.2.21}$$

for sufficiently large $|\boldsymbol{x}|$. Then there exists a unique decomposition

$$\boldsymbol{A}(\boldsymbol{x}) = \boldsymbol{B}(\boldsymbol{x}) + \boldsymbol{D}(\boldsymbol{x}), \tag{0.2.22}$$

$$\text{where} \quad \operatorname{div} \boldsymbol{B}(\boldsymbol{x}) = 0 \quad \text{is source-free} \tag{0.2.23}$$

$$\text{and} \quad \operatorname{curl} \boldsymbol{D}(\boldsymbol{x}) = \boldsymbol{0} \quad \text{is vortex-free} \tag{0.2.24}$$

and $\boldsymbol{B}, \boldsymbol{D} \to \boldsymbol{0}$ for $|\boldsymbol{x}| \to \infty$. Furthermore, we have the following explicit expressions

$$\boldsymbol{D}(\boldsymbol{x}) = \frac{1}{4\pi} \int d^3y \operatorname{div} \boldsymbol{A}(\boldsymbol{y}) \frac{\boldsymbol{x} - \boldsymbol{y}}{|\boldsymbol{x} - \boldsymbol{y}|^3} \tag{0.2.25}$$

$$\boldsymbol{B}(\boldsymbol{x}) = \frac{1}{4\pi} \int d^3y \operatorname{curl} \boldsymbol{A}(\boldsymbol{y}) \wedge \frac{\boldsymbol{x} - \boldsymbol{y}}{|\boldsymbol{x} - \boldsymbol{y}|^3}. \tag{0.2.26}$$

Here the vortex- and source-free parts are given in terms of the source density $\operatorname{div} \boldsymbol{A}$ and the vortex density $\operatorname{curl} \boldsymbol{A}$. The right-hand side of (0.2.25) is the gradient of a scalar potential while (0.2.26) is the curl of a vector potential. This is seen in the course of the following

Proof. We first prove (0.2.25). From (0.2.24) and Corollary 5 it follows that there exists a potential $\Phi(\boldsymbol{x})$ such that

$$\boldsymbol{D}(\boldsymbol{x}) = \operatorname{grad} \Phi(\boldsymbol{x}). \tag{0.2.27}$$

To determine Φ, we take the divergence of this equation

$$\operatorname{div} \boldsymbol{D} = \operatorname{div} \operatorname{grad} \Phi = \triangle \Phi = \operatorname{div} \boldsymbol{A}. \tag{0.2.28}$$

This is an inhomogeneous Laplace equation for Φ. By Theorem 1 the solution is given by

$$\Phi(\boldsymbol{x}) = -\frac{1}{4\pi} \int d^3y \, \frac{\operatorname{div} \boldsymbol{A}(\boldsymbol{y})}{|\boldsymbol{x} - \boldsymbol{y}|}. \tag{0.2.29}$$

We must show that this integral is actually convergent. The singularity for $\boldsymbol{y} \to \boldsymbol{x}$ is integrable, so that only $|\boldsymbol{y}| \to \infty$ must be considered. Let $a > |\boldsymbol{x}|$, then

$$\int\limits_{|y| \geq a} d^3y \, \frac{1}{|\boldsymbol{x} - \boldsymbol{y}|} \operatorname{div} \boldsymbol{A}(\boldsymbol{y}) = \int\limits_{|y| \geq a} d^3y \operatorname{div}_y \Big(\frac{1}{|\boldsymbol{x} - \boldsymbol{y}|} \boldsymbol{A}(\boldsymbol{y}) \Big)$$

$$+ \int d^3y \, \frac{(\boldsymbol{x} - \boldsymbol{y}) \cdot \boldsymbol{A}(\boldsymbol{y})}{|\boldsymbol{x} - \boldsymbol{y}|^3}. \tag{0.2.30}$$

By Gauss' theorem the first integral is a surface integral with an integrand which is $O(y^{-2-\varepsilon})$. Thus, for $a \to \infty$, the integral tends to 0. The integrand of the second volume integral is $O(y^{-3-\varepsilon})$, so that this integral also goes to 0. This proves the existence of (0.2.29). Substituting (0.2.29) into (0.2.27) we obtain the desired expression

$$\boldsymbol{D}(\boldsymbol{x}) = -\frac{1}{4\pi} \operatorname{grad}_x \int d^3y \, \frac{\operatorname{div} \boldsymbol{A}(\boldsymbol{y})}{|\boldsymbol{x} - \boldsymbol{y}|}$$

$$= \frac{1}{4\pi} \int d^3y \, \frac{(\boldsymbol{x} - \boldsymbol{x}) \operatorname{div} \boldsymbol{A}(\boldsymbol{y})}{|\boldsymbol{x} - \boldsymbol{y}|^3}. \tag{0.2.31}$$

Let us consider the difference

$$\boldsymbol{B}(\boldsymbol{x}) = \boldsymbol{A}(\boldsymbol{x}) - \boldsymbol{D}(\boldsymbol{x}) = \operatorname{curl} \boldsymbol{a}'(\boldsymbol{x}), \tag{0.2.32}$$

where $\boldsymbol{a}'(\boldsymbol{x})$ is a vector potential which must now be determined. Such a vector potential is never unique: If $\boldsymbol{a}(\boldsymbol{x})$ is another vector potential, we have $\operatorname{curl}(\boldsymbol{a}' - \boldsymbol{a}) = \boldsymbol{0}$ and it follows from Corollary 5 that

$$\boldsymbol{a}' - \boldsymbol{a} = \operatorname{grad} \chi. \tag{0.2.33}$$

Therefore, a vector potential is only determined up to a gradient (gauge freedom or gauge transformation). We can choose this gradient so that

$$\operatorname{div} \boldsymbol{a} = 0. \tag{0.2.34}$$

Indeed, we must only determine χ in (0.2.33) from

$$\operatorname{div} \boldsymbol{a}' = \triangle \chi, \tag{0.2.35}$$

and this can be uniquely solved for χ by Theorem 1. (Then the gauge is fixed.) From (0.2.32) we now obtain

$$\operatorname{curl} \boldsymbol{B} = \operatorname{curl}\operatorname{curl} \boldsymbol{a}' = \operatorname{curl}\operatorname{curl} \boldsymbol{a} = \operatorname{grad}\operatorname{div} \boldsymbol{a} - \triangle \boldsymbol{a},$$

using (0.1.26). Taking (0.2.34) into account, we have

$$\triangle \boldsymbol{a} = -\operatorname{curl} \boldsymbol{B} = -\operatorname{curl} \boldsymbol{A}$$

which leads to the desired vector potential

$$\boldsymbol{a}(\boldsymbol{x}) = \frac{1}{4\pi} \int d^3y \, \frac{\operatorname{curl} \boldsymbol{A}(\boldsymbol{y})}{|\boldsymbol{x} - \boldsymbol{y}|}. \tag{0.2.36}$$

The existence of this integral can be shown as above with help of Corollary 2. Hence,

$$\begin{aligned} \boldsymbol{B}(\boldsymbol{x}) = \operatorname{curl} \boldsymbol{a}(\boldsymbol{x}) &= \frac{1}{4\pi} \operatorname{curl}_x \int d^3y \, \frac{\operatorname{curl} \boldsymbol{A}(\boldsymbol{y})}{|\boldsymbol{x} - \boldsymbol{y}|} \\ &= \frac{1}{4\pi} \int d^3y \operatorname{grad}_x \frac{1}{|\boldsymbol{x} - \boldsymbol{y}|} \wedge \operatorname{curl} \boldsymbol{A}(\boldsymbol{y}) \\ &= \frac{1}{4\pi} \int d^3y \operatorname{curl} \boldsymbol{A}(\boldsymbol{y}) \wedge \frac{\boldsymbol{x} - \boldsymbol{y}}{|\boldsymbol{x} - \boldsymbol{y}|^3}. \end{aligned} \tag{0.2.37}$$

Finally uniqueness must be proved. If $\boldsymbol{A} = \boldsymbol{B}' + \boldsymbol{D}'$ is another decomposition, then $\operatorname{curl} \boldsymbol{A} = \operatorname{curl} \boldsymbol{B}' = \operatorname{curl} \boldsymbol{B}$. This implies $\operatorname{curl}(\boldsymbol{B}' - \boldsymbol{B}) = \boldsymbol{0}$. In addition, we have $\operatorname{div}(\boldsymbol{B}' - \boldsymbol{B}) = 0$, hence

$$\triangle(\boldsymbol{B}' - \boldsymbol{B}) = (\operatorname{grad}\operatorname{div} - \operatorname{curl}\operatorname{curl})(\boldsymbol{B}' - \boldsymbol{B}) = \boldsymbol{0}.$$

It follows from Corollary 6 that $\boldsymbol{B}' = \boldsymbol{B}$, and in the same way $\boldsymbol{D}' = \boldsymbol{D}$. □

Theorem 4 is the basis of electrodynamics: A stationary electromagnetic field consists of an electric field $\boldsymbol{E}$ plus a magnetic field $\boldsymbol{B}$. The former is the vortex-free part, $\operatorname{curl} \boldsymbol{E} = \boldsymbol{0}$, the latter is source-free, $\operatorname{div} \boldsymbol{B} = 0$. Only the electric field has sources, namely the electric charges. Magnetic charges (monopoles) have never been found.

0.3 Problems

1. Prove: Let K_2 be a compact, oriented, 2-dimensional smooth surface in $\mathbb{R}^3$ with boundary ∂K_2 and $\Phi(\boldsymbol{x}) \in C^1(K_2)$. Then

$$\int_{K_2} d\boldsymbol{\sigma} \wedge \operatorname{grad} \Phi = \int_{\partial K_2} \Phi \, d\boldsymbol{s}. \tag{0.3.1}$$

2. Which of the following vector fields are source-free or vortex-free?
   a) $\boldsymbol{A}(\boldsymbol{x}) = (2x_1 - 2x_2, -2x_1 + 4x_2 - 3x_3, -3x_2 + 6x_3)$.
   b)
   $$\boldsymbol{A}(\boldsymbol{x}) = \boldsymbol{a} \wedge \frac{\boldsymbol{x}}{r^3}, \quad \boldsymbol{a} = \text{const.}, \quad r = |\boldsymbol{x}|.$$
   c)
   $$\boldsymbol{A}(\boldsymbol{x}) = \frac{\boldsymbol{a}}{r + \alpha}, \quad \alpha > 0.$$

3. Calculate for the vector fields in Problem 2 a potential, a vector potential or a potential for the vortex-free part, respectively.
4. Verify that (0.2.3) is a solution of (0.2.1).
5. Prove the mean value theorem in the following form: Let $\Phi(\boldsymbol{x})$ be harmonic in the sphere $K_R = \{\boldsymbol{y} \mid |\boldsymbol{y} - \boldsymbol{y}| \leq R\}$, then

$$\Phi(\boldsymbol{x}) = \frac{3}{4\pi R^3} \int_{K_R} \Phi(\boldsymbol{y})\, d^3y. \tag{0.3.2}$$

1. Electrostatics in Vacuum

Let us start with the discussion of an important question of methodology, how one defines basic quantities in physics. Here philosophers beg permission to speak, sometimes. Fortunately, physics is in a very good situation (perhaps the best possible one) regarding its fundamental quantities: **Every really basic quantity is defined by a measuring procedure**. This point of view was already taken by Poincaré in 1905, when he said about mechanics: "The English teach mechanics as an experimental science; on the continent it is taught always more or less as a deductive and *a priori* science. The English are right, no doubt." *(H.Poincaré, Science and Hypothesis, Dover Publication, New York 1952)*. Two different measuring processes define two different quantities. The equivalence of two measuring procedures must be checked and may be an important result, as for example the equivalence of gravitational and inertial mass. The defining measuring procedure need not agree with the most accurate method to measure the quantity in the lab, because the latter (which is used to define the unit) may change with improvements of experimental techniques.

With such operational definitions experimental physics can start. But with measurements as the only tool, we get a purely empirical science. It consists of lists of measuring data. That is not yet physics. To order the data, we associate to every physical quantity a mathematical object. The mapping on a mathematical structure must be such that all observations are correctly described. If an exception is found, the mapping must be changed. This mapping of the physical reality on a mathematical structure is the ultimate goal of physics. To find the correct mapping one proceeds as follows. The experimental studies can be summarized in certain empirical laws (as Coulomb's law for example). By physical and mathematical analysis of the empirical laws (i.e. making use of general physical principles as the relativity principle, and applying mathematical theorems) one is led to general natural laws (as Maxwell's equations for example) and to a physical theory. Such a theory is only fruitful if it makes predictions that can be checked by new experiments (as Maxwell's theory predicts radiation). The more different predictions a theory makes, the more different observations it successfully describes, the better. This is the falsifybility criterion of K.Popper: A physical theory can never be proven, it can only be falsified, and the easier this is, that means the more tests there are, the better is the theory. If a theory stands all very different experimental tests, the whole construction is checked in a much more subtle way as if an empirical law is checked by repeating experiments of the more or less same kind again and again. This interplay of experiment and theory is typical for the physical method. Needless to say that the actual historical development does not follow this logical scheme (neither did Maxwell find his equations by using relativity, nor did he look for an experimental confirmation of electromagnetic radiation when he had become director of the Cavendish lab). History is never logical, but a basic physical course must be so. We therefore will follow the route just sketched.

With some care we can even go one step further. It is usually said that mathematics is the science with the most sound basis. The introduction of mathematical objects by a system of axioms has some similarity with the definition of physical quantities by a measuring procedure. Its main purpose is also to be operational, it makes things going because one can prove theorems from the axioms. But within mathematics there are serious limitations to check the whole construction: K.Goedel and others have shown that in any sufficiently complicated axiom system there are theorems which cannot be proved to be true. Physicists, on the other hand, are able to check their constructions by experiments. The interesting point is that all the open parts of mathematics, where unprovable theorems occur, are used in the physical theories, they are even essential components of those theories. In a way the experiments check these mathematical components, too. That means, the fact that nature can be successfully described by an ingenious web of mathematics and physics gives a qualitatively new kind of truth also to mathematics. Perhaps by this combination of methods one gets the highest degree of intellectual certainty which human intelligence can gain. Mathematicians may repel being embraced in that way. But this is a sentimental and not a logical objection. Some parts of mathematics have no direct connection with physics. Then the only possible check of truth is their internal consistency, as far as the latter can really be checked. Remarkably enough, that portion of mathematics without relevance to physics seems to decrease with time. The signs in science point towards unification.

After this philosophical excursion we come to our subject.

1.1 Electric Charge, Field Strength and the Equations of Electrostatics

The measuring procedure for defining electric charge q and field strength $\boldsymbol{E}(\boldsymbol{x})$ is the study of the forces between "electrically charged bodies". Let us take a big body and "charge" it by some definite procedure, for example by connecting it with a battery. Taking another small charged test body, we can then measure the force $\boldsymbol{K}(\boldsymbol{x})$ on this test body, which is at the position $\boldsymbol{x}$, at rest with respect to the big body. We know of course from mechanics how to measure forces. We write this force as a product of a space independent factor q, depending only on the test body, and a space dependent factor $\boldsymbol{E}(\boldsymbol{x})$ which is independent of the test body:

$$\boldsymbol{K}(\boldsymbol{x}) = q\boldsymbol{E}(\boldsymbol{x}). \tag{1.1.1}$$

We claim that **this equation defines the charge** q **and the field strength** $\boldsymbol{E}$ and, in addition, contains further empirical results. This might sound strange: how can one equation define two quantities, and still contain additional information ? Here is the answer:

1. **Definition of charge**: To compare the charges of two different test bodies, we bring them (one after the other) at the same place $\boldsymbol{x}$ and measure the forces on them. According to (1.1.1) we have
$$\frac{|q_1|}{|q_2|} \stackrel{\text{def}}{=} \frac{|\boldsymbol{K}_1|}{|\boldsymbol{K}_2|}, \tag{1.1.2}$$
and we say that $\operatorname{sgn} q_1 = \operatorname{sgn} q_2$, if $\boldsymbol{K}_1$ and $\boldsymbol{K}_2$ have the same direction. In addition we find by such measurements that $\boldsymbol{K}_1$ and $\boldsymbol{K}_2$ are always parallel or antiparallel and that the ratio (1.1.2) is independent of $\boldsymbol{x}$ and of the big body. Since this ratio is a property of the test body alone, it defines its charge. q is a scalar quantity which can be positive or negative. The unit is fixed by specifying some normal body. The best would be an electron or a proton. In practise one measures the few elementary charges on an oil droplet (Millikan's experiment). The historical and by now legal unit is 1 Coulomb (C) $= 6.242 \times 10^{18}$ elementary charges.
2. **Definition of the electric field strength** $\boldsymbol{E}(\boldsymbol{x})$: We bring one and the same test body at different places. Then
$$\frac{|\boldsymbol{E}(\boldsymbol{x}_1)|}{|\boldsymbol{E}(\boldsymbol{x}_2)|} \stackrel{\text{def}}{=} \frac{|\boldsymbol{K}(\boldsymbol{x}_1)|}{|\boldsymbol{K}(\boldsymbol{x}_2)|} \tag{1.1.3}$$
and the direction of $\boldsymbol{E}(\boldsymbol{x})$ is defined by the direction of $\boldsymbol{K}(\boldsymbol{x})$ if $q > 0$. In addition one finds that the ratio (1.1.3) is independent of the choice of the (small) test body. It is an attribute of space which exists also without the test body. In fact, we will see much later that the electric field can manifest itself not only by forces on test bodies (also by radiation, for example).

The same discussion is necessary at the beginning of mechanics, where Newton's first law

$$\boldsymbol{K}(\boldsymbol{x}) = m\boldsymbol{b}(\boldsymbol{x}) \tag{1.1.4}$$

simultaneously defines mass m and force $\boldsymbol{K}(\boldsymbol{x})$, assuming that accelerations $\boldsymbol{b}(\boldsymbol{x})$ can be measured. But in electrostatics we must be careful: If the test bodies are too big (or better too strongly charged), they influence the field generated by the big body. Therefore, strictly speaking, in (1.1.1) one has to take a microscopic test body and we have defined the electric field of macroscopic (strongly charged) bodies only. We shall return to this point at the end of Sect.1.4.

It is now possible to investigate the electric field in simple geometric situations. For this purpose one uses the concept of point charges which means that the extension of the charged bodies is small compared to their distances. As a consequence, such a configuration of point charges is completely described if the charges and their positions are specified, there is no further degree of freedom. In the same way mass points are used in mechanics. Coulomb found in 1785 that the force between a point charge q_1 at $\boldsymbol{y}$ and another point charge q at $\boldsymbol{x}$ is given by

$$\boldsymbol{K}(\boldsymbol{x},\boldsymbol{y}) = \frac{1}{4\pi\varepsilon_0}\frac{qq_1}{|\boldsymbol{x}-\boldsymbol{y}|^2}\frac{\boldsymbol{x}-\boldsymbol{y}}{|\boldsymbol{x}-\boldsymbol{y}|}. \tag{1.1.5}$$

This is **Coulomb's law** which obviously satisfies actio = reactio

$$\boldsymbol{K}(\boldsymbol{x},\boldsymbol{y}) = -\boldsymbol{K}(\boldsymbol{y},\boldsymbol{x}). \tag{1.1.6}$$

Since we have already fixed the unit of charge (C), the natural constant ε_0 in (1.1.5) is also fixed

$$\varepsilon_0 = \frac{10^7}{4\pi c^2}\frac{\mathrm{C}^2}{\mathrm{newton}\cdot \mathrm{s}^2}. \tag{1.1.7}$$

c is the light velocity in vacuum. The reason for this funny choice of units will become clear later when we consider radiation (Sect.3.3).

The system of quantities and units just introduced is the official international SI system. Its fundamental units are the metre (m), the kilogram (kg), the second (s) and the ampere (A) (besides the kelvin, mole and candela which are of no importance here). The metre is defined as the length travelled by light in vacuum during a time interval of 1/299 792 458 of a second. This fixes the light velocity

$$c = 299\,792\,458\,\mathrm{m/s}. \tag{1.1.8}$$

The second is the duration of 9 192 631 770 periods of the radiation emitted in the transition between the two hyperfine levels of the ground state of caesium 133. The kilogram is the mass of the prototype at Paris. Ampère is the unit of current corresponding to the unit of charge 1 C = 1 A·sec. Its precise experimental definition is the following: The current 1 A is flowing in two straight parallel wires of infinite length and negligible circular cross-section, placed 1 m apart in vacuum, if the force $2\cdot10^{-7}$ newton per m of length is produced between them. We shall discuss this definition in detail at the end of Sect.3.1 (3.1.41).

Other systems of quantities are obtained by giving the constant ε_0 (1.1.7) another value. In atomic physics one often uses the Gauss' or electrostatic system with

$$4\pi\varepsilon_0 = 1.$$

It has the advantage that one usually has to write less factors, for example Coulomb's law reads

$$|\boldsymbol{K}| = \frac{q^s q_1^s}{r^2}. \tag{1.1.9}$$

Then the unit of charge is solely expressed by mechanical units: 1esu = $1\mathrm{g}^{1/2}\mathrm{cm}^{3/2}\mathrm{sec}^{-1}$ and similarly for all other electric units. However, the factor 4π in Coulomb's law has an obvious mathematical reason (Theorem 1 (0.2.3)) which should not be obscured by the choice of units. In fact, in Gauss' system 4π appears in Maxwell's equations instead, which is the much worse place for it. One could avoid this by choosing $\varepsilon_0 = 1$, which is the Heaviside - Lorentz system. We shall sometimes use it for simplification of the notation. But otherwise we shall find ε_0 very practical when we go over to electrodynamics in matter in Chap.4. For these reasons we will use the SI system. Due to (1.1.1) the unit of the electric field strength is 1 newton/C = 1 Volt/m, with the voltage unit that will appear at the end of this section.

Coulomb's law can immediately be generalized to several point charges q_j at places $\boldsymbol{y}_j$. The force of these charges on a test point charge q at $\boldsymbol{x}$ is additive

$$\boldsymbol{K}(\boldsymbol{x}) = \frac{q}{4\pi\varepsilon_0}\sum_j \frac{q_j}{|\boldsymbol{x}-\boldsymbol{y}_j|^3}(\boldsymbol{x}-\boldsymbol{y}_j). \tag{1.1.10}$$

From (1.1.1) we get the electric field generated by those point charges

$$\boldsymbol{E}(\boldsymbol{x}) = \frac{1}{4\pi\varepsilon_0}\sum_j \frac{q_j}{|\boldsymbol{x}-\boldsymbol{y}_j|^3}(\boldsymbol{x}-\boldsymbol{y}_j). \tag{1.1.11}$$

We shall often make the transition from discrete point charges q_j to a continuous charge distribution $\varrho(\boldsymbol{x})$. This is achieved by the substitutions

$$q_j \to \varrho(\boldsymbol{y})\, d^3y, \quad \sum_j \to \int d^3y. \tag{1.1.12}$$

Then $\varrho(\boldsymbol{y})$ is the charge density, i.e. the charge per unit volume. The electric field (1.1.11) of a continuous charge distribution is now given by

$$\boldsymbol{E}(\boldsymbol{x}) = \frac{1}{4\pi\varepsilon_0}\int \varrho(\boldsymbol{y})\frac{\boldsymbol{x}-\boldsymbol{y}}{|\boldsymbol{x}-\boldsymbol{y}|^3}\, d^3y. \tag{1.1.13}$$

Similarly, from (1.1.1) we obtain the force density $\boldsymbol{k}(\boldsymbol{x})$ on a continuous charge distribution

$$\boldsymbol{k}(\boldsymbol{x}) = \varrho(\boldsymbol{x})\boldsymbol{E}(\boldsymbol{x}). \tag{1.1.14}$$

The equations (1.1.11) and (1.1.13) are typical action-at-a-distance laws: a portion of charge at place $\boldsymbol{y}$ generates an electric field at another place $\boldsymbol{x}$. In the spirit of field theory we want to replace such laws by local field equations: we look for partial differential equations for $\boldsymbol{E}(\boldsymbol{x})$. Remembering Theorem 4 (0.2.25) of the last chapter, we immediately see that $\boldsymbol{E}(\boldsymbol{x})$ (1.1.13) is vortex-free

$$\operatorname{curl} \boldsymbol{E}(\boldsymbol{x}) = \boldsymbol{0}, \tag{1.1.15}$$

and that its source density is given by

$$\operatorname{div} \boldsymbol{E}(\boldsymbol{x}) = \frac{\varrho}{\varepsilon_0}. \tag{1.1.16}$$

The latter equation is **Gauss' law**. According to Theorem 4, the two equations (1.1.15-16) are equivalent to (1.1.13), these are the **fundamental field equations of electrostatics**.

Let us apply Gauss' theorem to (1.1.16): For any three-dimensional region K_3 we have

$$\int\limits_{\partial K_3} \boldsymbol{E} \cdot d\boldsymbol{\sigma} = \frac{1}{\varepsilon_0} \int\limits_{K_3} \varrho(\boldsymbol{x})\, d^3x. \tag{1.1.17}$$

The left side is the electric flux through the closed surface ∂K_3, the right side is the total charge inside ∂K_3 (apart from ε_0). This is the **theorem of flux, the integral form of Gauss' law** (1.1.16). It states that the total charge in a region can be determined by only measuring the field strength at the boundaries.

From (1.1.15) and Corollary 5 (0.1.45) we know that $\boldsymbol{E}(\boldsymbol{x})$ is a potential field

$$\boldsymbol{E}(\boldsymbol{x}) = -\operatorname{grad} V(\boldsymbol{x}). \tag{1.1.18}$$

The minus sign is the usual convention. Using Theorem 4 again (0.2.29), the potential can be read off immediately from (1.1.13)

$$V(\boldsymbol{x}) = \frac{1}{4\pi\varepsilon_0} \int d^3x\, \frac{\varrho(\boldsymbol{y})}{|\boldsymbol{x} - \boldsymbol{y}|}. \tag{1.1.19}$$

By Theorem 1 (0.2.3) it is the solution of the **potential equation**

$$\Delta V(\boldsymbol{x}) = -\frac{\varrho(\boldsymbol{x})}{\varepsilon_0}. \tag{1.1.20}$$

These relations can easily be specialized to discrete point charges by taking the charge density

$$\varrho(\boldsymbol{y}) = \sum_j q_j \delta(\boldsymbol{y} - \boldsymbol{y}_j). \tag{1.1.21}$$

Then (1.1.19) gives

$$V(\boldsymbol{x}) = \frac{1}{4\pi\varepsilon_0} \sum_j \frac{q_j}{|\boldsymbol{x} - \boldsymbol{y}_j|} \tag{1.1.22}$$

in agreement with (1.1.11).

For an arbitrary electric field $\boldsymbol{E}(\boldsymbol{x})$ the potential $V(\boldsymbol{x})$ is determined by Corollary 5 (0.1.44)

$$\int_{0}^{\boldsymbol{x}} \boldsymbol{E}\cdot d\boldsymbol{s} = -\int_{0}^{\boldsymbol{x}} \operatorname{grad} V(\boldsymbol{x})\, d\boldsymbol{s} = V(\boldsymbol{0}) - V(\boldsymbol{x}). \tag{1.1.23}$$

The potential difference on the right side is called the voltage. This is an important measurable quantity. Its unit is

$$1\frac{\text{newton}\cdot\text{m}}{\text{C}} = 1\text{Volt}. \tag{1.1.24}$$

Using 1 C = 1 A s, this implies 1 N m = 1 V A s = 1 Watt s, which is the unit of mechanical work or energy. Hence, in the SI system electrical energy is very simply related to mechanical energy.

1.2 Multipole Expansion

We now want to discuss the electric field and the potential of a general charge distribution $\varrho(\boldsymbol{x})$. It is given by (1.1.19)

$$V(\boldsymbol{x}) = \frac{1}{4\pi\varepsilon_0}\int d^3y\, \frac{\varrho(\boldsymbol{y})}{|\boldsymbol{x}-\boldsymbol{y}|}. \tag{1.2.1}$$

We assume that the charge distribution is concentrated in a compact region. Expanding the denominator in (1.2.1) for large $|\boldsymbol{x}| = r$ by means of the binomial series we shall obtain

$$\frac{1}{|\boldsymbol{x}-\boldsymbol{y}|} = \frac{1}{r}\Big(1 - 2\frac{\boldsymbol{x}\cdot\boldsymbol{y}}{r^2} + \frac{y^2}{r^2}\Big)^{-\frac{1}{2}}$$

$$= \frac{1}{r}\Big[1 - \frac{1}{2}\Big(-2\frac{\boldsymbol{x}\cdot\boldsymbol{y}}{r^2} + \frac{y^2}{r^2}\Big) + \frac{3}{8}\Big(-2\frac{\boldsymbol{x}\cdot\boldsymbol{y}}{r^2} + \frac{y^2}{r^2}\Big)^2 + \dots\Big]$$

$$= \frac{1}{r} + \frac{\boldsymbol{x}\cdot\boldsymbol{y}}{r^3} + \frac{1}{2}\frac{3(\boldsymbol{x}\cdot\boldsymbol{y})^2 - y^2r^2}{r^5} + \mathrm{O}\Big(\frac{y^3}{r^4}\Big). \tag{1.2.2}$$

The series is convergent for $r > y$ (see (1.2.17)). Substituting this into (1.2.1) we get the multipole expansion

$$V(\boldsymbol{x}) = \frac{1}{4\pi\varepsilon_0}\Big[\frac{1}{r}\int d^3y\, \varrho(\boldsymbol{y}) + \frac{\boldsymbol{x}}{r^3}\cdot\int d^3y\, \boldsymbol{y}\varrho(\boldsymbol{y}) +$$

$$+ \frac{1}{2}\frac{x_jx_k}{r^5}\int d^3y\, (3y_jy_k - y^2\delta_{jk})\varrho(\boldsymbol{y}) + \mathrm{O}\Big(\frac{y^3}{r^4}\Big)\Big]. \tag{1.2.3}$$

The integral in the first term is the total charge Q, the second integral is the **dipole moment** $\boldsymbol{P}$, in the third term appears the **quadrupole moment**

Q_{jk} etc. The multipoles so defined depend on the choice of the origin of the coordinate system. The l th multipole has an x-dependence that decreases $\sim 1/r^{l+1}$. The order l has a geometric meaning which will be discussed below. If the charge vanishes $Q = 0$, we have $V \sim 1/r^2$ and the surface integrals

$$\int\limits_{\partial K_3} \frac{\partial V}{\partial x_j}\, d\sigma \to 0 \tag{1.2.4}$$

vanish for $r \to \infty$. The multipole expansion (1.2.3) can also be written in terms of distributive densities as follows

$$V(\boldsymbol{x}) = \frac{1}{4\pi\varepsilon_0}\int \frac{d^3y}{|\boldsymbol{x}-\boldsymbol{y}|}\Big[Q\delta(\boldsymbol{y}) + \operatorname{div}_{y}\boldsymbol{P}\delta(\boldsymbol{y}) + \ldots\Big]. \tag{1.2.5}$$

The divergence in the second term operating on $\delta(\boldsymbol{y})$, by definition, has to be shifted to $1/|\boldsymbol{x}-\boldsymbol{y}|$ by formal partial integration.

By computing the gradient of (1.2.3) we get the electric field:

$$\boldsymbol{E}(\boldsymbol{x}) = \frac{1}{4\pi\varepsilon_0}\Big[\frac{\boldsymbol{x}}{r^3}Q - \operatorname{grad}\frac{\boldsymbol{x}\cdot\boldsymbol{P}}{r^3} + \ldots\Big]. \tag{1.2.6}$$

Using (0.1.22-25) we shall obtain:

$$\begin{aligned}\operatorname{grad}\Big(\boldsymbol{x}\cdot\frac{\boldsymbol{P}}{r^3}\Big) &= (\boldsymbol{x}\cdot\operatorname{grad})\frac{\boldsymbol{P}}{r^3} + \Big(\frac{\boldsymbol{P}}{r^3}\cdot\operatorname{grad}\Big)\boldsymbol{x} + \boldsymbol{x}\wedge\operatorname{curl}\frac{\boldsymbol{P}}{r^3}\\ &= -\Big(\boldsymbol{x}\cdot 3\frac{\boldsymbol{x}}{r^5}\Big)\boldsymbol{P} + \frac{\boldsymbol{P}}{r^3} + \boldsymbol{x}\wedge\Big(\operatorname{grad}\frac{1}{r^3}\wedge\boldsymbol{P}\Big)\\ &= -2\frac{\boldsymbol{P}}{r^3} + 3\Big(\frac{\boldsymbol{x}}{r^5}\wedge\boldsymbol{P}\Big)\wedge\boldsymbol{x} = -\frac{3\boldsymbol{x}(\boldsymbol{P}\cdot\boldsymbol{x}) - r^2\boldsymbol{P}}{r^5}.\end{aligned} \tag{1.2.7}$$

In

$$\boldsymbol{E}(\boldsymbol{x}) = \frac{1}{4\pi\varepsilon_0}\Big[\frac{\boldsymbol{x}}{r^3}Q + \frac{3\boldsymbol{x}(\boldsymbol{P}\cdot\boldsymbol{x}) - r^2\boldsymbol{P}}{r^5} + \ldots\Big] \tag{1.2.8}$$

the first term is the field of a point charge (1.1.11), whereas the second term is a **dipole field**. For later use we want to write this result in spherical coordinates

$$x_1 = r\sin\vartheta\cos\varphi, \quad x_2 = r\sin\vartheta\sin\varphi, \quad x_3 = r\cos\vartheta. \tag{1.2.9}$$

Choosing the 3-axis parallel to $\boldsymbol{P}$, we have for the radial component

$$\begin{aligned}E_r \overset{\text{def}}{=} E\cdot\frac{\boldsymbol{x}}{r} &= \frac{1}{4\pi\varepsilon_0}\Big[\frac{Q}{r^2} + \frac{3r^2\cos\vartheta - r^2\cos\vartheta}{r^5}P + \ldots\Big]\\ &= \frac{1}{4\pi\varepsilon_0}\Big[\frac{Q}{r^2} + \frac{2P\cos\vartheta}{r^3}\Big]\end{aligned} \tag{1.2.10}$$

and for the longitudinal and azimuthal components

$$E_\vartheta \stackrel{\text{def}}{=} (E_1 \cos\varphi + E_2 \sin\varphi)\cos\vartheta - E_3 \sin\vartheta \tag{1.2.11}$$

$$= \frac{1}{4\pi\varepsilon_0}\frac{P\sin\vartheta}{r^3} \tag{1.2.12}$$

$$E_\varphi \stackrel{\text{def}}{=} -E_1 \sin\varphi + E_2\cos\varphi = 0. \tag{1.2.13}$$

The higher multipoles are best studied in spherical coordinates

$$\boldsymbol{x} = (r, \vartheta, \varphi), \quad \boldsymbol{y} = (r', \vartheta', \varphi'). \tag{1.2.14}$$

Using

$$\cos\alpha = \frac{\boldsymbol{x}\cdot\boldsymbol{y}}{rr'} = \cos\vartheta\cos\vartheta' + \sin\vartheta\sin\vartheta'\cos(\varphi - \varphi'), \tag{1.2.15}$$

we expand

$$\frac{1}{|\boldsymbol{x}-\boldsymbol{y}|} = (r^2 + r'^2 - 2rr'\cos\alpha)^{-\frac{1}{2}} = \frac{1}{r}\left(1 - 2\frac{r'}{r}\cos\alpha + \frac{r'^2}{r^2}\right)^{-\frac{1}{2}} \tag{1.2.16}$$

for $r' < r$

$$= \frac{1}{r}\sum_{l=0}^{\infty}\left(\frac{r'}{r}\right)^l P_l(\cos\alpha). \tag{1.2.17}$$

The P_l are polynomials of degree l, the so-called **Legendre polynomials**. The defining equation (1.2.17) implies the following explicit representation (Problem 4)

$$P_l(z) = \frac{1}{2^l l!}\frac{d^l}{dz^l}(z^2 - 1)^l. \tag{1.2.18}$$

In addition, the so-called associated Legendre functions play an important rôle

$$P_l^m(z) = \frac{1}{2^l l!}(1 - z^2)^{-\frac{m}{2}}\left(\frac{d}{dz}\right)^{l-m}(z^2-1)^l, \tag{1.2.19}$$

where $m = -l, -l+1, \ldots l$, and $P_l^0(z) = P_l(z)$. They are connected with the **spherical harmonics**

$$Y_l^m(\vartheta,\varphi) = (-)^m\sqrt{\frac{2l+1}{4\pi}\frac{(l-m)!}{(l+m)!}}\,P_l^m(\cos\vartheta)e^{im\varphi}. \tag{1.2.20}$$

We will study them in some detail in the next section where we prove the following addition theorem (1.3.63)

$$\sum_{m=-l}^{l} Y_l^m(\vartheta',\varphi')^* Y_l^m(\vartheta,\varphi) = \frac{2l+1}{4\pi}P_l(\cos\alpha). \tag{1.2.21}$$

The importance of this equation lies in the fact that on the left side the coordinates of $\boldsymbol{x}$ and $\boldsymbol{y}$ are separated in products of two spherical harmonics.

Now, substituting (1.2.21) into (1.2.17), we can write the potential (1.2.1) in the following form

$$V(\boldsymbol{x}) = \frac{1}{4\pi\varepsilon_0} 4\pi \sum_{l=0}^{\infty} \sum_{m=-l}^{l} \frac{Y_l^m(\vartheta,\varphi)}{(2l+1)r^{l+1}} \int d^3y \, Y_l^m(\vartheta',\varphi')^* \varrho(\boldsymbol{y}) r'^l \tag{1.2.22}$$

$$\stackrel{\text{def}}{=} \frac{1}{\varepsilon_0} \sum_{l,m} \frac{q_l^m}{2l+1} \frac{Y_l^m(\vartheta,\varphi)}{r^{l+1}}. \tag{1.2.23}$$

Here the integral in (1.2.22) defines the **spherical multipole moments** q_l^m. The order l of the multipole agrees with the order of the corresponding spherical harmonics (which generate a representation D^l of the rotation group). This is the geometric meaning of l. It follows from (1.2.20) that for an axial symmetrical charge distribution only multipoles with $m = 0$ occur. The multipole expansion expresses the potential as a power series in r^{-1}. At great distances from the charges, only the lowest multipoles are important.

1.3 Boundary Value Problems and Eigenfunction Expansion

In this section we calculate the electric field in a certain region of space, provided the potential at the boundary is given. Let K be a compact region with a smooth boundary ∂K and assume that the potential $V(\boldsymbol{x})$ on ∂K is known. We seek the potential for all $\boldsymbol{x}$ in K. Supposing that there are no charges inside K, V satisfies the homogeneous Laplace equation

$$\triangle V(\boldsymbol{x}) = 0. \tag{1.3.1}$$

This is the **inner Dirichlet problem for the Laplace equation**. The solution is unique: If V' is a second solution, then $V - V'$ is harmonic in K and $= 0$ on ∂K; by Corollary 7 it must vanish.

We solve the problem in spherical coordinates (r, ϑ, φ) (1.2.9). This is very appropriate for spherical geometry, but the method is general and applies to other geometric situations as well. We have to express the Laplace operator

$$\triangle = -\boldsymbol{p}^2, \tag{1.3.2}$$

$$\boldsymbol{p} \stackrel{\text{def}}{=} \frac{1}{i}\text{grad} = \frac{1}{i}\frac{\partial}{\partial \boldsymbol{x}} \tag{1.3.3}$$

in spherical coordinates. This can be done by a tedious calculation with the chain rule. We prefer a more physical reasoning that has a quantum mechanical background, because $\boldsymbol{p}$ (1.3.3) is the quantum mechanical momentum operator. i is the complex unit. We introduce the radial unit vector

$$e = \frac{x}{r} \tag{1.3.4}$$

and decompose p into a radial part and the rest orthogonal to e

$$p = e(e \cdot p) - e \wedge (e \wedge p). \tag{1.3.5}$$

This elementary algebraic identity remains also valid with the gradient operator p, because it stands at the right end, so that the differentiation does not operate on e. The radial part p_r can be written as follows

$$p_r = \frac{e}{i}\Big(\frac{x_j}{r}\frac{\partial}{\partial x_j}\Big) = \frac{e}{i}\Big(\frac{\partial x_j}{\partial r}\frac{\partial}{\partial x_j}\Big) = \frac{1}{i}\frac{x}{r}\frac{\partial}{\partial r}. \tag{1.3.6}$$

The second term in (1.3.5) contains

$$e \wedge p = \frac{1}{r} x \wedge p \overset{\text{def}}{=} \frac{1}{r} L, \tag{1.3.7}$$

which is the angular momentum operator in quantum mechanics

$$L = \frac{1}{i}(x_2\partial_3 - x_3\partial_2,\ x_3\partial_1 - x_1\partial_3,\ x_1\partial_2 - x_2\partial_1). \tag{1.3.8}$$

This operator must only depend on the angles. Indeed, using spherical coordinates (1.2.9) we get

$$L = \frac{1}{i}(-\sin\varphi\partial_\vartheta - \cos\varphi\cot\vartheta\partial_\varphi,\ \cos\varphi\partial_\vartheta - \sin\varphi\cot\vartheta\partial_\varphi,\ \partial_\varphi). \tag{1.3.9}$$

For p (1.3.5) we now have

$$p = \frac{1}{i}\frac{x}{r}\partial_r - \frac{x}{r^2} \wedge L, \tag{1.3.10}$$

and, operating once more with p (1.3.3) from the left, we arrive at

$$\begin{aligned} p^2 &= -\Big(\frac{3}{r}\partial_r - \frac{r^2}{r^3}\partial_r + \frac{x}{r}\cdot\frac{\partial}{\partial x}\frac{\partial}{\partial r}\Big) - p \cdot \Big(\frac{x}{r^2} \wedge L\Big) \\ &= -\Big(\frac{2}{r}\partial_r + \partial_r^2\Big) - \Big(p \wedge \frac{x}{r^2}\Big) \cdot L. \end{aligned} \tag{1.3.11}$$

In the vector product

$$p \wedge \frac{x}{r^2} = (p \wedge x)\frac{1}{r^2} = -(x \wedge p)\frac{1}{r^2} = -L\frac{1}{r^2} = -\frac{1}{r^2}L \tag{1.3.12}$$

we may commute the factors because the derivatives always act on coordinates different from the second factor. This leads to the desired final result for the **Laplace operator in spherical coordinates**

$$\boldsymbol{p}^2 = -\left(\partial_r^2 + \frac{2}{r}\partial_r\right) + \frac{\boldsymbol{L}^2}{r^2}$$

$$= -\frac{1}{r}\frac{\partial^2}{\partial r^2} r \cdot + \frac{\boldsymbol{L}^2}{r^2} = -\triangle\,. \tag{1.3.13}$$

The dot always indicates the place of the function, the operator is operating on. The square of the angular momentum is easily calculated from (1.3.9)

$$\boldsymbol{L}^2 = -\Big[\frac{1}{\sin\vartheta}\partial_\vartheta(\sin\vartheta\partial_\vartheta\cdot) + \frac{1}{\sin^2\vartheta}\partial_\varphi^2\Big]. \tag{1.3.14}$$

The Laplace equation for $V = V(r, \vartheta, \varphi)$ now reads

$$\triangle V = \frac{1}{r}\frac{\partial^2}{\partial r^2}(rV) - \frac{1}{r^2}\boldsymbol{L}^2 V = 0. \tag{1.3.15}$$

Since $\boldsymbol{L}^2$ only operates on the angles, the variables can be separated by a product ansatz

$$V = u(r)Y(\vartheta, \varphi), \tag{1.3.16}$$

$$\frac{1}{r}\left(\frac{\partial^2}{\partial r^2} ru\right)Y - \frac{u}{r^2}\boldsymbol{L}^2 Y = 0: \tag{1.3.17}$$

$$\frac{r}{u}\frac{\partial^2}{\partial r^2} ru(r) = \frac{1}{Y}\boldsymbol{L}^2 Y = \lambda. \tag{1.3.18}$$

Here the l.h.s. depends only on r whereas the r.h.s. is depending on ϑ, φ only, thus λ must be a constant. This leads to two separated eigenvalue problems

$$r\frac{d^2}{dr^2} ru(r) = \lambda\, u(r) \tag{1.3.19}$$

$$\boldsymbol{L}^2 Y = \lambda Y. \tag{1.3.20}$$

The first one is immediately solved by a power law

$$u(r) = r^l, \tag{1.3.21}$$

which gives

$$\lambda = l(l+1). \tag{1.3.22}$$

Choosing $l = 0, 1, 2, \ldots$, we get a complete system of solutions, regular at $r = 0$ for the inner Dirichlet problem. For the outer problem we have to take $l = 0, -1, -2, \ldots$.

To solve the second eigenvalue problem (1.3.20)

$$\frac{1}{\sin\vartheta}\frac{\partial}{\partial\vartheta}\sin\vartheta\frac{\partial Y}{\partial\vartheta} + \frac{1}{\sin^2\vartheta}\frac{\partial^2 Y}{\partial\varphi^2} = -l(l+1)Y, \tag{1.3.23}$$

we make again a product ansatz

$$Y(\vartheta, \varphi) = P(\vartheta)Q(\varphi) \tag{1.3.24}$$

and separate the variables

$$\frac{\sin\vartheta}{P}\frac{d}{d\vartheta}\sin\vartheta\frac{dP}{d\vartheta}+l(l+1)\sin^2\vartheta=-\frac{1}{Q}\frac{d^2Q}{d\varphi^2}=\text{const.}\stackrel{\text{def}}{=}\mu^2. \tag{1.3.25}$$

Here the second eigenvalue problem can simply be solved

$$\frac{d^2Q}{d\varphi^2}=-\mu^2 Q,\quad Q(\varphi)=e^{\pm i\mu\varphi}. \tag{1.3.26}$$

We choose $\mu=0,1,2,\dots$ in order to have a single-valued function $Q(\varphi)=Q(\varphi+2\pi)$

$$Q(\varphi)=e^{im\varphi},\quad m=0,\pm1,\pm2,\dots \tag{1.3.27}$$

This is a complete basis in the Hilbert space $L^2[0,2\pi]$ as one knows from the theory of Fourier series.

There remains to solve the ordinary differential equation

$$\sin\vartheta\frac{d}{d\vartheta}\sin\vartheta\frac{dP}{d\vartheta}+[l(l+1)\sin^2\vartheta-m^2]P=0. \tag{1.3.28}$$

We now make the ansatz

$$P(\vartheta)=\sin^{-m}\vartheta\, P_{lm}(\cos\vartheta) \tag{1.3.29}$$

and write a prime for the derivative $d/d\cos\vartheta$:

$$\sin^{4-m}\vartheta\, P''_{lm}+2(m-1)\sin^{2-m}\vartheta\cos\vartheta\, P'_{lm}+[l(l+1)-m^2+m]\sin^{2-m}\vartheta\, P_{lm}=0. \tag{1.3.30}$$

Introducing $\cos\vartheta=z$, this is equivalent to

$$(1-z^2)P''_{lm}+2(m-1)zP'_{lm}+[l(l+1)-m^2+m]P_{lm}=0. \tag{1.3.31}$$

Let us first consider the special case $m=l$:

$$(1-z^2)P''_{ll}+2(l-1)zP'_{ll}+2lP_{ll}=0, \tag{1.3.32}$$

with the solution

$$P_{ll}(z)=(1-z^2)^l. \tag{1.3.33}$$

Next we differentiate (1.3.31) once more with respect to z:

$$(1-z^2)P'''_{lm}+2(m-2)zP''_{lm}+[l(l+1)-(m-1)^2+m-1]P'_{lm}=0. \tag{1.3.34}$$

This equation is identical with (1.3.31) with m replaced by $m-1$, hence

$$P_{lm-1}(z)=\frac{d\,P_{lm}(z)}{d\,z},\quad\text{and} \tag{1.3.35}$$

$$P_{lm}(z)=\left(\frac{d}{dz}\right)^{l-m}(1-z^2)^l \tag{1.3.36}$$

due to (1.3.33). Since this vanishes for $m<-l$, we must have

$$m=-l,-l+1,\dots+l. \tag{1.3.37}$$

For $l = 0, 1, 2, \ldots$ all powers of z are obtained which shows the completeness of the solutions. For $m = 0$ we have the Legendre polynomials (1.2.18)

$$P_{l0}(z) = \frac{d^l}{dz^l}(1-z^2)^l = \text{const} \cdot P_l(z). \tag{1.3.38}$$

Furthermore, introducing the associated Legendre functions (1.2.19)

$$\begin{aligned} P_l^m(\cos\vartheta) &\stackrel{\text{def}}{=} \text{const} \cdot \sin^{-m}\vartheta P_{lm}(\cos\vartheta) \\ &= \text{const} \cdot (1-z^2)^{-m/2}\left(\frac{d}{dz}\right)^{l-m}(1-z^2)^l, \end{aligned} \tag{1.3.39}$$

the eigenfunctions (1.3.24) are given by the **spherical harmonics**

$$Y_l^m(\vartheta, \varphi) = \text{const} \cdot e^{im\varphi} P_l^m(\cos\vartheta). \tag{1.3.40}$$

The normalization factor will be fixed below. According to (1.3.16), the total solution of the Laplace equation is equal to

$$V_l^m(r, \vartheta, \varphi) = \text{const} \cdot r^l Y_l^m(\vartheta, \varphi). \tag{1.3.41}$$

We will discuss some important properties of the spherical harmonics, first their orthogonality. Since V_l^m are solutions of the Laplace equation, Green's theorem implies

$$\int\limits_{K_3} (V_l^m \bigtriangleup V_{l'}^{m'} - V_{l'}^{m'} \bigtriangleup V_l^m) d^3x = \int\limits_{\partial K_3} \left(V_l^m \frac{\partial V_{l'}^{m'}}{\partial n} - V_{l'}^{m'} \frac{\partial V_l^m}{\partial n}\right) d\sigma = 0. \tag{1.3.42}$$

Taking for K_3 the full unit sphere, we have

$$\frac{\partial V_l^m}{\partial n} = \frac{\partial V_l^m}{\partial r} = \frac{l}{r} V_l^m. \tag{1.3.43}$$

For $r = 1$ we find from (1.3.42)

$$\int (Y_l^m l' Y_{l'}^{m'} - Y_{l'}^{m'} l Y_l^m) d\Omega = 0, \tag{1.3.44}$$

where $d\Omega = d\cos\vartheta d\varphi$ is the surface element on the unit sphere S_2. Thus

$$\int Y_l^m Y_{l'}^{m'} d\Omega = 0, \quad l \neq l', \tag{1.3.45}$$

and this is also true for $l = l'$, $m \neq m'$, because then the integral over φ gives 0. Since P_l^m is real, we have

$$Y_l^{m\,*} = Y_l^{-m}, \tag{1.3.46}$$

and, therefore, we get the orthogonality

$$\int Y_l^{m\,*} Y_{l'}^{m'}\, d\Omega = 0, \quad l \neq l', \quad \text{or} \quad m \neq m'. \tag{1.3.47}$$

The constant in (1.3.40) is fixed by the normalization condition:

$$\int |Y_l^m|^2 d\Omega = 1, \tag{1.3.48}$$

which gives the factor in (1.2.20). The sign is conventional. We note the special values

$$Y_l^m(0,0) = 0, \quad \text{if} \quad m \neq 0 \tag{1.3.49}$$

$$Y_l^0(0,0) = \sqrt{\frac{2l+1}{4\pi}}. \tag{1.3.50}$$

The **spherical harmonics form a complete orthonormal system on the unit sphere** S_2. Consequently, any quadratically integrable function $f(\vartheta, \varphi) \in L^2(S_2)$ can be expanded as follows

$$f(\vartheta, \varphi) = \sum_{l=0}^{\infty} \sum_{m=-l}^{l} C_l^m Y_l^m(\vartheta, \varphi), \tag{1.3.51}$$

where the coefficients are obtained by multiplication with Y_l^m, integrating over S_2 and using orthonormality (1.3.47, 48)

$$C_l^m = \int Y_l^m(\vartheta, \varphi)^* f(\vartheta, \varphi) d\Omega. \tag{1.3.52}$$

The solutions

$$V_l^m = r^l Y_l^m, \tag{1.3.53}$$

on the other hand, are orthogonal over the full sphere of arbitrary radius, for example over K_1 with $r = 1$:

$$\int_{K_1} d^3x\, V_l^{m\,*} V_{l'}^{m'} \stackrel{\text{def}}{=} (V_l^m,\, V_{l'}^{m'})$$

$$= \int_0^1 dr\, r^2 \int d\Omega\, r^{l+l'} Y_l^{m\,*} Y_{l'}^{m'} = \frac{\delta_{ll'}\delta_{mm'}}{l+l'+3}. \tag{1.3.54}$$

Here we have introduced the notation of the L^2 scalar product. The normalized solutions

$$\Psi_l^m = \sqrt{2l+3}\, V_l^m \tag{1.3.55}$$

form a complete orthonormal system over $L^2(K_1)$

$$(\Psi_l^m,\, \Psi_{l'}^{m'}) = \delta_{ll'}\delta_{mm'}. \tag{1.3.56}$$

Consequently, any function $f(\boldsymbol{x}) \in L^2(K_1)$ can be expanded as follows

$$f(\boldsymbol{x}) = \sum_{lm} (\Psi_l^m,\, f)\Psi_l^m$$

$$= \sum_{lm} \Psi_l^m(\boldsymbol{x}) \int \Psi_l^m(\boldsymbol{y})^* f(\boldsymbol{y})\, d^3y \tag{1.3.57}$$

$$= \int d^3y\, \delta(\boldsymbol{x} - \boldsymbol{y}) f(\boldsymbol{y}). \tag{1.3.58}$$

The last distributive relation makes sense for all test functions $f \in C_0^\infty(K_1)$. It implies the so-called **completeness relation**

$$\sum_l (2l+3) r^l r'^l \sum_{m=-l}^{l} Y_l^m(\vartheta', \varphi')^* Y_l^m(\vartheta, \varphi) = \delta(\boldsymbol{x} - \boldsymbol{y}). \tag{1.3.59}$$

The sum on the left side converges only in the sense of distributions (in $S'(K_1)$).

We use (1.3.59) to prove the addition theorem. The completeness relation holds independent of the choice of the polar axis. Consequently the inner sum

$$\sum_{m=-l}^{l} Y_l^m(\vartheta', \varphi')^* Y_l^m(\vartheta, \varphi) = F(\alpha) \tag{1.3.60}$$

must depend on the angle α between $\boldsymbol{x}$ and $\boldsymbol{y}$, only, where

$$\cos\alpha = \cos\vartheta \cos\vartheta' + \sin\vartheta \sin\vartheta' \cos(\varphi - \varphi'). \tag{1.3.61}$$

To determine $F(\alpha)$ we choose $\boldsymbol{y}$ as the polar axis, then $\vartheta' = 0$ and only $m = 0$ contributes. Since $\vartheta = \alpha$, we find

$$Y_l^0(0,0)^* Y_l^0(\alpha, \varphi) = \frac{2l+1}{4\pi} P_l(\cos\alpha) = F(\alpha). \tag{1.3.62}$$

This leads to the **addition theorem**

$$\sum_{m=-l}^{l} Y_l^m(\vartheta', \varphi')^* Y_l^m(\vartheta, \varphi) = \frac{2l+1}{4\pi} P_l(\cos\alpha). \tag{1.3.63}$$

Let us now return to the **inner Dirichlet problem for the sphere**. We want to construct the solution of $\triangle V = 0$ inside the unit sphere K_1 with

$$V(\boldsymbol{x})\Big|_{r=1} = V_0(\vartheta, \varphi) \tag{1.3.64}$$

given. $V(\boldsymbol{x})$ has an eigenfunction expansion (1.3.51)

$$V(\boldsymbol{x}) = \sum_{l=0}^{\infty} r^l \sum_{m=-l}^{l} C_l^m Y_l^m(\vartheta, \varphi). \tag{1.3.65}$$

For $r = 1$ we have

$$V_0(\vartheta, \varphi) = \sum_{l,m} C_l^m Y_l^m(\vartheta, \varphi), \tag{1.3.66}$$

which allows us to determine the unknown coefficients by orthonormality (1.3.47, 48)

$$C_l^m = \int Y_l^m(\vartheta', \varphi')^* V_0(\vartheta', \varphi')\, d\Omega'. \tag{1.3.67}$$

Inserting this into (1.3.65) leads to the desired solution $V(\boldsymbol{x})$. We choose the polar axis in the direction of $\boldsymbol{x}$, then $\vartheta = \varphi = 0$ and only $m = 0$ contributes. Since (1.3.50)

$$Y_l^0(0,0) = \sqrt{\frac{2l+1}{4\pi}}, \quad Y_l^0(\vartheta, \varphi) = \sqrt{\frac{2l+1}{4\pi}} P_l(\cos\vartheta),$$

we get

$$V(r,0,0) = \sum_l r^l \frac{2l+1}{4\pi} \int_0^{2\pi} d\varphi' \int_0^{\pi} d\vartheta' \,\sin\vartheta'\, P_l(\cos\vartheta') V_0(\vartheta', \varphi'). \tag{1.3.68}$$

Here the summation can be carried out. Let us differentiate the equation (1.2.17)

$$\frac{1}{\sqrt{1 - 2rz + r^2}} = \sum_l r^l P_l(z) \tag{1.3.69}$$

with respect to r

$$-\frac{1}{2} \frac{-2z + 2r}{(1 - 2rz + r^2)^{3/2}} = \sum_{l=0}^{\infty} l r^{l-1} P_l(z) \tag{1.3.70}$$

and multiply by $2r$. Then the sum in (1.3.68) is equal to

$$\begin{aligned} \sum_{l=0}^{\infty} (2l+1) r^l P_l(z) &= \frac{2rz - 2r^2}{(1 - 2rz + r^2)^{3/2}} + \frac{1}{(1 - 2rz + r^2)^{1/2}} \\ &= \frac{1 - r^2}{(1 - 2rz + r^2)^{3/2}}, \end{aligned} \tag{1.3.71}$$

hence

$$V(r,0,0) = \frac{1 - r^2}{4\pi} \int \frac{V_0(\vartheta', \varphi')}{(1 - 2r\cos\vartheta' + r^2)^{3/2}}\, d\Omega'. \tag{1.3.72}$$

Finally, since the polar axis is arbitrary, the solution can be written in the following general form

$$V(\boldsymbol{x}) = \frac{1 - |\boldsymbol{x}|^2}{4\pi} \int_{|\boldsymbol{y}|=1} \frac{V_0(\boldsymbol{y})}{|\boldsymbol{x} - \boldsymbol{y}|^3}\, d\Omega_y. \tag{1.3.73}$$

This is **Poisson's integral formula**. It expresses the potential in the sphere by the boundary values.

1.4 Green's Functions

The method of Green's functions is the most elegant way to solve boundary value problems, supposed that one can actually calculate the Green's function. We start from the potential of a continuous charge distribution in infinite space

$$V(\boldsymbol{x}) = \frac{1}{4\pi} \int d^3y' \, \frac{\varrho(\boldsymbol{y}')}{|\boldsymbol{x} - \boldsymbol{y}'|}, \tag{1.4.1}$$

where we have set $\varepsilon_0 = 1$ (Heaviside system) to simplify the notation. For a unit point charge at $\boldsymbol{y}$ we have to substitute $\varrho(\boldsymbol{y}') \to \delta(\boldsymbol{y}' - \boldsymbol{y})$, which gives the potential

$$V(\boldsymbol{x}) = \frac{1}{4\pi|\boldsymbol{x} - \boldsymbol{y}|} \stackrel{\text{def}}{=} G(\boldsymbol{x}, \boldsymbol{y}). \tag{1.4.2}$$

This is a distributive solution of the Laplace equation (compare (0.2.11))

$$-\Delta_x \, G(\boldsymbol{x}, \boldsymbol{y}) = \delta(\boldsymbol{x} - \boldsymbol{y}), \tag{1.4.2}$$

where $\boldsymbol{y}$ is fixed. **If the inhomogeneous term in the differential equation is a δ-distribution, one** usually **calls the solution a Green's function, or a fundamental solution.**

We wish to take over this method to problems with boundaries. Let $\Omega \subset \mathbb{R}^3$ be an open region (not necessaryly compact). We look for a (distributive) solution satisfying

$$-\Delta_x \, G(\boldsymbol{x}, \boldsymbol{y}) = \delta(\boldsymbol{x} - \boldsymbol{y}), \quad \text{for} \quad \boldsymbol{x} \in \Omega \tag{1.4.4}$$

$$G(\boldsymbol{x}, \boldsymbol{y})\Big|_{\boldsymbol{x} \in \partial\Omega} = 0. \tag{1.4.5}$$

The first equation means that we have a unit point charge at $\boldsymbol{x}$ inside Ω, and the second requires that the potential on the surface $\partial\Omega$ is equal to 0; physically speaking, this is a conducting surface that is grounded. The Green's function G is the potential of this arrangement. We shall see below, that if G is known, we can solve the general boundary value problem.

Let us first try to construct Green's functions for various geometric situations. Without any boundary the point charge at $\boldsymbol{y}$ would produce the potential

$$G_0(\boldsymbol{x}, \boldsymbol{y}) = \frac{1}{4\pi|\boldsymbol{x} - \boldsymbol{y}|}. \tag{1.4.6}$$

We now try to place further point charges $e_k(\boldsymbol{y})$ outside Ω, where the Laplace equation (1.4.4) need not hold, in such a way that the boundary condition (1.4.5) is fulfilled. We make the following ansatz

$$G(\boldsymbol{x}, \boldsymbol{y}) = \frac{1}{4\pi|\boldsymbol{x} - \boldsymbol{y}|} + \sum_k \frac{e_k(\boldsymbol{y})}{4\pi|\boldsymbol{x} - \boldsymbol{y}_h^*|}, \quad \boldsymbol{y}_k^* \notin \Omega. \tag{1.4.7}$$

The point charges $e_k(\boldsymbol{y})$ are the so-called **mirror or image charges**. They must be chosen and arranged in such a way that their potential compensates that of the unit charge on the boundary $\partial\Omega$. This is not possible for regions of arbitrary shape, the region must have a certain symmetry. We want to illustrate the method in the following examples.

Example 1. Half-space $x_3 > 0$: In this simplest case one must only place a negative image charge at the mirror place $\boldsymbol{y}^* = (y_1, y_2, -y_3)$

$$G(\boldsymbol{x}, \boldsymbol{y}) = \frac{1}{4\pi|\boldsymbol{x}-\boldsymbol{y}|} - \frac{1}{4\pi|\boldsymbol{x}-\boldsymbol{y}^*|}, \tag{1.4.8}$$

in order to satisfy the boundary condition (1.4.5).

Example 2. Quarter-space $x_3 > 0$, $x_1 > 0$: Now we have three mirror points

$$\boldsymbol{y}_1^* = (-y_1, y_2, y_3), \quad \boldsymbol{y}_2^* = (-y_1, y_2, -y_3), \quad \boldsymbol{y}_3^* = (y_1, y_2, -y_3). \tag{1.4.9}$$

The mirror charges must have alternating sign

$$G(\boldsymbol{x}, \boldsymbol{y}) = \frac{1}{4\pi|\boldsymbol{x}-\boldsymbol{y}|} - \frac{1}{4\pi|\boldsymbol{x}-\boldsymbol{y}_1^*|} + \frac{1}{4\pi|\boldsymbol{x}-\boldsymbol{y}_2^*|} - \frac{1}{4\pi|\boldsymbol{x}-\boldsymbol{y}_3^*|}. \tag{1.4.10}$$

Example 3. Sphere $|\boldsymbol{x}| > R$ or $|\boldsymbol{x} < R$: In this case the mirror point is given by

$$\frac{|\boldsymbol{y}^*|}{R} = \frac{R}{|\boldsymbol{y}|}, \quad \text{or} \tag{1.4.11}$$

$$\boldsymbol{y}^* = \frac{R^2}{|\boldsymbol{y}|^2}\boldsymbol{y}. \tag{1.4.12}$$

This is the transformation by reciprocal radii or inversion at the surface of the sphere $\partial\Omega$ (Fig.1).

It follows from (1.4.11) that the triangles $(\boldsymbol{0}, \boldsymbol{y}, \boldsymbol{x})$ and $(\boldsymbol{0}, \boldsymbol{x}, \boldsymbol{y}^*)$ are similar. Thus

$$\frac{|\boldsymbol{x}-\boldsymbol{y}^*|}{|\boldsymbol{x}-\boldsymbol{y}|} = \frac{R}{|\boldsymbol{y}|}, \quad \text{and}$$

$$\frac{1}{4\pi|\boldsymbol{x}-\boldsymbol{y}|} - \frac{1}{4\pi|\boldsymbol{x}-\boldsymbol{y}^*|}\frac{R}{|\boldsymbol{y}|} = 0. \tag{1.4.13}$$

Remarkably enough, the last factor in the second term is independent of $\boldsymbol{x} \in \partial\Omega$. Hence, it can be taken as the magnitude $e(\boldsymbol{y})$ of the image charge and the method works:

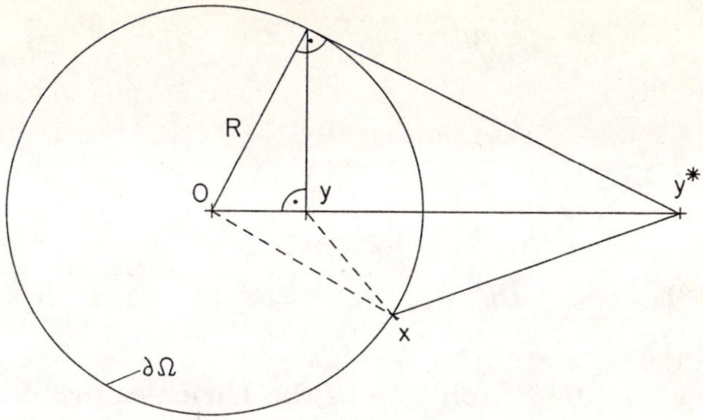

Fig. 1. Transformation by reciprocal radii

$$G(\boldsymbol{x},\boldsymbol{y}) = \frac{1}{4\pi|\boldsymbol{x}-\boldsymbol{y}|} - \frac{1}{4\pi|\boldsymbol{x}-\boldsymbol{y}^*|}\frac{R}{|\boldsymbol{y}|}$$

$$= \frac{1}{4\pi|\boldsymbol{x}-\boldsymbol{y}|} - \frac{R|\boldsymbol{y}|}{4\pi\big|\boldsymbol{x}|\boldsymbol{y}|^2 - \boldsymbol{y}R^2\big|}, \tag{1.4.14}$$

where (1.4.12) has been used. Since the inversion is an involution $(\boldsymbol{y}^*)^* = \boldsymbol{y}$, the same Green's function (1.4.14) can be used for the outer problem.

Next we study properties of the Green's functions. It follows from (1.4.7) (and it is generally true), that

$$G(\boldsymbol{x},\boldsymbol{y}) = \frac{1}{4\pi|\boldsymbol{x}-\boldsymbol{y}|} + h(\boldsymbol{x},\boldsymbol{y}), \tag{1.4.15}$$

where $h(\boldsymbol{x},\boldsymbol{y})$ is a harmonic function in $\boldsymbol{x} \in \Omega$. Note that the image charges are outside Ω. This Eq.(1.4.15) implies uniqueness of the Green's function: h is a harmonic function in $\boldsymbol{x} \in \Omega$ with boundary condition

$$h(\boldsymbol{x},\boldsymbol{y})\Big|_{\boldsymbol{x}\in\partial\Omega} = -\frac{1}{4\pi|\boldsymbol{x}-\boldsymbol{y}|}. \tag{1.4.16}$$

By Corollary 7, h is unique (this corollary has been proven only for compact regions, but it holds also in the non-compact situation, assuming appropriate decrease at infinity). Finally, the Green's function is symmetric

$$G(\boldsymbol{x},\boldsymbol{y}) = G(\boldsymbol{y},\boldsymbol{x}). \tag{1.4.17}$$

We do not give a general proof of this property, but check it for the sphere: Since

$$\left|\boldsymbol{x}y - \frac{R^2}{y}\boldsymbol{y}\right|^2 = x^2y^2 - 2R^2(\boldsymbol{x}\cdot\boldsymbol{y}) + R^4$$
$$= \left|\frac{R^2}{x}\boldsymbol{x} - \boldsymbol{y}x\right|^2,$$

we have

$$\frac{y}{|\boldsymbol{x}y^2 - \boldsymbol{y}R^2|} = \frac{1}{|\boldsymbol{x}y - \boldsymbol{y}R^2/y|} = \frac{1}{|\boldsymbol{y}x - \boldsymbol{x}R^2/x|} = \frac{x}{|\boldsymbol{y}x^2 - \boldsymbol{x}R^2|}. \tag{1.4.18}$$

This implies the symmetry (1.4.17) of (1.4.14).

Let us now turn to the **solution of the Dirichlet problem**. We want to determine the solution of

$$-\triangle V(\boldsymbol{x}) = \varrho(\boldsymbol{x}), \quad \boldsymbol{x} \in \Omega \tag{1.4.19}$$

with prescribed boundary values

$$V(\boldsymbol{x})\Big|_{\boldsymbol{x}\in\partial\Omega} = V_0(\boldsymbol{x}). \tag{1.4.20}$$

For $u \in C^2(\Omega)$ we write down Green's theorem

$$\int_\Omega (V(\boldsymbol{y}) \triangle u(\boldsymbol{y}) - u(\boldsymbol{y}) \triangle V(\boldsymbol{y}))d^3y = \int_{\partial\Omega} \Big(V\frac{\partial u}{\partial n_y} - u\frac{\partial V}{\partial n_y}\Big) d\sigma_y. \tag{1.4.21}$$

First we choose

$$u(\boldsymbol{y}) = \frac{1}{4\pi|\boldsymbol{x}-\boldsymbol{y}|}, \quad \boldsymbol{x} \in \Omega, \quad \text{fixed.} \tag{1.4.22}$$

We again exclude the point $\boldsymbol{x}$ by a small sphere which finally goes to 0, as in the proof of Theorem 1 (0.2.7) (or we use the distributive relation (0.2.11)). Then we get

$$-V(\boldsymbol{x}) + \int_\Omega \frac{\varrho(\boldsymbol{y})}{4\pi|\boldsymbol{x}-\boldsymbol{y}|}\, d^3y = \frac{1}{4\pi}\int_{\partial\Omega} \Big(V_0(\boldsymbol{y})\frac{\partial}{\partial n_y}\frac{1}{|\boldsymbol{x}-\boldsymbol{y}|}$$
$$-\frac{1}{|\boldsymbol{x}-\boldsymbol{y}|}\frac{\partial V}{\partial n_y}\Big) d\sigma_y. \tag{1.4.23}$$

Let us now choose $u(\boldsymbol{y}) = h(\boldsymbol{x},\boldsymbol{y})$ in (1.4.21), where h is the harmonic function occurring in (1.4.15):

$$\int_\Omega h(\boldsymbol{x},\boldsymbol{y})\varrho(\boldsymbol{y})\, d^3y = \int_{\partial\Omega} \Big(V_0(\boldsymbol{y})\frac{\partial h(\boldsymbol{x},\boldsymbol{y})}{\partial n_y} - h(\boldsymbol{x},\boldsymbol{y})\frac{\partial V}{\partial n_y}\Big) d\sigma_y. \tag{1.4.24}$$

Adding this to the foregoing equation we arrive at

$$-V(\boldsymbol{x}) + \int\limits_{\Omega} G(\boldsymbol{x},\boldsymbol{y})\varrho(\boldsymbol{y})\, d^3y = \int\limits_{\partial\Omega} \Big(V_0(\boldsymbol{y}) \frac{\partial}{\partial n_y} G(\boldsymbol{x},\boldsymbol{y})$$

$$-G(\boldsymbol{x},\boldsymbol{y}) \frac{\partial V}{\partial n_y} \Big) d\sigma_y. \qquad (1.4.25)$$

Since G vanishes on $\partial\Omega$, the unknown derivative of V in the last term disappears. The solution of the Dirichlet problem is thus given by

$$V(\boldsymbol{x}) = \int\limits_{\Omega} G(\boldsymbol{x},\boldsymbol{y})\varrho(\boldsymbol{y})\, d^3y - \int\limits_{\partial\Omega} V_0(\boldsymbol{y}) \frac{\partial}{\partial n_y} G(\boldsymbol{x},\boldsymbol{y})\, d\sigma_y. \qquad (1.4.26)$$

The first (volume) integral represents the potential of the charges inside Ω. The second (surface) integral gives the contribution of the charges on the surface $\partial\Omega$, which are described by the boundary values $V_0(\boldsymbol{y})$ of the potential. The latter can be easily measured, whereas it is difficult to determine the charge density on the surface directly.

As an example we consider the **inner Dirichlet problem for the sphere** again. We have to calculate

$$\frac{\partial G(\boldsymbol{x},\boldsymbol{y})}{\partial n_y}\Big|_{\partial\Omega} = \frac{\partial}{\partial y}\Big(\frac{1}{4\pi|\boldsymbol{x}-\boldsymbol{y}|} - \frac{Ry}{4\pi|\boldsymbol{x}y^2 - \boldsymbol{y}R^2|} \Big)\Big|_{y=R}$$

$$= \frac{1}{4\pi}\frac{\partial}{\partial y}\Big(\frac{1}{\sqrt{x^2 - 2xy\cos\vartheta + y^2}} - \frac{R}{\sqrt{x^2y^2 - 2xyR^2\cos\vartheta + R^4}} \Big)\Big|_{y=R}$$

$$= \frac{1}{4\pi}\Big(\frac{x\cos\vartheta - y}{|\boldsymbol{x}-\boldsymbol{y}|^3} + R\frac{x^2y - xR^2\cos\vartheta}{R^3\sqrt{x^2 - 2xy\cos\vartheta + y^2}^{\,3}} \Big)\Big|_{y=R} = \frac{x^2 - R^2}{4\pi R|\boldsymbol{x}-\boldsymbol{y}|^3}. \qquad (1.4.27)$$

Thus

$$V(\boldsymbol{x}) = \int G(\boldsymbol{x},\boldsymbol{y})\varrho(\boldsymbol{y})\, d^3y + \frac{R^2 - x^2}{4\pi R} \int\limits_{|y|=R} \frac{V_0(\boldsymbol{y})}{|\boldsymbol{x}-\boldsymbol{y}|^3}\, d\Omega_y, \qquad (1.4.28)$$

for $|\boldsymbol{x}| \le R$. The first term is the potential generated by the charge distribution ϱ and the second term is Poisson's integral (1.3.73).

To get the solution of the outer problem, we must only substitute $\partial/\partial n_y$ by $-\partial/\partial n_y$ which leads to a different sign in the last term:

$$V(\boldsymbol{x}) = \int\limits_{\Omega} G(\boldsymbol{x},\boldsymbol{y})\varrho(\boldsymbol{y}) d^3y + \frac{x^2 - R^2}{4\pi R} \int\limits_{|y|=R} \frac{V_0(\boldsymbol{y})}{|\boldsymbol{x}-\boldsymbol{y}|^3}\, d\Omega_y. \qquad (1.4.29)$$

This solution holds for $|\boldsymbol{x}| \ge R$. As an example let us consider a grounded conducting sphere ($V_0 = 0$) with a point charge q outside at $\boldsymbol{y}_0$:

$$\varrho(\boldsymbol{x}) = q\delta(\boldsymbol{x} - \boldsymbol{y}_0). \qquad (1.4.30)$$

Then the potential for $|\boldsymbol{x}| \geq R$ is given by

$$V(\boldsymbol{x}) = \frac{q}{4\pi\varepsilon_0|\boldsymbol{x}-\boldsymbol{y}_0|} - \frac{qR|\boldsymbol{y}_0|}{4\pi\varepsilon_0|\boldsymbol{x}y_0^2 - \boldsymbol{y}_0R^2|}, \tag{1.4.31}$$

where we use SI units from now on.

The first term in (1.4.31) is the potential of the point charge and the second term the potential of the image charge. But in reality there is no image charge, the potential inside the conducting sphere vanishes. Instead, some **charge is induced on the conducting surface** ∂K_R, although the potential vanishes there ($V_0 = 0$). To show this we calculate a surface charge density $s(\boldsymbol{y})$ with the property that

$$V_2(\boldsymbol{x}) \stackrel{\text{def}}{=} \int\limits_{|\boldsymbol{y}|=R} \frac{s(\boldsymbol{y})}{4\pi\varepsilon_0|\boldsymbol{x}-\boldsymbol{y}|} d\sigma_y = \begin{cases} -\dfrac{qRy_0}{4\pi\varepsilon_0|\boldsymbol{x}y_0^2 - \boldsymbol{y}_0R^2|}, & \text{if } |\boldsymbol{x}| > R \\ -\dfrac{q}{4\pi\varepsilon_0|\boldsymbol{x}-\boldsymbol{y}_0|}, & \text{if } |\boldsymbol{x}| < R. \end{cases} \tag{1.4.32}$$

The last condition guarantees that the total potential vanishes for $|\boldsymbol{x}| < R$. To determine $s(\boldsymbol{y})$ we note that the potential V_2 is continuous at $x = R$, but its normal derivative makes a jump: For $\boldsymbol{x}_0 \in \partial K_R$ on the boundary we have

$$\Big(\frac{\partial}{\partial n}\Big|_e - \frac{\partial}{\partial n}\Big|_i\Big) \int \frac{s(\boldsymbol{y})}{4\pi\varepsilon_0|\boldsymbol{x}-\boldsymbol{y}|}\, d\sigma_y\Big|_{\boldsymbol{x}=\boldsymbol{x}_0} = \frac{\partial V_2}{\partial n}\Big|_e - \frac{\partial V_2}{\partial n}\Big|_i$$

$$= \frac{\partial V(\boldsymbol{x})}{\partial n}\Big|_{x\downarrow R} = \frac{\partial}{\partial x}\Big(\frac{q}{4\pi\varepsilon_0|\boldsymbol{x}-\boldsymbol{y}_0|} - \frac{qRy_0}{4\pi\varepsilon_0|\boldsymbol{x}y_0^2 - \boldsymbol{y}_0R^2|}\Big)\Big|_{x=R=x_0}, \tag{1.4.33}$$

where e stands for external and i for internal. The last expression can be calculated as Poisson's integral above (1.4.27)

$$= q\frac{y_0^2 - R^2}{4\pi\varepsilon_0 R|\boldsymbol{y}_0 - \boldsymbol{x}_0|^3}. \tag{1.4.34}$$

This jump of the normal derivative is directly related to the surface charge density. This follows from the theorem of flux (1.1.17), as we are now going to show.

We consider a cylinder Z over a small plane surface F which includes a piece of the conducting surface (Gauss' cylinder, see the later Fig.14 (4.3.19)). According to (1.1.17) we have

$$\int\limits_{\partial Z} \boldsymbol{E}\cdot d\boldsymbol{\sigma} = \frac{1}{\varepsilon_0}\int\limits_Z \varrho(\boldsymbol{x})\, d^3x = \frac{1}{\varepsilon_0}\int\limits_F s(\boldsymbol{x})\, d\sigma. \tag{1.4.35}$$

In the limit where the height of the cylinder goes to 0, we find

$$\int\limits_F (E_e^n - E_i^n)\, d\sigma = \frac{1}{\varepsilon_0}\int\limits_F s(\boldsymbol{x})\, d\sigma. \tag{1.4.36}$$

Since F is arbitrary we obtain a **jump of the normal components of the electric field**

$$E_e^n - E_i^n = \frac{1}{\varepsilon_0} s = -\frac{\partial V}{\partial n}\bigg|_e + \frac{\partial V}{\partial n}\bigg|_i, \tag{1.4.37}$$

which **is given by the surface charge density** s/ε_0. Applying this to (1.4.34), we find the surface charge density on the sphere

$$s(\boldsymbol{x}_0) = q \frac{R^2 - y_0^2}{4\pi R |\boldsymbol{y}_0 - \boldsymbol{x}_0|^3}. \tag{1.4.38}$$

This surface charge is actually responsable for the modification of the potential of the point charge q in (1.4.31).

A similar consideration can be made with a small closed curve ∂K_2 around a piece of the surface (Stokes' contour, Fig.12). Using Stokes' theorem this time and (1.1.15), we have

$$\int_{K_2} \operatorname{curl} \boldsymbol{E} \cdot d\boldsymbol{\sigma} = \int_{\partial K_2} \boldsymbol{E} \cdot d\boldsymbol{s} = 0. \tag{1.4.39}$$

If the contour shrinks to 0, we conclude that the **tangential component of the electric field**

$$E_e^t - E_i^t = 0 \tag{1.4.40}$$

is continuous. We shall return to these boundary conditions in Sect.4.3 (see (4.3.18) and Fig.12).

We add some remarks on the problem of the conducting sphere just considered. The force of the grounded sphere on the point charge q can be calculated by integrating the forces of the induced charges on the sphere (1.4.38) (Problem 11). The same force would produce a real mirror charge without the conducting sphere, because the electric field at the place of the point charge q is the same in both cases. If we add the potential of another auxiliary point charge at the center of the sphere to (1.4.31), we get the potential of a point charge q outside of an isolated conducting sphere of total charge Q (Problem 12)

$$V(\boldsymbol{x}) = \frac{1}{4\pi\varepsilon_0}\left[\frac{q}{|\boldsymbol{x} - \boldsymbol{y}_0|} - \frac{Rq}{y_0}\frac{1}{|\boldsymbol{x} - \boldsymbol{y}_0^*|} + \left(Q + \frac{Rq}{y_0}\right)\frac{1}{|\boldsymbol{x}|}\right]. \tag{1.4.41}$$

The force on the point charge is then given by

$$K = q|\boldsymbol{E}| = \frac{1}{4\pi\varepsilon_0}\frac{q}{y_0^2}\left[-Q + qR^3 \frac{2y_0^2 - R^2}{(y_0^2 - R^2)^2 y_0}\right], \tag{1.4.42}$$

where the first term in (1.4.41) (the self-field) gives no contribution. We notice that this is not just the product of the point charge q times a factor that is independent of q, as we had provisionally said in the discussion of

(1.1.1). In the limit $|Q| \gg |q|$, the last term in (1.4.42) can be neglected and the simple picture is true. Otherwise the splitting of $\boldsymbol{K} = q\boldsymbol{E}$ is more complicated, i.e. $\boldsymbol{E}$ depends on q. In this case the precise meaning of $\boldsymbol{E}(\boldsymbol{x})$ cannot be understood without the complete field theory of electrostatics.

The concept of surface charge density can be used to establish a general method for solving the Dirichlet problem which we describe for the interested reader. Let us consider the potential

$$V_1(\boldsymbol{x}) = \frac{1}{4\pi\varepsilon_0} \int\limits_{S_0} \frac{s(\boldsymbol{y})}{|\boldsymbol{x}-\boldsymbol{y}|}\, d\sigma_y, \tag{1.4.43}$$

generated by a surface charge density $s(\boldsymbol{y})$ that is concentrated on a closed smooth surface S_0. We compute its normal derivative on S_0 using

$$\frac{\partial}{\partial n}\frac{1}{|\boldsymbol{x}-\boldsymbol{y}|} = -\frac{(\boldsymbol{x}-\boldsymbol{y})\cdot\boldsymbol{n}}{|\boldsymbol{x}-\boldsymbol{y}|^3} = \frac{\cos\varphi_{xy}}{|\boldsymbol{x}-\boldsymbol{y}|^2},$$

where φ_{xy} is the angle between $\boldsymbol{y}-\boldsymbol{x}$ and $\boldsymbol{n}$. The potential

$$V_0(\boldsymbol{x}) \overset{\text{def}}{=} \varepsilon_0 \frac{\partial V_1(\boldsymbol{x})}{\partial n} = \frac{1}{4\pi}\int\limits_{S_0} \frac{\cos\varphi_{xy}}{|\boldsymbol{x}-\boldsymbol{y}|^2}\, s(\boldsymbol{y})\, d\sigma_y \tag{1.4.44}$$

is a solution of the Laplace equation for $\boldsymbol{x}$ inside or outside S_0. On the surface S_0 there is a discontinuity, given by (1.4.37). In addition, $V_0(\boldsymbol{x})$ (1.4.44) has finite values on S_0 without performing any limit. We therefore write (1.4.37) in the following form

$$\begin{aligned} s(\boldsymbol{x}) = \varepsilon_0 \frac{\partial V_1}{\partial n}\Big|_i - \frac{1}{4\pi}\int\limits_{S_0} \frac{\cos\varphi_{xy}}{|\boldsymbol{x}-\boldsymbol{y}|^2}\, s(\boldsymbol{y})\, d\sigma_y \\ + \frac{1}{4\pi}\int\limits_{S_0} \frac{\cos\varphi_{xy}}{|\boldsymbol{x}-\boldsymbol{y}|^2}\, s(\boldsymbol{y})\, d\sigma_y - \varepsilon_0 \frac{\partial V_1}{\partial n}\Big|_e, \end{aligned} \tag{1.4.45}$$

where $\boldsymbol{x}$ is on the surface S_0. Since this equation is completely symmetric under exchange of the "interior" with the "exterior", the values of $V_0(\boldsymbol{x})$ (1.4.44) on S_0 must lie half-way between the inner and outer limit:

$$\frac{1}{2}s(\boldsymbol{x}) = \varepsilon_0 \frac{\partial V_1}{\partial n}\Big|_i - \frac{1}{4\pi}\int\limits_{S_0} \frac{\cos\varphi_{xy}}{|\boldsymbol{x}-\boldsymbol{y}|^2}\, s(\boldsymbol{y})\, d\sigma_y. \tag{1.4.46}$$

(This result can be verified by a more detailed investigation of (1.4.44)). The first term on the r.h.s. of (1.4.46) is the limit of the potential (1.4.44) for $\boldsymbol{x} \to S_0$ from inside. In the inner Dirichlet problem, these boundary values are given

$$\varepsilon_0 \frac{\partial V_1}{\partial n}\Big|_i = V_0(\boldsymbol{x}), \quad \boldsymbol{x} \in S_0.$$

Then (1.4.46) becomes an integral equation

$$2\pi s(\boldsymbol{x}) + \int\limits_{S_0} \frac{\cos\varphi_{xy}}{|\boldsymbol{x}-\boldsymbol{y}|^2}\, s(\boldsymbol{y})\, d\sigma_y = V_0(\boldsymbol{x}), \quad \boldsymbol{x} \in S_0 \tag{1.4.47}$$

for the unknown surface charge density $s(\boldsymbol{x})$ (Fredholm integral equation). If this is solved, the potential $V_0(\boldsymbol{x})$ for arbitrary $\boldsymbol{x}$ follows from (1.4.44).

1.5 Energy of the Electric Field

We consider an electric field

$$\boldsymbol{E}(\boldsymbol{x}) = -\operatorname{grad} V(\boldsymbol{x}) \tag{1.5.1}$$

that vanishes at infinity and we choose the potential such that $V(\infty) = 0$. The force on a point charge q is equal to

$$\boldsymbol{K}(\boldsymbol{x}) = q\boldsymbol{E}(\boldsymbol{x}) = -q \operatorname{grad} V(\boldsymbol{x}). \tag{1.5.2}$$

The mechanical work necessary to bring the charge from infinity to the place $\boldsymbol{x}$ is then given by

$$W_q = -\int\limits_{\infty}^{\boldsymbol{x}} \boldsymbol{K}\cdot d\boldsymbol{s} = -q\int\limits_{\infty}^{\boldsymbol{x}} \boldsymbol{E}\cdot d\boldsymbol{s} = qV(\boldsymbol{x}). \tag{1.5.3}$$

We assume that $V(\boldsymbol{x})$ is produced by other point charges q_j

$$V(\boldsymbol{x}) = \frac{1}{4\pi\varepsilon_0}\sum_j \frac{q_j}{|\boldsymbol{x}-\boldsymbol{x}_j|}, \tag{1.5.4}$$

then

$$W_q = \frac{1}{4\pi\varepsilon_0}\sum_j \frac{qq_j}{|\boldsymbol{x}-\boldsymbol{x}_j|}. \tag{1.5.5}$$

This is the energy of the charge q in the field of the other charges q_j. The **total energy W of a system of point charges** is therefore given by

$$W = \frac{1}{4\pi\varepsilon_0}\sum_{j<k} \frac{q_j q_k}{|\boldsymbol{x}_j-\boldsymbol{x}_k|} = \frac{1}{4\pi\varepsilon_0}\frac{1}{2}\sum_{j\neq k} \frac{q_j q_k}{|\boldsymbol{x}_j-\boldsymbol{x}_k|}. \tag{1.5.6}$$

The constraint in the summation prevents double counting, furthermore, the self-energy for $j = k$ is excluded. The last expression can immediately be taken over to a continuous charge distribution $\varrho(\boldsymbol{x})$:

$$W = \frac{1}{4\pi\varepsilon_0}\frac{1}{2}\int d^3x \int d^3y\, \frac{\varrho(\boldsymbol{x})\varrho(\boldsymbol{y})}{|\boldsymbol{x}-\boldsymbol{y}|}. \tag{1.5.7}$$

Here the self-energy is included, but this is irrelevant because the singularity at $\boldsymbol{x} = \boldsymbol{y}$ is integrable for continuous $\varrho(\boldsymbol{x})$, so that the contribution of coinciding points is negligible. But (1.5.7) is not true for point charges $\varrho \sim \delta(\boldsymbol{x} - \boldsymbol{x}_j)$, where we must use the original expression (1.5.6).

The expression (1.5.7) describes the energy as an "action over the distance" between the various parts of the charge distribution. In the spirit of field theory we would like to express W by the electric field $\boldsymbol{E}(\boldsymbol{x})$ that is generated by the charges. Since the potential is equal to

$$V(\boldsymbol{x}) = \frac{1}{4\pi\varepsilon_0} \int d^3y \, \frac{\varrho(\boldsymbol{y})}{|\boldsymbol{x} - \boldsymbol{y}|}, \tag{1.5.8}$$

we get

$$\begin{aligned} W &= \frac{1}{2} \int d^3x \, \varrho(\boldsymbol{x}) V(\boldsymbol{x}) \\ &= \frac{\varepsilon_0}{2} \int d^3x \, V \operatorname{div} \boldsymbol{E}, \end{aligned} \tag{1.5.9}$$

where we have used Gauss' law (1.1.16). We rewrite the integrand as follows

$$V \operatorname{div} \boldsymbol{E} = \operatorname{div}(V\boldsymbol{E}) - \boldsymbol{E} \cdot \operatorname{grad} V = \operatorname{div}(V\boldsymbol{E}) + \boldsymbol{E}^2. \tag{1.5.10}$$

By Gauss' theorem the divergence term is equal to a surface integral

$$\int_{\partial K_r} V\boldsymbol{E} \cdot d\boldsymbol{\sigma} \to 0, \quad r \to \infty, \tag{1.5.11}$$

which goes to 0 if the surface gets infinitely big. This follows from the multipole expansion because for a finite total charge Q we have

$$V \sim \frac{Q}{r}, \quad |\boldsymbol{E}| \sim \frac{Q}{r^2}, \quad \int_{\partial K_r} V\boldsymbol{E} \cdot d\boldsymbol{\sigma} \sim \frac{Q^2}{r^3} r^2 \to 0$$

for $r \to \infty$. Then we arrive at the desired expression

$$W = \frac{\varepsilon_0}{2} \int d^3x \, \boldsymbol{E}^2(\boldsymbol{x}), \tag{1.5.12}$$

where we can interpret

$$u(\boldsymbol{x}) = \frac{\varepsilon_0}{2} \boldsymbol{E}^2(\boldsymbol{x}) \tag{1.5.13}$$

as the **energy density of the electric field**. The energy is now attributed to the electric field that is produced by the charges, whereas in (1.5.6, 7) it sits in the relative spatial configuration of the charges.

We want to apply these concepts to the situation of the **capacitor**. We consider two conductors with constant potentials V_1, V_2 with a potential field $V(\boldsymbol{x})$ between them. Let $V^{(1)}(\boldsymbol{x})$ be the potential for the boundary condition $V_1 = 1$, $V_2 = 0$, and $V^{(2)}(\boldsymbol{x})$ the potential for $V_1 = 0$, $V_2 = 1$, then

$$V(\boldsymbol{x}) = V_1 V^{(1)}(\boldsymbol{x}) + V_2 V^{(2)}(\boldsymbol{x}). \tag{1.5.14}$$

The corresponding field energy is given by

$$W = \frac{\varepsilon_0}{2} \int |\mathrm{grad}\, V|^2 d^3x. \tag{1.5.15}$$

Here, only the exterior Ω of the conductors contributes to the volume integral, because in the interior we have $V = \text{const}$ and $\boldsymbol{E} = \boldsymbol{0}$. Hence

$$W = \frac{\varepsilon_0}{2} \Big[V_1^2 \int_\Omega |\mathrm{grad}\, V^{(1)}|^2 d^3x + 2V_1V_2 \int_\Omega \mathrm{grad}\, V^{(1)} \cdot \mathrm{grad}\, V^{(2)} d^3x$$

$$+ V_2^2 \int_\Omega |\mathrm{grad}\, V^{(2)}|^2 d^3x. \tag{1.5.16}$$

To evaluate this we use Gauss' theorem

$$\int_\Omega \mathrm{div}\,(V^{(1)} \mathrm{grad}\, V^{(1)}) d^3x = \int_\Omega V^{(1)} \triangle V^{(1)} d^3x + \int_\Omega |\mathrm{grad}\, V^{(1)}|^2 d^3x$$

$$= \int_{\partial\Omega} V^{(1)} \frac{\partial V^{(1)}}{\partial n_x} d\sigma_x. \tag{1.5.17}$$

The first volume integral vanishes, because $\triangle V^{(1)} = 0$. In the surface integral there is no contribution from infinity, provided the total charge on the conductors is zero

$$Q_2 = -Q_1 \stackrel{\text{def}}{=} -Q. \tag{1.5.18}$$

This is again a consequence of the multipole expansion (1.2.4). There remains the contribution of the surface $\partial\Omega_1$ of the conductor 1

$$V_1 V^{(1)} \frac{\partial V^{(1)}}{\partial n_x}\Big|_{\partial\Omega_1} = \frac{\partial V}{\partial n_x}\Big|_{\partial\Omega_1} = \frac{1}{\varepsilon_0} s_1(\boldsymbol{x}), \tag{1.5.19}$$

which has been expressed by the surface charge density according to (1.4.37). Thus

$$V_1 \int_\Omega |\mathrm{grad}\, V^{(1)}|^2 d^3x = \frac{1}{\varepsilon_0} Q_1 \tag{1.5.20}$$

and an analogous result follows for the last term in (1.5.16).

For the mixed term we get

$$2\int \operatorname{grad} V^{(1)} \cdot \operatorname{grad} V^{(2)} d^3x = \int \operatorname{div}\left(V^{(1)} \operatorname{grad} V^{(2)}\right) + \int \operatorname{div}\left(V^{(2)} \operatorname{grad} V^{(1)}\right)$$

$$- \int V^{(1)} \triangle V^{(2)} - \int V^{(2)} \triangle V^{(1)}$$

$$= \int\limits_{\partial\Omega_1} V^{(1)} \operatorname{grad} V^{(2)} \cdot d\boldsymbol{\sigma} + \int\limits_{\partial\Omega_2} V^{(2)} \operatorname{grad} V^{(1)} \cdot d\boldsymbol{\sigma}. \tag{1.5.21}$$

Multiplying this by $V_1 V_2$ we get

$$\frac{Q_2}{\varepsilon_0} + \frac{Q_1}{\varepsilon_0} = 0,$$

so that this term vanishes due to (1.5.18). Thus

$$W = \frac{1}{2}(V_1 Q_1 + V_2 Q_2) = \frac{Q}{2}(V_1 - V_2) = \frac{Q}{2} U, \tag{1.5.22}$$

where the potential difference or the voltage appears.

The charges and the potentials on the conductors are not independent. To find a relation between them, we consider

$$Q_1 = \int\limits_{\partial\Omega_1} s_1(\boldsymbol{x}) d\sigma = -\varepsilon_0 \int\limits_{\partial\Omega_1} \frac{\partial V}{\partial n} d\sigma$$

$$= -\varepsilon_0 V_1 \int\limits_{\partial\Omega_1} \frac{\partial V^{(1)}}{\partial n} d\sigma - \varepsilon_0 V_2 \int\limits_{\partial\Omega_1} \frac{\partial V^{(2)}}{\partial n} d\sigma \stackrel{\text{def}}{=} C_{11} V_1 + C_{12} V_2, \tag{1.5.23}$$

and similarly

$$Q_2 = C_{21} V_1 + C_{22} V_2. \tag{1.5.24}$$

Here the coefficients

$$C_{jk} = -\varepsilon_0 \int\limits_{\partial\Omega_j} \frac{\partial V^{(k)}}{\partial n} d\sigma \tag{1.5.25}$$

are purely geometric quantities called mutual capacities. They are determined by the solutions $V^{(1)}$, $V^{(2)}$ of the two normalized boundary value problems and, therefore, are independent of the actual charges or potentials on the conductors. From the two equations

$$Q_1 + Q_2 = 0 = (C_{11} + C_{21}) V_1 + (C_{12} + C_{22}) V_2$$

$$U = V_1 - V_2 \tag{1.5.26}$$

we find

$$V_1 = \frac{C_{12} + C_{22}}{C_{11} + C_{12} + C_{21} + C_{22}} U$$

$$V_2 = -\frac{C_{11} + C_{21}}{C_{11} + C_{12} + C_{21} + C_{22}} U. \tag{1.5.27}$$

This gives the desired relation between charge and voltage

$$Q_1 = Q = \frac{C_{11}C_{22} - C_{12}C_{21}}{C_{11} + C_{12} + C_{21} + C_{22}} U \stackrel{\text{def}}{=} CU. \tag{1.5.28}$$

Here the quantity C is the **capacitance of the arrangement of the two conductors**. It has been entirely expressed by the geometric quantities C_{jk} (1.5.25). Using this in (1.5.22) we see that the **electrostatic energy in the capacitor**

$$W = \frac{C}{2} U^2 \tag{1.5.29}$$

depends quadratically on the voltage.

1.6 Problems

1. Calculate the electric field of two point charges q and $-q$ with distance d between them. Expand for $|\boldsymbol{x}| \gg d$ and interpret the leading contribution.
2. Determine the potential $V(\boldsymbol{x})$ and the field strength $\boldsymbol{E}(\boldsymbol{x})$ of a spherically symmetric charge distribution $\varrho(r)$, $r = |\boldsymbol{x}|$. Discuss the result with help of the theorem of flux. Specialize to a homogeneously charged sphere with radius r_0 and total charge Q.
3. The electrostatic potential of a hydrogen atom in the ground state is given by

$$V(\boldsymbol{x}) = \frac{q}{4\pi\varepsilon_0} \frac{e^{-2r/r_0}}{r} \left(1 + \frac{r}{r_0}\right), \tag{1.6.1}$$

where q is the elementary charge and r_0 is Bohr's radius. Calculate the corresponding charge distribution and discuss it. Verify directly and with help of the theorem of flux that the total charge is 0.
4. Legendre polynomials: From the defining relation

$$\frac{1}{\sqrt{1 - 2xz + x^2}} = \sum_{l=0}^{\infty} x^l P_l(z) \tag{1.6.1}$$

prove:
a) the recursion relations

$$(n+1)P_{n+1} - (2n+1)zP_n + nP_{n-1} = 0 \tag{1.6.2}$$

$$P'_{n+1} - 2zP'_n + P'_{n-1} = P_n \tag{1.6.3}$$

$$zP'_n - nP_n - P'_{n-1} = 0 \tag{1.6.4}$$

$$P'_{n+1} - (2n+1)P_n - P'_{n-1} = 0. \tag{1.6.5}$$

b) the explicit representations

$$P_n(z) = \sum_{k=0}^{[n/2]} (-)^k \frac{1 \cdot 3 \cdot \ldots (2n-2k-1)}{2^k (n-2k)!} z^{n-2k}, \tag{1.6.6}$$

where

$$\left[\frac{n}{2}\right] = \begin{cases} \frac{n}{2}, & n \text{ even} \\ \frac{n-1}{2}, & n \text{ odd} \end{cases}$$

$$P_n(z) = \frac{1}{2^n n!} \frac{d^n}{dz^n} (z^2-1)^n. \tag{1.6.7}$$

c) the orthogonality

$$\int_{-1}^{1} dz\, P_n(z) P_m(z) = \frac{2}{2n+1} \delta_{nm}, \tag{1.6.8}$$

(Hint: integrate $1/\sqrt{1-2xz+x^2}\sqrt{1-2yz+y^2}$)

d) the differential equation

$$\frac{d}{dz}\left[(1-z^2)\frac{dP_l(z)}{dz}\right] + l(l+1)P_l = 0. \tag{1.6.9}$$

4. Ellipsoidal coordinates: Let $a_1 \geq a_2 \geq a_3$ be the three semi-axis of an ellipsoid. The cubic equation

$$\sum_{j=1}^{3} \frac{x_j}{a_j^2 + u} = 1 \tag{1.6.9}$$

has three real roots $-a_1^2 \leq u_1 \leq -a_2^2 \leq u_2 \leq -a_3^2 \leq u_3$. They define the ellipsoidal coordinates $u_k, k = 1, 2, 3$.

a) Show that

$$x_j^2 = \frac{\prod_i (u_i + a_j^2)}{\prod_{i \neq j} (a_i^2 - a_j^2)}. \tag{1.6.10}$$

Hint: Use and prove the identity

$$1 - \frac{x_1^2}{a_1^2+u} - \frac{x_2^2}{a_2^2+u} - \frac{x_3^2}{a_3^2+u} = \frac{(u-u_1)(u-u_2)(u-u_3)}{(a_1^2+u)(a_2^2+u)(a_3^2+u)}.$$

b) Calculate the metric tensor $g = (g_{ik})$

$$g_{ik} = \sum_l \frac{\partial x_l}{\partial u_i} \frac{\partial x_l}{\partial u_k} \tag{1.6.11}$$

and its inverse $g^{ik} = (g^{-1})^{ik}$. With the result

$$g_{ik} = \delta_{ik} \frac{\prod_{j \neq k} |u_k - u_j|}{4R_k^2}, \tag{1.6.12}$$

$$R_k^2 = \prod_j |u_k + a_j^2| \tag{1.6.13}$$

you find the Laplace operator in ellipsoidal coordinates

$$\triangle = \frac{1}{\sqrt{|g|}} \sum_{ik} \frac{\partial}{\partial u_i} \left(\sqrt{|g|} g^{ik} \frac{\partial}{\partial u_k} \right) \tag{1.6.14}$$

$$= 4 \sum_k \frac{R_k}{\prod_{j \neq k} |u_k - u_j|} \frac{\partial}{\partial u_k} R_k \frac{\partial}{\partial u_k}, \tag{1.6.15}$$

where $|g| = \det g$.

5. Calculate the potential outside of the conducting ellipsoid $u_1 = 0$ with constant potential V_0. The result is an elliptic integral of the first kind. What is the capacitance $C = Q/V_0$ (Q = charge on the ellipsoid) ? Show that

$$V \to \frac{Q}{4\pi\varepsilon_0 r} \quad \text{for} \quad r \to \infty. \tag{1.6.16}$$

Hint: Use the relation

$$\sum_k u_k = \sum_j x_j^2 - \sum_j a_j^2 \tag{1.6.17}$$

which follows from Problem 4.

6. Toroidal coordinates: In cylinder coordinates (ϱ, φ, z) one substitutes ϱ and z by α, β, where

$$z + i\varrho = c \cot \frac{\alpha - i\beta}{2} \tag{1.6.18}$$

and $c = \text{const} > 0$. (α, β, φ) are the toroidal coordinates.

   a) Calculate z, ϱ as functions of α, β. For which α, β goes $r^2 = \varrho^2 + z^2$ to infinity ?
   b) Show that the equation $\beta = \text{const} = \beta_1$ defines a torus $z^2 + (\varrho - b)^2 = a^2$, determine a and b. For which β one gets the interior and the exterior of the torus, respectively ?

7. Express the derivatives $\partial/\partial\varrho$ and $\partial/\partial z$ in cylinder coordinates by $\partial/\partial\alpha$, $\partial/\partial\beta$ in toroidal coordinates.
8. Transform the Laplace equation in cylinder coordinates into the following form in toroidal coordinates:

$$\frac{\partial}{\partial\alpha} \frac{\sinh\beta}{\cosh\beta - \cos\beta} \frac{\partial V}{\partial\alpha} + \frac{\partial}{\partial\beta} \frac{\sinh\beta}{\cosh\beta - \cos\alpha} \frac{\partial V}{\partial\beta}$$
$$+ \frac{1}{\sinh\beta(\cosh\beta - \cos\alpha)} \frac{\partial^2 V}{\partial\varphi^2} = 0. \tag{1.6.19}$$

9. Calculate the potential outside of a conducting torus with constant potential V_0. What is the capacitance $C = Q/V_0$ of the torus ? Hint: Substitute
$$V(\alpha, \beta) = \sqrt{\cosh\beta - \cos\alpha U(\alpha, \beta)}. \tag{1.6.20}$$
In the resulting differential equation make the product ansatz
$$U(\alpha, \beta) = U_n(\beta)\cos n\alpha, \quad n = 0, 1, 2, \ldots \tag{1.6.21}$$
and transform it by means of the substitution $s = \cosh\beta$ into the Legendre equation (1.6.9). For the determination of the coefficients use the Fourier series (*A.Erdelyi, Higher Transcendental Functions, Mac Graw-Hill (1953), Vol.1, p.166, eq.3)*
$$\frac{1}{\sqrt{x - \cos\alpha}} = \frac{2^{3/2}}{\pi}\Big[\tfrac{1}{2}Q_{-\frac{1}{2}}(x) + \sum_{n=1}^{\infty} Q_{n-\frac{1}{2}}(x)\cos n\alpha\Big], \tag{1.6.22}$$
where $Q_m(x)$ are the Legendre functions of second kind.
10. Verify by a limiting process that Poisson's integral
$$V(\boldsymbol{x}) = \frac{R^2 - x^2}{4\pi r} \int\limits_{|y|=R} \frac{V_0(\boldsymbol{y})}{|\boldsymbol{x} - \boldsymbol{y}|^3}\, d\Omega_y \tag{1.6.23}$$
satisfies the boundary condition $V = V_0$ on the sphere.
11. Calculate the force of a grounded conducting sphere of radius R on a charge q outside the sphere, by integrating the forces of the induced surface charges. Show that the same force would be produced by the fictive image charge.
12. Determine the potential of a point charge q outside of an isolated conducting sphere with total charge Q. Calculate the force on the charge q and discuss when it is attractive and repulsive.
13. Calculate the capacitances $C = Q/U$ of the following arrangements of conductors:
    a) A sphere of radius R in infinite space.
    b) Two concentric spheres (inner radius R_1, outer radius R_2).
    c) Two coaxial cylinders (inner radius R, outer radius $R + d$, length $L \gg d$), corrections from the ends are neglected.
    d) Two parallel plates (surface F, distance $d \ll \sqrt{F}$), corrections from the boundaries are neglected.
    e) Two parallel capacitors with capacitances C_1 and C_2.
    f) Two capacitors in series with capacitances C_1 and C_2.
14. **Two-dimensional potential theory**: Let $z = x + iy$ and $f(z) = V(x, y) + iW(x, y)$ be an analytic function of z. Show by means of the Cauchy-Riemann differential equations
$$\frac{\partial V}{\partial x} = \frac{\partial W}{\partial y}, \quad \frac{\partial V}{\partial y} = -\frac{\partial W}{\partial x}, \tag{1.6.24}$$

that $V(x, y)$ and $W(x, y)$ are solutions of the two-dimensional Laplace equation. What boundary value problems are solved by the following analytic functions ? Calculate the corresponding potentials :

a) $f(z) = -q \log z$.

b)

$$f(z) = \frac{2}{i\pi} V_0 \log \frac{1+z}{1-z}.$$

c) Which $f(z)$ corresponds to the boundary value problem of two parallel straight lines, with boundary conditions $V = 0$ on $y = 0$ and $V = V_0$ on $y = d$?

d) What is $f(z)$ for two oppositely charged wires perpendicular to the xy-plane with distance $2d$ between them ?

15. One two-dimensional boundary value problem can be mapped onto another one by a (conformal) mapping $z = g(z')$, where g is an analytic function.

a) What boundary value problem is obtained by applying the mapping $z = \log(z' + a)$ on 14c) ? Calculate the potential.

b) Which problem is obtained from 14b) by the mapping

$$z = \exp i \frac{\pi}{d} z' \quad ?$$

Calculate the potential.

16. The analytic function $f(z) = -q \log z$ solves the two-dimensional boundary value problem with boundary conditions $V = 0$ on $y = 0$, $x > 0$ and $V = V_0$ on $y = 0$, $x < 0$. This can be transformed by the conformal mapping

$$z' = \frac{a}{\pi} \left(\log z + \frac{1}{2} - \frac{1}{2} z^2 \right) \tag{1.6.25}$$

into a plate condenser.

a) Determine the surface charge densities on the interior and exterior sides of the lower plate ($V = 0$) in a parameter representation. What is the result at large distance from the end of the plate ?

b) Calculate the total charge (interior and exterior) contained in one unit length measured from the end of the plate. Determine the boundary corrections for the capacity. (Hint: Use besides the imaginary part of $f(z)$ also the real part.)

2. The Relativity Principles and Maxwell's Equations

In the last chapter on electrostatics all charges had to be at rest. Then the forces between them are mediated by the electric field. Now we consider moving charges. Here one observes additional forces which, in the spirit of field theory, are attributed to an additional field, the magnetic field. This field can also be present without any moving macroscopic charge, as for example in the neighborhood of an iron magnet. This shows again the importance of the field concept; a description of only the forces between moving charges would be a very restricted one.

But a serious problem arises here. Motion is a relative thing. Instead of moving a charge, the observer may move. This moving observer then would no longer observe the electrostatic field of the charge but partly also a magnetic field. For this reason we must carefully study the problem of how the physical laws transform under a change of the reference frame. This is special relativity theory, and, without it, electrodynamics cannot really be understood. We will discuss in the first two sections what we need of relativity theory in the following. In Sect.2.1 we start by analysing the problem, what is the correct transformation law for going from one reference system to another. It is not the Galilean transformation of Newtonian mechanics, but the Lorentz transformation. The next task is to write the basic physical equations in a form independent of the reference frame. The tool for doing this is the tensor calculus. Since we do not suppose that the reader is familiar with it, we give a brief but self-contained introduction in Sect.2.2, which only requires some basic facts of linear algebra and of differential and integral calculus. This tool then enables us in Sect.2.3 to deduce from the Lorentz force law the basic electromagnetic field tensor. The fundamental equations of electrodynamics (Maxwell's equations in vacuum) then follow essentially by transforming electrostatics to a moving reference system (Sect.2.4). In this way the very basic nature of these equations is evident. In the remainder of this chapter we discuss general properties of the equations, in particular the conservation laws, leaving the applications to the following chapter.

2.1 A Review of Special Relativity

Until now we have only studied time-independent problems. From now on time and space coordinates will play the same rôle according to the concept of field theory. It is then convenient to combine them into the four-dimensional real space $\mathbb{R}^4$ of space-time points $x = (x^0, x^1, x^2, x^3) = (x^\mu), x^0 = ct$. Here the velocity of light c has been introduced into the time component in order to have the same dimension in all four components of x. At this point we need not know what light really is. By the introduction of c we have only reduced the measuring procedure for time to that of space, or vice versa. This is in fact the point of view in metrology today (see (1.1.8)). The possibility of measuring time by studying the propagation of light signals is an important concept in relativity. Throughout we use the convention that greek indices assume the values 0,1,2,3, whereas latin indices are used for the spatial values 1,2,3. Specifying the position $\boldsymbol{x}$ of a physical object as a function of time t, defines a curve in $\mathbb{R}^4$. The light rays going out from the origin move on the light-cone

$$c^2t^2 - |\boldsymbol{x}|^2 = 0. \tag{2.1.1}$$

This double-cone consists of the past-cone $t < 0$ and the future-cone $t > 0$. All coordinates so far are understood with respect to a fixed basis e_μ which defines the frame of reference. A change of the frame from one observer to another is given by a change of the basis **and** by a corresponding change of the coordinates, because the real physical object to be described remains the same. This is the so-called passive point of view that we will always adopt.

In special relativity theory one only considers linear coordinate transformations

$$x'^\mu = \Lambda^\mu{}_\nu x^\nu, \tag{2.1.2}$$

where Λ is a real 4×4-matrix. If, in addition, we add constant terms a^μ in (2.1.2), we include translations of the origin of the coordinate system. We always assume the convention of summing over double upper and lower indices. The reason for using these upper and lower indices will be explained in the following section. In this section we will often omit the summation indices and, for example, write the linear coordinate transformation in (2.1.2) simply by

$$x \longrightarrow x' = \Lambda x. \tag{2.1.3}$$

The basis of relativity theory is given by two principles. The first one, simply called **relativity principle, says that all reference frames which are in relative uniform motion with respect to each other, are physically equivalent.** Such frames are called Lorentz frames. This is a very democratic principle because it means that all physical laws are the same for the various observers. However, it is a quite general postulate which is also true in ordinary (Galilean invariant) mechanics. The second principle is much more specific. It is the **principle of constant velocity of light** which **says that the light velocity is the same in all Lorentz frames, independent of the motion of the light source.**

This second principle is far from being obvious. It is in fact paradoxical because one is used to add velocities if an observer moves relative to the source. However, the principle has a firm experimental basis, supplied by the Michelson-Morley experiment. Its idea is the following: If c were depending on the observers velocity, then the light velocity in the direction of motion of the earth and perpendicular to it would be a little different. Such a difference can be very accurately measured by an interferometer with two perpendicular arms, one in the direction and the other perpendicular to the earth motion. Since no difference has been found, we must accept the principle of constant light velocity.

We are now going to work out the consequences of this principle. In view of (2.1.1) it can be expressed as follows: If

$$(x^0)^2 - \boldsymbol{x}^{\,2} = 0$$

is true in one frame of reference, then this also holds in another frame

$$(x'^0)^2 - \boldsymbol{x}'^{2} = 0.$$

It is convenient to write the quadratic forms appearing here as

$$Q(x) = x^T g\, x \tag{2.1.4}$$

$$Q'(x) = Q(\Lambda x) = x^T \Lambda^T g\, \Lambda\, x, \tag{2.1.5}$$

where

$$g = (g_{\mu\nu}) = \begin{pmatrix} 1 & 0 & 0 & 0 \\ 0 & -1 & 0 & 0 \\ 0 & 0 & -1 & 0 \\ 0 & 0 & 0 & -1 \end{pmatrix} \tag{2.1.6}$$

is the fundamental metric tensor. T means the transposed vector or matrix. Both forms (2.1.4, 5) vanish for fixed $\boldsymbol{x}$ if $x^0 = \pm|\boldsymbol{x}|$, hence

$$Q'(x) = \lambda(x^0 - |\boldsymbol{x}|)(x^0 + |\boldsymbol{x}|) = \lambda((x^0)^2 - \boldsymbol{x}^{\,2}) = \lambda\, Q(x).$$

The case $\lambda \neq 1$ corresponds to a change of units which we disregard. Then we arrive at

$$x^T \Lambda^T g\, \Lambda\, x = x^T g\, x$$

for all $x \in \mathbb{R}^4$, or

$$\Lambda^T g\, \Lambda = g. \tag{2.1.7}$$

We emphasize that we have used the condition of constant $x^2 = x'^2$ only for light rays ($x^2 = 0$). **All transformations satisfying (2.1.7) are called Lorentz transformations.** They obviously form a group, the Lorentz group $\mathcal{L}$. Equation (2.1.7) suggests the introduction of the indefinite scalar product

$$(x,y) = x^T g\, y = x^0 y^0 - \boldsymbol{x}\cdot\boldsymbol{y} = x^0y^0 - x^1y^1 - x^2y^2 - x^3y^3$$
$$= x^\nu g_{\nu\mu} y^\mu . \qquad (2.1.8)$$

It is invariant under Lorentz transformations

$$(x', y') = (\Lambda x, \Lambda y) = (\Lambda x)^T g\, \Lambda y = x^T \Lambda^T g\, \Lambda y = x^T g\, y = (x,y).$$

The four-dimensional real vector space with scalar product (2.1.8) is called Minkowski space $\mathbb{M}$. Lorentz transformations are the congruency transformations of $\mathbb{M}$. The elements of $\mathbb{M}$ are called points or (four) vectors.

There are three classes of vectors in $\mathbb{M}$: (i) time-like vectors x with $x^2 > 0$, (ii) space-like vectors y with $y^2 < 0$ and (iii) light-like vectors z with $z^2 = 0$. Each class is mapped into itself under Lorentz transformations because x^2 remains constant. We shall often find that functions of a four-vector x behave differently for time-like or space-like x. A three-dimensional surface S in $\mathbb{M}$ is called time-like or space-like if any tangent vector to S is time-like or space-like, respectively. Two disjoint sets X, Y of points are space-like separated if every vector $x - y, x \in X, y \in Y$ is space-like. Then it is impossible to connect the points $x \in X$ and $y \in Y$ in a causal way, for instants by light signals. On the other hand, if $x - y$ is time-like, then the two points are causally connected. This causal structure of Minkowski space will be important later.

Equation (2.1.7) implies det $\Lambda = \pm 1$ for all $\Lambda \in \mathcal{L}$. Examples with determinant $= -1$ are time-reflection T and space-reflection P (parity transformation)

$$T = \begin{pmatrix} -1 & 0 & 0 & 0 \\ 0 & 1 & 0 & 0 \\ 0 & 0 & 1 & 0 \\ 0 & 0 & 0 & 1 \end{pmatrix}, \qquad P = \begin{pmatrix} 1 & 0 & 0 & 0 \\ 0 & -1 & 0 & 0 \\ 0 & 0 & -1 & 0 \\ 0 & 0 & 0 & -1 \end{pmatrix}. \qquad (2.1.9)$$

The Lorentz transformations Λ with det $\Lambda = +1$ form the subgroup

$$\mathcal{L}_+ = SO(1,3)$$

of $\mathcal{L}$. It is a special pseudo-orthogonal group. The defining equation (2.1.7) means that the rows and columns of a Lorentz matrix $\Lambda^\mu{}_\nu$ are orthogonal with respect to the Minkowski scalar product (2.1.8), for example

$$\Lambda^0{}_\mu \Lambda^0{}_\nu - \sum_{j=1}^{3} \Lambda^j{}_\mu \Lambda^j{}_\nu = \begin{cases} 0, & \text{for } \mu \neq \nu \\ 1, & \text{for } \mu = \nu = 0\,. \\ -1 & \text{for } \mu = \nu \neq 0 \end{cases} \qquad (2.1.10)$$

Taking $\mu = \nu = 0$,we have

$$(\Lambda^0{}_0)^2 - \sum_{j=1}^{3} (\Lambda^j{}_0)^2 = 1$$

and therefore

$$(\Lambda^0{}_0)^2 \geq 1 \quad \text{i.e.} \quad \Lambda^0{}_0 \geq 1 \quad \text{or} \quad \Lambda^0{}_0 \leq -1.$$

For $\Lambda^0{}_0 \geq 1$, the direction of time is not reversed. The subgroup

$$\mathcal{L}_+^\uparrow = \{\Lambda \in \mathcal{L}_+ | \Lambda^0{}_0 \geq 1\} \tag{2.1.11}$$

is the **proper Lorentz group**. Only this group is an exact symmetry group of physics (neglecting gravitation), because parity and time-reversal (2.1.9) are not conserved in weak interactions.

We now study some important subgroups of the proper Lorentz group $\mathcal{L}_+^\uparrow$. First we consider a Lorentz matrix $\Lambda \in \mathcal{L}_+^\uparrow$ with the following structure

$$\Lambda(R) = \begin{pmatrix} 1 & 0 \\ 0 & R_3 \end{pmatrix}, \tag{2.1.12}$$

where R_3 is a real 3×3 matrix. Equation (2.1.7) implies $R_3^T R_3 = 1$, which means that R_3 is a 3-dimensional rotation $\in SO(3)$. Time is not transformed in (2.1.12). This is obviously a subgroup of the Lorentz group. Another subgroup is constituted by the **Lorentz boosts**, for example

$$\Lambda(\chi) = \begin{pmatrix} \cosh\chi & 0 & 0 & \sinh\chi \\ 0 & 1 & 0 & 0 \\ 0 & 0 & 1 & 0 \\ \sinh\chi & 0 & 0 & \cosh\chi \end{pmatrix}. \tag{2.1.13}$$

This is a special Lorentz transformation along the 3-axis

$$x'^0 = x^0 \cosh\chi + x^3 \sinh\chi$$

$$x'^3 = x^0 \sinh\chi + x^3 \cosh\chi. \tag{2.1.14}$$

Every $\Lambda \in \mathcal{L}_+^\uparrow$ can be generated by taking products of these special transformations (2.1.12) and (2.1.13). Let us study the motion of the new spatial origin $\boldsymbol{x}' = \mathbf{0}$ in (2.1.14). For $x'^3 = 0$ in we get

$$x^3 = c\frac{\sinh\chi}{\cosh\chi} t = c \tanh\chi\, t. \tag{2.1.15}$$

This is a uniform motion with relative velocity

$$v = c\tanh\chi < c \tag{2.1.16}$$

in the 3-direction. Substituting

$$\cosh\chi = \frac{1}{\sqrt{1-\tanh^2\chi}} = \frac{1}{\sqrt{1-\frac{v^2}{c^2}}} \stackrel{\text{def}}{=} \gamma \tag{2.1.17}$$

$$\sinh\chi = \frac{\tanh\chi}{\sqrt{1-\tanh^2\chi}} = \gamma\frac{v}{c}$$

into (2.1.14), we get the **Lorentz transformation from one inertial frame to another, moving with relative velocity v in the 3-direction**:

$$t' = \gamma\Big(t - \frac{v}{c^2}x^3\Big), \quad x'^3 = \gamma\Big(x^3 - vt\Big). \tag{2.1.18}$$

Such frames which are at rest or in uniform motion with respect to each other with relative velocities $v < c$, are the only allowed reference frames in special relativity. They are called Lorentz frames or inertial frames, because Newton's basic law of inertia is true in these coordinate systems. The main problem to be studied in the following sections is how the laws of nature, in particular electrodynamics, behave under a change of the reference frame.

2.2 Tensors in Minkowski Space

The main result of the foregoing discussion of the relativity principles can be summarized as follows: If the frame of reference is changed by a Lorentz transformation, the physical laws in the two frames must, nevertheless, look the same. From now on we work with proper Lorentz transformations $\Lambda \in \mathcal{L}_+^\uparrow$, only. Reflections will be considered separately (see end of Sect.2.5). The frame-independence means mathematically that it must be possible to write the basic equations in a form independent of the reference system. The tool to do this is the tensor calculus.

Lorentz tensors are linear forms over Minkowski space. A real linear form F on Minkowski space $\mathbb{M}$ is a linear function from $\mathbb{M}$ to the real numbers $\mathbb{R}$. This means that F is a rule which assigns a real number $F(x)$ to each vector x in $\mathbb{M}$ with the following properties:

$$F(x+y) = F(x) + F(y), \quad x, y \in \mathbb{M}$$

$$F(\lambda x) = \lambda F(x), \quad \lambda \in \mathbb{R}.$$

A simple example is obtained by means of the scalar product (2.1.8)

$$F(x) = (f,\, x), \tag{2.2.1}$$

where f is a fixed vector in $\mathbb{M}$. It is proved in linear algebra that every linear form F is of this form (2.2.1). The linear forms on $\mathbb{M}$ obviously form a linear space. It is called the dual space and is denoted by $\mathbb{M}'$. According to (2.2.1), $\mathbb{M}'$ can be identified with $\mathbb{M}$. However, in tensor calculus it is convenient to distinguish between the two, as they play different rôles in the theory. We use the prime for this purpose. The value $F(x)$ of the form F on the vector x is also denoted by $\langle F, x\rangle = F(x)$.

Let e_ν, $\nu = 0, 1, 2, 3$ be a basis of $\mathbb{M}$, that means every vector $B \in \mathbb{M}$ has a unique representation

$$B = B^\nu e_\nu. \tag{2.2.2}$$

To e_ν there exists a dual basis e^μ in $\mathbb{M}'$, defined by

$$\langle e^\mu, e_\nu\rangle = \delta^\mu{}_\nu, \quad \mu, \nu = 0, 1, 2, 3, \tag{2.2.3}$$

where $\delta^\mu{}_\nu$ is the four-dimensional Kronecker-delta (=1 for $\mu = \nu$ and 0 otherwise). Then every element A' in $\mathbb{M}'$ has also a unique representation

$$A' = A'_\mu e^\mu. \tag{2.2.4}$$

The linear form A' now operates on $B \in \mathbb{M}$ as follows

$$\langle A', B\rangle = A'_\mu B^\mu = A'^\nu g_{\nu\mu} B^\mu \tag{2.2.5}$$

because of (2.2.3) and (2.1.8). This leads to the definition of covariant (A'_μ) and contravariant (A'^ν) components

$$A'_\mu = A'^\nu g_{\nu\mu} \tag{2.2.6}$$

and to the lowering of indices by means of the metric tensor g. If upper and lower indices are contracted in couples as in (2.2.5), we get a number. Writing the inverse matrix of g as

$$g_{\mu\nu} g^{\nu\lambda} = \delta_\mu{}^\lambda, \tag{2.2.7}$$

we find

$$g^{\mu\nu} = g_{\mu\nu} = g_{\nu\mu}.$$

Multiplying (2.2.6) with the inverse g^{-1} , we have lifted an index

$$A^\mu = A_\nu g^{\nu\mu} = g^{\mu\nu} A_\nu. \tag{2.2.8}$$

Vectors in $\mathbb{M}$ and linear forms in $\mathbb{M}'$ are called **tensors** of first rank. It should be stressed that they **are geometrical objects that do not depend on a coordinate system or basis**. This is the reason why they are so useful for writing physical equations in a form independent of the reference system. If we change the (dual) basis e^μ in $\mathbb{M}'$ by a proper Lorentz transformation

$$\tilde{e}^\mu = \Lambda^\mu{}_\nu e^\nu, \tag{2.2.9}$$

the basis e_ν in $\mathbb{M}$ must change so that (2.2.3) remains valid:

$$\langle \tilde{e}^\mu, \tilde{e}_\nu\rangle = \delta^\mu{}_\nu = \Lambda^\mu{}_\alpha \langle e^\alpha, \tilde{\Lambda}_\nu{}^\beta e_\beta\rangle = \Lambda^\mu{}_\alpha \tilde{\Lambda}_\nu{}^\alpha.$$

This implies that e_ν transforms with the inverse transposed matrix

$$\tilde{\Lambda}_\nu{}^\mu = ((\Lambda^{-1})^T)_\nu{}^\mu = (\Lambda^{-1})^\mu{}_\nu = ((\Lambda^T)^{-1})_\nu{}^\mu, \tag{2.2.10}$$

using the fact that inverting and transposing can be interchanged. The components of the tensor A' (2.2.4) must also change, so that A' remains the same:

$$A' = A'_\mu e^\mu = \tilde{A}'_\nu \tilde{e}^\nu = \tilde{A}'_\nu \Lambda^\nu{}_\mu e^\mu. \tag{2.2.11}$$

Hence

$$\tilde{A}'_\nu = ((\Lambda^T)^{-1})_\nu{}^\mu A'_\mu \tag{2.2.12}$$

transforms also with the inverse transposed matrix. The fact that A'_ν varies in the same way as the basis e_ν in $\mathbb{M}$ is the reason for the notion "covariant" components. On the other hand by lowering the index in (2.2.9) we find

$$\tilde{e}_\mu = \Lambda_\mu{}^\nu e_\nu. \tag{2.2.13}$$

Comparison with (2.2.10) gives us the following simpler expression for the inverse transposed matrix

$$((\Lambda^T)^{-1})_\nu{}^\mu = \Lambda_\nu{}^\mu = ((\Lambda^{-1})^T)_\nu{}^\mu = (\Lambda^{-1})^\mu{}_\nu. \tag{2.2.14}$$

This relation can also be obtained directly by writing indices in (2.1.7). So one must be careful in keeping in order upper and lower Lorentz indices (before TEX I had always printing problems here). It follows in the same way that the contravariant components B^ν in (2.2.2) transform with $\Lambda^\mu{}_\nu$ as the dual basis e^ν (2.2.9), i.e. contrary to the basis e_ν in $\mathbb{M}$. This is in accordance with (2.1.2) in the last section. The different transformation properties of covariant and contravariant components have the important consequence that **by contracting upper and lower indices, we get a Lorentz invariant**:

$$\tilde{A}_\mu \tilde{B}^\mu = \Lambda^\mu{}_\nu \Lambda_\mu{}^\lambda A^\nu B_\lambda = A^\nu B_\nu. \tag{2.2.15}$$

Next we consider bilinear forms T over $\mathbb{M}\times\mathbb{M}$. They assign a real number

$$T(A,B) = T_{\mu\nu} A^\mu B^\nu \tag{2.2.16}$$

to every pair of vectors in $\mathbb{M}$. This is called a covariant tensor of second rank. By lifting one index, we obtain a mixed tensor $T^\mu{}_\nu$. An example of this is the Lorentz transformation

$$\tilde{A}^\mu = \Lambda^\mu{}_\nu A^\nu. \tag{2.2.17}$$

A contravariant tensor of second rank $T^{\mu\nu}$ transforms the covariant components A_ν of a four-vector into the contravariant components of another four-vector:

$$B^\mu = T^{\mu\nu} A_\nu. \tag{2.2.18}$$

Since we know how the components A^μ, B^ν transform, the transformation of $T_{\mu\nu}$ follows as above (2.2.11) from the basis-independence of the tensor:

$$T_{\mu\nu} A^\mu B^\nu = \tilde{T}_{\alpha\beta} \tilde{A}^\alpha \tilde{B}^\beta = \tilde{T}_{\alpha\beta} \Lambda^\alpha{}_\mu \Lambda^\beta{}_\nu A^\mu B^\nu,$$

hence

$$\tilde{T}_{\alpha\beta} = \Lambda_\alpha{}^\mu \Lambda_\beta{}^\nu T_{\mu\nu}, \tag{2.2.19}$$

where (2.2.14) has been used. In the same way, the transformation properties of arbitrary higher and mixed tensors can be written down. The **components of a tensor T of rank k transform with k Lorentz matrices** Λ with upper and lower indices, arranged in such a way that **every upper index of T is contracted with a lower index of a Λ and vice versa.** The position of the indices shows whether one has an inverse transposed matrix (2.2.14) or an ordinary Lorentz matrix.

Now we consider vectors and tensors with space and time-dependent components, like $A^\mu(x)\,, T^{\mu\nu}(x)$. These objects are called **vector and tensor fields.** Under the change of the reference frame by a Lorentz transformation $\Lambda^\mu{}_\nu$

$$\tilde{x}^\alpha = \Lambda^\alpha{}_\mu x^\mu \tag{2.2.20}$$

these **fields transform two-fold.** First there is the usual transformation of the tensor components just discussed, but in addition, the argument x must be transformed to the new reference frame $x \to \tilde{x}$, for example

$$\tilde{A}^\mu(\tilde{x}) = \Lambda^\mu{}_\nu A^\nu(x)$$

$$\tilde{T}^{\mu\nu}(\tilde{x}) = \tilde{T}^{\mu\nu}(\Lambda x) = \Lambda^\mu{}_\alpha \Lambda^\nu{}_\beta T^{\alpha\beta}(x). \tag{2.2.21}$$

We always assume the so-called passive point of view, where the reference frame is changed, whereas the (physical) objects remain what they are. As discussed above, the change of the reference system then consists of a change of the basis and the corresponding transformations of the tensor components. We shall sometimes be careless with the language, we simply speak of transformation of a tensor field which really means the transformation of its components and a corresponding transformation of the basis. All these transformation properties are called Lorentz (or Poincaré) covariance.

Tensor fields are differentiable with respect to x, if the increments can be linearly approximated:

$$\lim_{\varepsilon \to 0} \frac{1}{\varepsilon}[T(x+\varepsilon y) - T(x)] = \langle DT(x), y \rangle \quad , \quad y \in \mathbb{M}. \tag{2.2.22}$$

Since this is a linear form on $\mathbb{M}$, differentiation increases the covariant degree of a tensor field by one. This follows also by the chain rule from (2.2.20)

$$\partial_\mu = \frac{\partial}{\partial x^\mu} = \frac{\partial \tilde{x}^\alpha}{\partial x^\mu} \frac{\partial}{\partial \tilde{x}^\alpha} = \Lambda^\alpha{}_\mu \tilde{\partial}_\alpha,$$

hence the **derivative**

$$\tilde{\partial}_\alpha = \Lambda_\alpha{}^\mu \partial_\mu \tag{2.2.23}$$

transforms indeed **with the inverse transposed matrix** (2.2.14). In components we write

$$(DT)_\lambda^{\mu\nu\cdots} = \frac{\partial T^{\mu\nu\cdots}(x)}{\partial x^\lambda} = T^{\mu\nu\cdots}{}_{,\lambda} \,.$$

We give some important examples :

1) A scalar field $\phi(x)$ is a tensor field of rank 0. It transforms trivially under Lorentz transformation

$$\phi'(x') = \phi'(\Lambda x) = \phi(x).$$

Since for a scalar field no basis is needed, the prime of ϕ' is sometimes omitted. By differentiation

$$(D\phi(x))_\mu = \frac{\partial \phi(x)}{\partial x^\mu} = \partial_\mu \phi \tag{2.2.24}$$

we get a covariant vector field, the **gradient**.

2) Let $A^\mu(x)$ be a contravariant vector field. Differentiating it, we obtain the mixed second rank tensor

$$(DA(x))^\mu{}_\nu = \frac{\partial A^\mu(x)}{\partial x^\nu} = A^\mu{}_{,\nu} \,.$$

If this is contracted, we obtain the scalar field

$$A^\mu{}_{,\mu} = \operatorname{div} A(x), \tag{2.2.25}$$

which is the **divergence** of $A(x)$.

3) If we differentiate a covariant vector field $A_\mu(x)$

$$(DA(x))_{\mu\nu} = \frac{\partial A_\mu(x)}{\partial x^\nu} = A_{\mu,\nu}$$

and form the antisymmetric combination

$$A_{\nu,\mu} - A_{\mu,\nu} = (\operatorname{curl} A)_{\mu\nu}, \tag{2.2.26}$$

we get the **curl** of $A(x)$. It is an antisymmetric second rank tensor. These are the four-dimensional generalizations of (0.1.12-14).

4) We now take the contravariant components of $\operatorname{grad}\phi$

$$(\operatorname{grad}\phi)^\mu = g^{\mu\nu} \frac{\partial \phi}{\partial x^\nu} \tag{2.2.27}$$

and form the divergence according to 2) above :

$$g^{\mu\nu}\frac{\partial^2\phi}{\partial x^\mu \partial x^\nu} = \frac{\partial^2\phi}{\partial (x^0)^2} - \frac{\partial^2\phi}{\partial (x^1)^2} - \frac{\partial^2\phi}{\partial (x^2)^2} - \frac{\partial^2\phi}{\partial (x^3)^2} = \partial_\mu \partial^\mu \phi. \quad (2.2.28)$$

This gives the **wave operator which obviously is Lorentz invariant.**

Finally we mention the integral theorems in four dimensions which we have to use later. The Lebesgue measure on $\mathbb{R}^4$

$$d^4x = dx^0 dx^1 dx^2 dx^3 \quad (2.2.29)$$

is invariant under Lorentz transformations Λ because $|\det \Lambda| = 1$. We almost only need the following simple form of Gauss' theorem : Let $A^\mu(x)$ be a continuously differentiable contravariant vector field defined on a region G in Minkowski space with smooth boundary ∂G, and let A^μ vanish on ∂G. Then we have

$$\int_G \operatorname{div} A(x) d^4x = \int_G \frac{\partial A^\mu(x)}{\partial x^\mu} d^4x = 0. \quad (2.2.30)$$

This theorem immediately extends to tensor fields : Let $a \in \mathbb{M}$ be an arbitrary constant vector. Then, given a differentiable tensor field $T_{\mu\nu}(x)$ in G vanishing on ∂G , Gauss' theorem applied to the vector field $T_{\mu\nu}(x)a^\nu$ leads to

$$\int_G \partial^\mu T_{\mu\nu}(x) d^4x \, a^\nu = 0 \quad (2.2.31)$$

for arbitrary $a \in \mathbb{M}$. Therefore

$$\int_G \partial^\mu T_{\mu\nu}(x) d^4x = 0. \quad (2.2.32)$$

Partial integration is another consequence of Gauss' theorem :

$$\int_G d^4x \partial_\mu (A^\mu(x) g(x)) = 0 = \int_G d^4x A^\mu(x) \partial_\mu g(x) + \int_G d^4x (\partial_\mu A^\mu) g. \quad (2.2.33)$$

2.3 Lorentz Force and the Electromagnetic Field Tensor

In Sect.1.1 the electric field was defined by the force

$$\boldsymbol{K} = q\boldsymbol{E} \quad (2.3.1)$$

on a charged test body **which is at rest**. If the test body is in motion and the force still is the same, we say that only the electric field $\boldsymbol{E}(t, \boldsymbol{x})$ is present, where $t, \boldsymbol{x}$ are time and position of the test charge. But in general the force will be different and depends on the velocity $\boldsymbol{v}$ of the test body. This velocity dependent part is due to the magnetic field and the total force is given by the **Lorentz force law**

$$\boldsymbol{K}_L(t,\boldsymbol{x}) = q(\boldsymbol{E}(t,\boldsymbol{x}) + \boldsymbol{v} \wedge \boldsymbol{B}(t,\boldsymbol{x})), \tag{2.3.2}$$

where the test charge q is at the place $\boldsymbol{x}$ at time t. Similarly to the discussion in Sect.1.1, **this equation defines the magnetic field strength $\boldsymbol{B}$ and contains additional experimental information**, for example that $\boldsymbol{K}_L$ depends linearly on $\boldsymbol{v}$. It is only important that this defines a measuring procedure for $\boldsymbol{B}$ in principle. In reality one measures magnetic fields differently, for example by magnetic resonance (as we will discuss at the end of Sect.3.2). Altogether, the **Lorentz force law defines the charge and the electric and magnetic fields simultaneously**.

However there is a serious difficulty with this definition. The velocity $\boldsymbol{v}$ of the test body is measured with respect to the reference frame of the observer. The situation changes drastically if the observer also moves, which is allowed according to the relativity principle. **Eq.(2.3.2) can only have a general meaning if it correctly accounts for such a change, that means, it must be possible to write it in a form independent of the reference frame, i.e. in tensor form**. Otherwise, the Lorentz force law (2.3.2) would not be a fundamental law of nature, because it would change from one observer to the other. This clearly shows that **it is impossible to understand electrodynamics without the basis of relativity theory**. It is our aim now to achieve this rewriting.

For our purpose we must find four-dimensional quantities (Minkowski tensors) for all quantities appearing in (2.3.2). This is simple in case of the velocity. The **four-velocity** is defined by

$$u^\mu = \frac{dx^\mu}{ds}, \tag{2.3.3}$$

where

$$ds^2 = (dx^0)^2 - (dx^1)^2 - (dx^2)^2 - (dx^3)^3 = dx^\mu dx_\mu \tag{2.3.4}$$

is the Lorentz invariant line element. Taking the square root, we obtain the so-called **proper time**

$$ds = dx^0\sqrt{1-\left(\frac{d\boldsymbol{x}}{dx^0}\right)^2} = dx^0\sqrt{1-\frac{\boldsymbol{v}^2}{c^2}} \stackrel{\text{def}}{=} \frac{dx^0}{\gamma} \tag{2.3.5}$$

of the moving test charge. The square root often occurs in relativistic equations and is usually abbreviated by $1/\gamma$. Since ds is a scalar, u^μ is indeed a four-vector. The usual 3-velocity in the laboratory frame is equal to

$$\boldsymbol{v} = \frac{d\boldsymbol{x}}{dt}. \tag{2.3.6}$$

Then we can write u^μ as follows

$$u^\mu = \gamma\Big(1, \frac{\boldsymbol{v}}{c}\Big). \tag{2.3.7}$$

It is normalized

$$u^2 = u^\mu u_\mu = \gamma^2\Big(1 - \frac{\boldsymbol{v}^2}{c^2}\Big) = 1, \tag{2.3.8}$$

but is dimensionless instead of having the dimension of a velocity. We might have multiplied (2.3.7) by c, but this is not the usual convention.

Next we consider the **four-momentum**

$$p^\mu = m_0 c u^\mu = m_0\gamma(c, \boldsymbol{v}), \tag{2.3.9}$$

where m_0 is the rest mass of the moving particle, which is a Lorentz invariant. The mass in the lab frame is given by

$$m = m_0\gamma = \frac{m_0}{\sqrt{1 - \frac{v^2}{c^2}}}, \tag{2.3.10}$$

this leads to the simple expression

$$p^\mu = m(c, \boldsymbol{v}). \tag{2.3.11}$$

The normalization of p

$$p^2 = p_\mu p^\mu = m_0^2 c^2 \tag{2.3.12}$$

has the correct dimension. From the four-momentum we get the **four-vector force**

$$K^\mu = \frac{dp^\mu}{ds} = \Big(c\frac{dm}{ds}, \gamma\frac{d}{dx^0}(m\boldsymbol{v})\Big). \tag{2.3.13}$$

Its spatial part

$$\boldsymbol{K} = \frac{\gamma}{c}\frac{d}{dt}(m\boldsymbol{v}) \stackrel{\text{def}}{=} \frac{\gamma}{c}\boldsymbol{K}_{\text{cl}} \tag{2.3.14}$$

defines the **classical force $\boldsymbol{K}_{\text{cl}}$** which is known from classical non-relativistic mechanics. In contrast to $\boldsymbol{K}$, $\boldsymbol{K}_{\text{cl}}$ has the correct physical dimension. It is this force which we have to identify with the Lorentz force (2.3.2), as we shall see below.

The meaning of the zeroth component K^0 in (2.3.13) is somewhat obscure. To understand this, we differentiate (2.3.12) with respect to s and get

$$\begin{aligned} 2p_\mu K^\mu &= 0 \qquad \text{or} \\ p^0K^0 - \boldsymbol{p}\cdot\boldsymbol{K} &= 0. \end{aligned} \tag{2.3.15}$$

This shows that

$$K^0 = \frac{1}{c}\boldsymbol{v}\cdot\boldsymbol{K} \tag{2.3.16}$$

is essentially the work per unit time (power) of the force which is transferred to the test charge. Hence

$$K^\mu = \Big(\frac{1}{c}\boldsymbol{v}\cdot\boldsymbol{K},\boldsymbol{K}\Big) = \frac{\gamma}{c}\Big(\frac{1}{c}\boldsymbol{v}\cdot\boldsymbol{K}_{\rm cl},\boldsymbol{K}_{\rm cl}\Big). \qquad (2.3.17)$$

As already said **the Lorentz force $\boldsymbol{K}_L$ must be identified with $\boldsymbol{K}_{\rm cl}$ not with $\boldsymbol{K}$, because otherwise we would not succeed in writing (2.3.2) in tensor form**. But with $\boldsymbol{K}_{\rm cl} = \boldsymbol{K}_L$ we get

$$\begin{aligned} K^\mu &= q\frac{\gamma}{c}\Big(\frac{1}{c}\boldsymbol{v}\cdot\boldsymbol{E},\boldsymbol{E}+\boldsymbol{v}\wedge\boldsymbol{B}\Big) \\ &= q\Big(\frac{1}{c}\boldsymbol{u}\cdot\boldsymbol{E},\frac{1}{c}u^0\boldsymbol{E}+\boldsymbol{u}\wedge\boldsymbol{B}\Big) \stackrel{\rm def}{=} qF^{\mu\nu}u_\nu. \end{aligned} \qquad (2.3.18)$$

It is essential that the four-velocity appears here, the three-vector $\boldsymbol{v}$ has disappeared completely. This would not be so if we had used $\boldsymbol{K}$ instead of $\boldsymbol{K}_{\rm cl}$. Since the l.h.s. K^μ is a four-vector and the r.h.s. depends linearly on the four-vector u_ν, **the linear transformation $F^{\mu\nu}$ must be a contravariant tensor of second rank** (see (2.2.10)). This field tensor $F^{\mu\nu}$ can now be read off from (2.3.18). Since the covariant components of u are equal to

$$u_\nu = (u^0,-\boldsymbol{u}), \qquad (2.3.19)$$

the **field tensor** has the following form

$$F^{\mu\nu} = \begin{pmatrix} 0 & -E^1/c & -E^2/c & -E^3/c \\ E^1/c & 0 & -B^3 & B^2 \\ E^2/c & B^3 & 0 & -B^1 \\ E^3/c & -B^2 & B^1 & 0 \end{pmatrix}. \qquad (2.3.20)$$

This is easy to verify because (2.3.18) is just the multiplication of the matrix (2.3.20) by (2.3.19), considered as a column matrix. This antisymmetric second rank tensor combines the electric and magnetic fields. Its transformation law follows from (2.2.8). A change of the reference frame by a Lorentz transformation

$$x' = \Lambda x, \quad x'^\mu = \Lambda^\mu{}_\nu x^\nu \qquad (2.3.21)$$

implies

$$F'^{\alpha\beta}(x') = \Lambda^\alpha{}_\mu \Lambda^\beta{}_\nu F^{\mu\nu}(x). \qquad (2.3.22)$$

We want to evaluate this for a Lorentz transformation in 1-direction

$$\Lambda^\alpha{}_\beta = \begin{pmatrix} \gamma & -\gamma v/c & 0 & 0 \\ -\gamma v/c & \gamma & 0 & 0 \\ 0 & 0 & 1 & 0 \\ 0 & 0 & 0 & 1 \end{pmatrix}. \qquad (2.3.23)$$

Using (2.3.20) we get

$$E'^1(x') = cF'^{10} = \gamma^2\Big(-\frac{v^2}{c^2}+1\Big)E^1 = E^1(x)$$

$$E'^2(x') = cF'^{20} = \gamma(E^2 - vB^3)(x)$$

$$E'^3 = \gamma(E^3 + vB^2), \quad B'^1 = B^2$$

$$B'^2(x') = \gamma\Big(B^2 + \frac{v}{c^2}E^3\Big)(x)$$

$$B'^3 = \gamma\Big(B^3 - \frac{v}{c^2}E^2\Big). \tag{2.3.24}$$

We notice that **electric and magnetic fields are partially transformed into each other**, therefore, **the identification of these fields depends on the reference frame.** Only the transverse fields perpendicular to the 1-direction are transformed. In first order in v we may drop $\gamma = 1 + O(v^2)$ and the result (2.3.24) can be simplified as follows

$$E'_x = E_x, \quad E'_y = E_y - vB_z, \quad E'_z = E_z + vB_y$$

$$B'_x = B_x, \quad B'_y = B_y + \frac{v}{c^2}E_z, \quad B'_z = B_z - \frac{v}{c^2}E_y, \tag{2.3.25}$$

where we denoted the components by x, y, z instead of 1,2,3.

2.4 Maxwell's Equations

The fundamental equation of electrostatics was Gauss' law

$$\operatorname{div} \boldsymbol{E} = \frac{1}{\varepsilon_0}\rho, \tag{2.4.1}$$

saying that the sources of the electric field are the electric charges. We must also know what the sources of the magnetic field are. Up to now magnetic charges (monopoles) have not been found experimentally, thus

$$\operatorname{div} \boldsymbol{B} = 0. \tag{2.4.2}$$

By the relativity principle **these two source equations must hold in any frame**, if they are really fundamental laws of nature. For (2.4.2) that means

$$\operatorname{div}'\boldsymbol{B}'(x') = \frac{\partial B'_x}{\partial x'} + \frac{\partial B'_y}{\partial y'} + \frac{\partial B'_z}{\partial z'} = 0. \tag{2.4.3}$$

We consider a Lorentz transformation in x-direction (2.3.23). Since the gradient transforms with the inverse transposed matrix, we have

$$\frac{\partial}{\partial t'} = \gamma\left(\frac{\partial}{\partial t} + v\frac{\partial}{\partial x}\right), \quad \frac{\partial}{\partial x'} = \gamma\left(\frac{\partial}{\partial x} + \frac{v}{c^2}\frac{\partial}{\partial t}\right)$$

$$\frac{\partial}{\partial y'} = \frac{\partial}{\partial y}, \quad \frac{\partial}{\partial z'} = \frac{\partial}{\partial z}. \tag{2.4.4}$$

The sign of the terms $\sim v$ has been changed by the inverse Lorentz matrix. Using (2.3.24) in (2.4.3), we find

$$0 = \gamma\left(\frac{\partial B_x}{\partial x} + \frac{v}{c^2}\frac{\partial B_x}{\partial t} + \frac{\partial B_y}{\partial y} + \frac{v}{c^2}\frac{\partial E_z}{\partial y} + \frac{\partial B_z}{\partial z} - \frac{v}{c^2}\frac{\partial E_y}{\partial z}\right). \tag{2.4.5}$$

To leading order $O(1)$ this is the original source equation (2.4.2), but to $O(v/c)$ we obtain the new equation

$$\frac{\partial B_x}{\partial t} + \frac{\partial E_z}{\partial y} - \frac{\partial E_y}{\partial z} = 0. \tag{2.4.6}$$

The terms containing $\boldsymbol{E}$ are just the x-component of curl $\boldsymbol{E}$. Since the direction of the Lorentz transformation is arbitrary, we conclude

$$\frac{\partial \boldsymbol{B}}{\partial t} + \operatorname{curl}\boldsymbol{E} = \boldsymbol{0}. \tag{2.4.7}$$

This is the **induction law**. The equations (2.4.2) and (2.4.7) are the **two homogeneous Maxwell equations**.

It remains to consider the Lorentz transformation of **Gauss' law** (2.4.1). Assuming that this is also a fundamental law of nature, it **must hold in any Lorentz frame**. That means, in particular, that it is not only true for stationary charges but also for moving charges. We do not yet know how the charge density $\rho(t, \boldsymbol{x})$ transforms under a change of the reference frame. If a charge is viewed from a moving frame, it becomes a current. The definition of current follows from that of charge: If the charge dQ moves through a surface F during time dt, we say that the current

$$J = \frac{\partial Q}{\partial t} \tag{2.4.8}$$

is flowing. Its unit is 1 Coul/sec = 1 Ampère. In general the strength of the current is not constant over the surface F. Then it must be expressed by a surface integral of a current density $\boldsymbol{j}(t, \boldsymbol{x})$

$$J = \int_F \boldsymbol{j}(t, \boldsymbol{x}) \cdot d\boldsymbol{\sigma}. \tag{2.4.9}$$

By definition the direction of the current density vector coincides with the direction of motion of the positive charges.

In order to understand how ρ and $\boldsymbol{j}$ must be combined into a four-dimensional quantity, it is convenient to discuss charge conservation. The temporal change of the total charge in a volume K_3 is given by the current through the boundary ∂K_3

$$\frac{d}{dt}\int\limits_{K_3} \rho(t,\boldsymbol{x})\,d^3x = -\int\limits_{\partial K_3} \boldsymbol{j}(t,\boldsymbol{x})\,d\boldsymbol{\sigma}. \qquad (2.4.10)$$

This expresses the fact that charge cannot be generated, nor destroyed in K_3. Using Gauss' theorem we conclude

$$\int\limits_{K_3}\Big(\frac{\partial\rho}{\partial t} + \operatorname{div}\boldsymbol{j}\Big)\,d^3x = 0,$$

and

$$\frac{\partial\rho}{\partial t} + \operatorname{div}\boldsymbol{j} = 0, \qquad (2.4.11)$$

because K_3 is arbitrary. This is the continuity equation which expresses charge conservation. It can immediately be written in four-dimensional form

$$\frac{\partial c\rho}{\partial x^0} + \frac{\partial j^1}{\partial x^1} + \frac{\partial j^2}{\partial x^2} + \frac{\partial j^3}{\partial x^3} = \partial_\mu j^\mu = 0. \qquad (2.4.12)$$

This shows that

$$j^\mu = (c\rho,\,\boldsymbol{j}) \qquad (2.4.13)$$

is a four-vector field, the **charge-current density four-vector**. It, therefore, transforms under a Lorentz transformation in x-direction (2.3.23) as follows

$$\rho' = \gamma\Big(\rho - \frac{v}{c^2}j_x\Big), \quad j'_x = \gamma(j_x - \rho v), \quad j'_y = j_y, \quad j'_z = j_z. \qquad (2.4.14)$$

Now we are ready to transform Gauss' law

$$\operatorname{div}'\boldsymbol{E}' - \frac{1}{\varepsilon_0}\rho' = 0:$$

$$0 = \gamma\Big(\frac{\partial E'_x}{\partial x'} + \frac{\partial E'_y}{\partial y'} + \frac{\partial E'_z}{\partial z'} - \frac{\rho'}{\varepsilon_0}\Big) = \gamma\Big(\frac{\partial E_x}{\partial x} + \frac{v}{c^2}\frac{\partial E_x}{\partial t}$$
$$+ \frac{\partial E_y}{\partial y} - v\frac{\partial B_z}{\partial y} + \frac{\partial E_z}{\partial z} + v\frac{\partial B_y}{\partial z} - \frac{\rho}{\varepsilon_0} + \frac{v}{c^2\varepsilon_0}j_x\Big). \qquad (2.4.15)$$

The leading order $O(1)$ is Gauss' law again. To $O(v/c)$ we get

$$\frac{1}{c^2}\frac{\partial E_x}{\partial t} - \frac{\partial B_z}{\partial y} + \frac{\partial B_y}{\partial z} + \frac{1}{c^2\varepsilon_0}j_x = 0. \qquad (2.4.16)$$

Here the terms containing the magnetic field are equal to $-(\operatorname{curl}\boldsymbol{B})_x$. Consequently, we get the following general equation

$$\frac{1}{c^2}\frac{\partial \boldsymbol{E}}{\partial t} = \operatorname{curl} \boldsymbol{B} - \mu_0 \boldsymbol{j}, \tag{2.4.17}$$

where

$$\mu_0 = \frac{1}{c^2 \varepsilon_0} = \frac{4\pi}{10^7} \frac{\mathrm{N}}{\mathrm{A}^2} \tag{2.4.18}$$

due to the previous convention (1.1.7). Gauss' law (2.4.1) and (2.4.17) constitute the **two inhomogeneous Maxwell's equations.**

It is clear from this derivation that it is possible to write **Maxwell's equations in tensor form** using the field tensor (2.3.20). In fact, the inhomogeneous equations read

$$\partial_\nu F^{\mu\nu} = -\mu_0 j^\mu. \tag{2.4.19}$$

For $\mu = 1$, for example, one easily checks

$$\frac{1}{c}\frac{\partial E^1}{\partial x^0} - \frac{\partial B^3}{\partial x^2} + \frac{\partial B^2}{\partial x^3} = \frac{1}{c^2}\frac{\partial E^1}{\partial t} - (\operatorname{curl} \boldsymbol{B})^1 = -\mu_0 j^1. \tag{2.4.20}$$

Concerning the homogeneous equations it is convenient to use the field tensor $F_{\mu\nu}$ with two lower (covariant) indices

$$F_{\mu\nu} = \begin{pmatrix} 0 & E^1/c & E^2/c & E^3/c \\ -E^1/c & 0 & -B^3 & B^2 \\ -E^2/c & B^3 & 0 & -B^1 \\ -E^3/c & -B^2 & B^1 & 0 \end{pmatrix}. \tag{2.4.21}$$

It differs from $F^{\mu\nu}$ (2.3.20) only in the signs of the temporal components. If we consider

$$\operatorname{div} \boldsymbol{B} = -\frac{\partial F_{23}}{\partial x^1} - \frac{\partial F_{31}}{\partial x^2} - \frac{\partial F_{12}}{\partial x^3} = 0, \tag{2.4.22}$$

we realize that the indices appear in cyclic order. Thus we conjecture the following form of the **homogeneous equations**

$$\frac{\partial F_{\mu\nu}}{\partial x^\lambda} + \frac{\partial F_{\nu\lambda}}{\partial x^\mu} + \frac{\partial F_{\lambda\mu}}{\partial x^\nu} = 0. \tag{2.4.23}$$

The left side is an antisymmetric covariant tensor of third rank. Such a tensor has only 4 non-vanishing components with the following index triples: (0,1,2), (0,1,3), (0,2,3) and (1,2,3). The case (1,2,3) was just considered, let us also check (0,1,2):

$$\frac{\partial F_{01}}{\partial x^2} + \frac{\partial F_{12}}{\partial x^0} + \frac{\partial F_{20}}{\partial x^1} = \frac{1}{c}\left(\frac{\partial E^1}{\partial x^2} - \frac{\partial E^2}{\partial x^1}\right) - \frac{1}{c}\frac{\partial B^3}{\partial t} = 0.$$

This is indeed the 3-component of the induction law and the other two components follow from the remaining triples.

Summing up, we have obtained Maxwell's equations from (i) the Lorentz force law (2.3.2), (ii) from the two source equations

$$\operatorname{div} \boldsymbol{B} = 0 \tag{2.4.24}$$

$$\operatorname{div} \boldsymbol{E} = \frac{\rho}{\varepsilon_0}. \tag{2.4.25}$$

and (iii) from Lorentz covariance of the theory. The Lorentz force defines the electric and magnetic fields and eqs. (2.4.24-25) specify their sources. Therefore, none of these equations can be omitted. In fact, if we change one of the three basic equations, we get a completely different theory. For example, if we assume that magnetic charges exist and the right side of (2.4.24) is some monopole density $\mu(t, \boldsymbol{x})$, the homogeneous Maxwell equations are no longer true. The resulting theory (Problem 1) is symmetric in $\boldsymbol{E}$ and $\boldsymbol{B}$. It does not allow the introduction of electromagnetic potentials, which will be discussed in Sect.3.4. This shows that one should not introduce potentials too early. This theory with magnetic monopoles has found considerable interest in modern times, because "grand unified gauge theories" predict such objects. Curiously enough, one monopole was even "observed" a single time, but this apparently was a flop. If we take the source equation (2.4.24) without the Lorentz force law, then also gravitation theory is a possible example. Its relativistic formulation by Einstein (general relativity) is very different from electrodynamics. If one assumes only a part of the Lorentz force law, then there exist other counter examples. This shows that the two source equations (2.4.24-25) together with the Lorentz force are really the basis of electrodynamics in vacuum. However, **all these equations remain rather meaningless without the relativity principle**. This is the indispensable frame of the theory, not merely an elegant formalism. The latter wrong opinion was sometimes expressed in the early days of relativity theory.

2.5 Discussion of Maxwell's Equations, Induction Law

Maxwell's equations consists of the two source equations

$$\operatorname{div} \boldsymbol{B}(t, \boldsymbol{x}) = 0 \tag{2.5.1}$$

$$\varepsilon_0 \operatorname{div} \boldsymbol{E}(t, \boldsymbol{x}) = \rho(t, \boldsymbol{x}), \tag{2.5.2}$$

which contain no time derivative and specify the source densities $\operatorname{div} \boldsymbol{B}$, $\operatorname{div} \boldsymbol{E}$ of the fields, and the two dynamical equations with time derivatives

$$\frac{\partial \boldsymbol{B}}{\partial t} = -\operatorname{curl} \boldsymbol{E} \tag{2.5.3}$$

$$\varepsilon_0 \frac{\partial \boldsymbol{E}}{\partial t} = \frac{1}{\mu_0} \operatorname{curl} \boldsymbol{B} - \boldsymbol{j}. \tag{2.5.4}$$

Eqs.(2.5.1) and (2.5.3) are the homogeneous equations, the other two equations, involving ρ and $\boldsymbol{j}$, are the inhomogeneous ones. We consider charge and current densities $\rho(t, \boldsymbol{x})$, $\boldsymbol{j}(t, \boldsymbol{x})$ as given and want to determine the fields $\boldsymbol{E}(t, \boldsymbol{x})$ and $\boldsymbol{B}(t, \boldsymbol{x})$ by solving the system of linear partial differential equations of first order.

Since the equations are first order in t, the initial value problem (Cauchy problem) is well posed if we know the fields at an initial time $t = 0$ in the whole space, i.e. $\boldsymbol{E}(0, \boldsymbol{x})$ and $\boldsymbol{B}(0, \boldsymbol{x})$ are given. But then the equations (2.5.3) and (2.5.4) alone are sufficient to calculate the fields for later times. **The source equations** (2.5.1-2) are not true equations of motion because they contain no time derivative. They **are constraints on the fields**. Here arises a consistency problem: If the source equations are satisfied by the initial data at $t = 0$, they must automatically be fulfilled for all $t > 0$ due to (2.5.3-4). To check this we differentiate (2.5.1) and use (2.5.3)

$$\frac{\partial}{\partial t}\text{div}\,\boldsymbol{B} = \text{div}\,\frac{\partial \boldsymbol{B}}{\partial t} = -\text{div}\,\text{curl}\,\boldsymbol{E} = 0.$$

Since this holds for all $\boldsymbol{x}$ and t, it follows indeed

$$\text{div}\,\boldsymbol{B}(t, \boldsymbol{x}) = 0,$$

if this is true at $t = 0$. Differentiating (2.5.2) and using (2.5.4) we get

$$\varepsilon_0 \text{div}\,\frac{\partial \boldsymbol{E}}{\partial t} - \frac{\partial \rho}{\partial t} = \frac{1}{\mu_0}\text{div}\,\text{curl}\,\boldsymbol{B} - \text{div}\,\boldsymbol{j} - \frac{\partial \rho}{\partial t}. \tag{2.5.5}$$

This vanishes if and only if the continuity equation (2.4.11) holds. Hence, charge conservation is a necessary and sufficient requirement for the consistency of Maxwell's equations.

We have said already that (2.5.3) is **Faraday's induction law**. In fact, if K_2 is a fixed two-dimensional compact surface, Stokes' theorem implies

$$\int_{K_2} \text{curl}\,\boldsymbol{E}\cdot d\boldsymbol{\sigma} = \int_{\partial K_2} \boldsymbol{E}\cdot d\boldsymbol{x} = -\int_{K_2} \frac{\partial \boldsymbol{B}}{\partial t}\cdot d\boldsymbol{\sigma}. \tag{2.5.6}$$

Here the temporal change of the magnetic flux Φ appears:

$$\frac{\partial}{\partial t}\int_{K_2} \boldsymbol{B}\cdot d\boldsymbol{\sigma} = \frac{d\Phi}{dt} = -\int_{\partial K_2} \boldsymbol{E}\cdot d\boldsymbol{x} = -U_i, \tag{2.5.7}$$

and it is equal to the voltage U_i induced along the contour ∂K_2. Such a voltage along a given path is also called electromotive force, because if the contour is a conducting wire, the electromotive force gives rise to a current flow through the wire. If the the conducting loop is opened and the two ends are connected with a voltmeter, it measures the voltage U_i. But the electromotive force is also present without any wire. This can be seen in the propagation of electromagnetic waves in vacuum (see Sect.3.3, 3.4).

The induction law in integral form (2.5.7) has been derived for a fixed time-independent surface K_2, only. But it is also true if $K_2(t)$ varies with time. In this case it follows most elegantly and in full generality from Stokes' theorem in 4-dimensional Minkowski space. This requires some knowledge of differential forms. Since we do not use this tool elsewhere in the book, we only make a small insertion for the interested reader, which may be skipped without harm for the further understanding.

The (covariant) field tensor (2.4.21) defines a 2-form

$$F = F_{\mu\nu} dx^\mu \wedge dx^\nu . \tag{2.5.8}$$

Then the homogeneous Maxwell's equations (2.4.23) assume the simple form

$$\begin{aligned} dF &= \sum_{\mu\nu\lambda} \frac{\partial F_{\mu\nu}}{\partial x^\lambda} dx^\lambda \wedge dx^\mu \wedge dx^\nu \\ &= \frac{1}{3} \sum_{\mu\nu\lambda} \left(\frac{\partial F_{\mu\nu}}{\partial x^\lambda} + \frac{\partial F_{\nu\lambda}}{\partial x^\mu} + \frac{\partial F_{\lambda\mu}}{\partial x^\nu} \right) dx^\lambda \wedge dx^\mu \wedge dx^\nu = 0, \end{aligned} \tag{2.5.9}$$

where d denotes the external derivative. If K_3 is a 3-dimensional compact region with 2-dimensional boundary ∂K_3, Stokes' theorem gives

$$\int_{K_3} dF = \int_{\partial K_3} F = 0. \tag{2.5.10}$$

Let now $K_2(t)$ be a compact 2-dimensional surface for all $t_1 \le t \le t_2$. For K_3 we choose the following set (Fig.2)

$$K_3 = \left\{ (t, \boldsymbol{x}) \in \mathbb{R}^4 \,\middle|\, t_1 \le t \le t_2, \quad \boldsymbol{x} \in K_2(t) \right\}.$$

Then the boundary ∂K_3 consists of three pieces: $K_2(t_1)$, $K_2(t_2)$ and the cylinder-like surface

$$C_2 = \left\{ (t, \boldsymbol{x}) \in \mathbb{R}^4 \,\middle|\, t_1 \le t \le t_2, \quad \boldsymbol{x} \in \partial K_2(t) \right\}.$$

The corresponding surface integrals in (2.5.10) are equal to

$$-2 \int_{K_2(t_2)} \boldsymbol{B} \cdot d\boldsymbol{\sigma} + 2 \int_{K_2(t_1)} \boldsymbol{B} \cdot d\boldsymbol{\sigma} + \int_{C_2} F = 0. \tag{2.5.11}$$

To compute the last integral, we assume that $\partial K_2(t)$ is given by an equation $\boldsymbol{x} = \boldsymbol{x}(t, s)$, where s measures the arc length along ∂K_2. Then

$$dx^j = \frac{\partial x^j}{\partial t} dt + \frac{\partial x^j}{\partial s} ds, \quad dx^0 = c\, dt,$$

and, by means of (2.4.21), we get

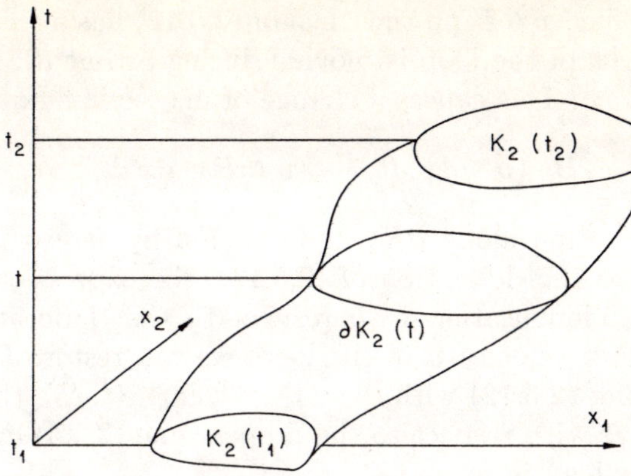

Fig. 2. General induction law in integral form

$$F_{\mu\nu}dx^\mu \wedge dx^\nu = 2F_{0j}dx^0 \wedge dx^j + F_{jk}dx^j \wedge dx^k$$

$$= 2E^j dt \wedge \left(\frac{\partial x^j}{\partial s}\, ds\right) - \varepsilon^{jkl}B^l\left(\frac{\partial x^j}{\partial t}\, dt + \frac{\partial x^j}{\partial s}\, ds\right) \wedge \left(\frac{\partial x^k}{\partial t}\, dt + \frac{\partial x^k}{\partial s}\, ds\right)$$

$$= 2\boldsymbol{E}\cdot\frac{\partial \boldsymbol{x}}{\partial s}\, dt \wedge ds - 2\varepsilon^{ljk}B^l\frac{\partial x^j}{\partial t}\frac{\partial x^k}{\partial s}\, dt \wedge ds.$$

This leads to

$$\int\limits_{C_2} = 2\int\limits_{t_1}^{t_2} dt \oint\limits_{\partial K_2(t)} ds\left(\boldsymbol{E}\cdot\frac{\partial \boldsymbol{x}}{\partial s} - \boldsymbol{B}\cdot\left[\frac{\partial \boldsymbol{x}}{\partial t}\wedge\frac{\partial \boldsymbol{x}}{\partial s}\right]\right)$$

$$= 2\int\limits_{t_1}^{t_2} dt \oint\limits_{\partial K_2(t)} ds\left(\boldsymbol{E} - \left[\boldsymbol{B}\wedge\frac{\partial \boldsymbol{x}}{\partial t}\right]\right)\cdot\frac{\partial \boldsymbol{x}}{\partial s}$$

$$= 2\int\limits_{t_1}^{t_2} dt \oint\limits_{\partial K_2(t)} d\boldsymbol{x}\left(\boldsymbol{E} + \frac{\partial \boldsymbol{x}}{\partial t}\wedge\boldsymbol{B}\right).$$

Substituting this into (2.5.11) and differentiating with respect to t_2, we arrive at

$$\frac{d}{dt}\int\limits_{K_2(t)} \boldsymbol{B}\cdot d\boldsymbol{\sigma} = -\oint\limits_{\partial K_2(t)} d\boldsymbol{x}\cdot\left(\boldsymbol{E} + \frac{\partial \boldsymbol{x}}{\partial t}\wedge\boldsymbol{B}\right). \tag{2.5.12}$$

This is the general induction law in integral form for time-dependent surfaces $K_2(t)$. Comparing with (2.5.7), we observe that the induced electric field is changed by a "motional field" $\boldsymbol{v}\wedge\boldsymbol{B}$ as in the Lorentz force law. Here $\boldsymbol{v}$ need not be a uniform constant velocity, it may change with t and along the loop $\partial K_2(t)$.

The motional field $\boldsymbol{v} \wedge \boldsymbol{B}$ on the r.h.s. of (2.5.12) has a simple geometric origin: If a piece $d\boldsymbol{x}$ of the loop is moving during a time dt, it sweeps over a surface $(\boldsymbol{v}\, dt) \wedge d\boldsymbol{x}$. This causes a change of magnetic flux

$$\boldsymbol{B} \cdot (\boldsymbol{v} \wedge d\boldsymbol{x})dt = -(\boldsymbol{v} \wedge \boldsymbol{B}) \cdot d\boldsymbol{x}\, dt,$$

which, after integration along ∂K_2 and division by dt gives the motional term. According to the derivation of (2.5.12), the expression was Lorentz invariant before differentiation with respect to the (laboratory) time dt. Assuming a uniform velocity $\boldsymbol{v}$ of the loop, we can restore Lorentz invariance by multiplying (2.5.12) with $\gamma = (1 - v^2/c^2)^{-1/2}$. We then get on the l.h.s. the derivative with respect to the proper time (2.3.5) $d\tau = dt/\gamma$. The integrand on the r.h.s.

$$\gamma(\boldsymbol{E} + \boldsymbol{v} \wedge \boldsymbol{B}) \stackrel{\text{def}}{=} \boldsymbol{E}'$$

is the spatial part of a four-vector (2.3.18). This is a pure electric field in the rest system of the loop. It has a direct physical meaning because it drives the induced current J which is flowing along the loop, if the latter is a conducting wire:

$$\frac{d\Phi}{d\tau} = -\oint_{\partial K_2} d\boldsymbol{x} \cdot \boldsymbol{E}' = -RJ. \tag{2.5.13}$$

Here Φ is the magnetic flux through K_2 and R is the resistivity of the loop circuit. As in (2.5.7), the temporal change of magnetic flux is equal to the induced voltage.

When Einstein struggled with relativity, he noticed an apparent lack of symmetry in the induction law, as was mentioned in the historical introduction: If a magnet is moving in the presence of a conducting loop at rest, the change of the magnetic flux Φ on the l.h.s. in (2.5.12) is different from zero. This induces an electric field $\boldsymbol{E}$, which gives rise to an induction current according to (2.5.13). The motional field vanishes since $\boldsymbol{v} = \boldsymbol{0}$. But if the magnet is stationary and the conducting loop is in motion, there is no electric field in the lab system. The current is now driven by the motional field or by the Lorentz force on the electrons in the wire. In both cases the induced current J must be the same, apart from a factor γ due to the different reference frames.

The induction law has an enormous practical importance. The change of the magnetic flux can be generated by moving a conducting loop in a constant magnetic field. This is the principle of the dynamo. Or the change of magnetic flux can be produced by changing the magnetic field in time (for example by an alternating current in a solenoid), leaving the loop (the secondary solenoid) unchanged. This is the principle of the transformer. Dynamos and transformers are the basic elements of the electric supply system: The alternating current produced in the dynamos of the power stations is transformed by transformers to very high voltage. In this form it can be transported by cable over large distances with minimal losses. Near the consumers the current is transformed to low voltage again by further small transformers.

The minus sign in the induction law prevents the possibility of a perpetuum mobile. Let us consider a current that forms a right screw, that means it flows along the direction of the fingers of the right fist, and produces a magnetic field in the direction of the thumb. Moving a conducting ring into this field and choosing the normal direction $\boldsymbol{d\sigma}$ parallel to $\boldsymbol{B}$, we have

$$\int_{K_2} \frac{\partial \boldsymbol{B}}{\partial t} \cdot \boldsymbol{d\sigma} > 0. \tag{2.5.13a}$$

Then (2.5.6) implies

$$\int_{\partial K_2} \boldsymbol{E} \cdot \boldsymbol{ds} < 0.$$

Thus the induced current forms a left screw and its magnetic field is opposite to the inducing field $\boldsymbol{B}$. Consequently, the motion of the ring into the field costs mechanical work. If it were the other way around, one would gain mechanical energy and in addition produce an induction current. This would be a perpetuum mobile. The induced current always weakens the inducing magnetic field, which is the so-called Lenz's rule. The induction law also shows a fundamental difference between time-independent and time-dependent electromagnetic fields. In electrostatics the electric field was vortex-free, $\operatorname{curl} \boldsymbol{E} = \boldsymbol{0}$. If the fields are time-dependent, this is no longer the case.

We turn now to (2.5.4). In the stationary, time-independent situation this is **Ampère's law**

$$\operatorname{curl} \boldsymbol{B} = \mu_0 \boldsymbol{j}. \tag{2.5.14}$$

For this reason we shall also refer to (2.5.4) as Ampère's law. In addition, by (2.5.1) the magnetic field is source-free. Then according to Theorem 4 (0.2.37) we can express $\boldsymbol{B}(\boldsymbol{x})$ by its vortex density (2.5.14) as follows

$$\boldsymbol{B}(\boldsymbol{x}) = \frac{\mu_0}{4\pi} \int d^3y \, \frac{\boldsymbol{j}(\boldsymbol{y}) \wedge (\boldsymbol{x} - \boldsymbol{y})}{|\boldsymbol{x} - \boldsymbol{y}|^3} = \operatorname{curl} \boldsymbol{A}(\boldsymbol{x}). \tag{2.5.15}$$

This is the **Biot-Savart law**. The vector potential $\boldsymbol{A}(\boldsymbol{x})$ can be read off from (2.5.15)

$$\boldsymbol{A}(\boldsymbol{x}) = \frac{\mu_0}{4\pi} \int d^3y \, \frac{\boldsymbol{j}(\boldsymbol{y})}{|\boldsymbol{x} - \boldsymbol{y}|}, \tag{2.5.16}$$

but is of course not unique (see (0.2.33)). The direction of the magnetic field (2.5.15) is given by the so-called right fist-rule: If the fingers of the right fist point in the direction of the current, then the thumb shows the direction of the magnetic field.

Finally we want to study the behavior of Maxwell's equations under reflections. We consider space reflection or parity $P : \boldsymbol{x} \to -\boldsymbol{x}$ and time inversion $T : t \to -t$ (2.1.9). It follows from the definition of the electromagnetic fields by the Lorentz force law (2.3.2) that $\boldsymbol{E}$ and $\boldsymbol{B}$ transform differently under these reflections. Let us first consider time reversal T: We know from mechanics that velocity $\boldsymbol{v}$ is odd $\boldsymbol{v} \to -\boldsymbol{v}$, thus acceleration $\boldsymbol{b}$ is even and so is force $\boldsymbol{K}$ by definition (1.1.4). Then (2.3.2) implies that $\boldsymbol{E}$ is even and $\boldsymbol{B}$ is odd

$$\boldsymbol{E}(t,\boldsymbol{x}) \to \boldsymbol{E}(-t,\boldsymbol{x}), \quad \boldsymbol{B}(t,\boldsymbol{x}) \to -\boldsymbol{B}(-t,\boldsymbol{x}). \tag{2.5.17}$$

For parity the behavior is the other way around: $\boldsymbol{v}$ is still odd, but $\boldsymbol{b}$ and $\boldsymbol{K}$ are also odd, thus $\boldsymbol{E}$ is odd and $\boldsymbol{B}$ is even:

$$\boldsymbol{E}(t,\boldsymbol{x}) \to -\boldsymbol{E}(t,-\boldsymbol{x}), \quad \boldsymbol{B}(t,\boldsymbol{x}) \to \boldsymbol{B}(t,-\boldsymbol{x}). \tag{2.5.18}$$

A vector like $\boldsymbol{B}$ that does not change sign under parity is called a pseudovector or axial vector. All other vector quantities just considered are true or polar vectors. Since temporal and spatial derivatives change sign under T and P, respectively, it is now easy to verify that **Maxwell's equations** (2.5.1-4) **are invariant under** T **and** P, provided $\rho(t,\boldsymbol{x})$ is a scalar and $\boldsymbol{j}(t,\boldsymbol{x})$ is a true vector like $\boldsymbol{v}$ (note that this is consistent with $\boldsymbol{j} = \rho\boldsymbol{v}$)

$$\boldsymbol{j}(t,\boldsymbol{x}) \to -\boldsymbol{j}(-t,\boldsymbol{x}), \quad \boldsymbol{j}(t,\boldsymbol{x}) = -\boldsymbol{j}(t,-\boldsymbol{x}). \tag{2.5.19}$$

This is in accordance with their definitions. We observe that the four-vector $j^\mu = (c\rho, \boldsymbol{j})$ (2.4.13) transforms under time inversion different from $x^\mu = (ct, \boldsymbol{x})$. For this reason the tensor calculus in Sect.2.2 was based on the proper Lorentz group, only.

2.6 Conservation Laws

We now investigate systematically the conservation laws of Maxwell's equations

$$\varepsilon_0 \boldsymbol{E}' = \frac{1}{\mu_0}\text{curl}\,\boldsymbol{B} - \boldsymbol{j} \tag{2.6.1}$$

$$\boldsymbol{B}' = -\text{curl}\,\boldsymbol{E} \tag{2.6.2}$$

$$\varepsilon_0 \text{div}\,\boldsymbol{E} = \rho \tag{2.6.3}$$

$$\text{div}\,\boldsymbol{B} = 0. \tag{2.6.4}$$

As before, the prime denotes the partial derivative with respect to time. We consider the charge density ρ and the current density $\boldsymbol{j}$ as given for all $(t,\boldsymbol{x})$. The current density can be expressed by the velocities of the charges

$$\boldsymbol{j} = \rho\boldsymbol{v}. \tag{2.6.5}$$

For given $\boldsymbol{j}(t, \boldsymbol{x})$, this is the definition of the velocity field, or vice versa. The electromagnetic field acts on the charges according to the Lorentz force law with the force density

$$\boldsymbol{k} = \rho \boldsymbol{E} + \boldsymbol{j} \wedge \boldsymbol{B}. \tag{2.6.6}$$

All these equations are necessary to check conservation of charge, energy, momentum and angular momentum.

Charge conservation follows easily from (2.6.3) and (2.6.1):

$$\frac{\partial \rho}{\partial t} = \varepsilon_0 \operatorname{div} \boldsymbol{E}' = -\operatorname{div} \boldsymbol{j}. \tag{2.6.7}$$

But we have emphasized before (2.5.5) that this is a necessary condition for consistency of Maxwell's equations. Strictly speaking, this conservation law must then be verified outside of Maxwell's theory in the equations that determine $\rho(t, \boldsymbol{x})$ and $\boldsymbol{j}(t, \boldsymbol{x})$. For example, if $\boldsymbol{j}$ arises from the flow of a plasma fluid (2.6.5), consisting of one kind of charged particles, then charge conservation is a consequence of mass conservation, guaranteed by the corresponding hydrodynamic equations. We will discuss this electro-hydrodynamics in more detail in Sect.3.6 (3.6.49).

The first true conservation law is **energy conservation**. To obtain this, we multiply (2.6.1) by $\boldsymbol{E}$ and (2.6.2) by $\boldsymbol{B}/\mu_0$ (scalar products) and add the two equations:

$$\varepsilon_0 \boldsymbol{E}' \cdot \boldsymbol{E} + \frac{1}{\mu_0} \boldsymbol{B}' \cdot \boldsymbol{B} = \frac{1}{\mu_0} (\boldsymbol{E} \cdot \operatorname{curl} \boldsymbol{B} - \boldsymbol{B} \cdot \operatorname{curl} \boldsymbol{E}) - \boldsymbol{j} \cdot \boldsymbol{E}. \tag{2.6.8}$$

On the left-hand side we have a total time derivative and on the right-hand side we obtained a divergence

$$\frac{1}{2} \frac{\partial}{\partial t} \Big(\varepsilon_0 \boldsymbol{E}^2 + \frac{1}{\mu_0} \boldsymbol{B}^2 \Big) = \frac{1}{\mu_0} \operatorname{div} \boldsymbol{B} \wedge \boldsymbol{E} - \boldsymbol{j} \cdot \boldsymbol{E}, \tag{2.6.9}$$

due to (0.1.21). This is the differential form of energy conservation. In fact,

$$u = \frac{1}{2} \Big(\varepsilon_0 \boldsymbol{E}^2 + \frac{1}{\mu_0} \boldsymbol{B}^2 \Big) \tag{2.6.10}$$

must be the energy density of the electromagnetic field, because we know already the expression (1.5.13) for a pure electric field energy from electrostatics. The last term

$$\boldsymbol{j} \cdot \boldsymbol{E} = \rho \boldsymbol{v} \cdot \boldsymbol{E} = \boldsymbol{k} \cdot \boldsymbol{v} \tag{2.6.11}$$

is the work per unit time and volume transferred to the charges (2.7.13). The divergence in (2.6.9) leads to a surface integral, upon integrating over a compact volume K_3 and using Gauss' theorem:

$$\frac{\partial}{\partial t} \frac{1}{2} \int_{K_3} \Big(\varepsilon_0 \boldsymbol{E}^2 + \frac{1}{\mu_0} \boldsymbol{B}^2 \Big) d^3x = \frac{1}{\mu_0} \int_{\partial K_3} (\boldsymbol{B} \wedge \boldsymbol{E}) \cdot d\boldsymbol{\sigma} - \int_{K_3} \boldsymbol{j} \cdot \boldsymbol{E} \, d^3x. \tag{2.6.12}$$

The surface integral must represent the total energy flow through the boundary ∂K_3. Thus

$$\boldsymbol{S} = \frac{1}{\mu_0} \boldsymbol{E} \wedge \boldsymbol{B}, \tag{2.6.13}$$

the so-called **Poynting vector**, must be the **energy-flow density of the electromagnetic field**. This energy flow through vacuum is a very new aspect of electromagnetic field theory. We will investigate it in much detail in the following chapter when we study radiation.

We now turn to **momentum conservation**. The temporal change of momentum of the charges is equal to the Lorentz force. We therefore start from (2.6.6) and substitute ρ from (2.6.3) and $\boldsymbol{j}$ from (2.6.1):

$$\boldsymbol{k} = \varepsilon_0 (\operatorname{div} \boldsymbol{E}) \boldsymbol{E} + \frac{1}{\mu_0} (\operatorname{curl} \boldsymbol{B}) \wedge \boldsymbol{B} - \varepsilon_0 \boldsymbol{E}' \wedge \boldsymbol{B}.$$

In the last term we insert the Poynting vector (2.6.13) and use the induction law (2.6.2)

$$\varepsilon_0 \boldsymbol{E}' \wedge \boldsymbol{B} = \varepsilon_0 \mu_0 \boldsymbol{S}' - \varepsilon_0 \boldsymbol{E} \wedge \boldsymbol{B}' = \varepsilon_0 \mu_0 \boldsymbol{S}' + \varepsilon_0 \boldsymbol{E} \wedge (\operatorname{curl} E).$$

The resulting equation

$$\boldsymbol{k} + \varepsilon_0 \mu_0 \boldsymbol{S}' = \varepsilon_0 (\operatorname{div} \boldsymbol{E}) \boldsymbol{E} + \frac{1}{\mu_0} (\operatorname{curl} \boldsymbol{B}) \wedge \boldsymbol{B} - \varepsilon_0 \boldsymbol{E} \wedge (\operatorname{curl} \boldsymbol{E})$$

can be written in a symmetrical form in $\boldsymbol{E}$ and $\boldsymbol{B}$

$$= \varepsilon_0 (\operatorname{div} \boldsymbol{E}) \boldsymbol{E} + \varepsilon_0 (\operatorname{curl} \boldsymbol{E}) \wedge \boldsymbol{E} + \frac{1}{\mu_0} (\operatorname{curl} \boldsymbol{B}) \wedge \boldsymbol{B} + \frac{1}{\mu_0} (\operatorname{div} \boldsymbol{B}) \boldsymbol{B}. \tag{2.6.14}$$

The last added term vanishes due to (2.6.4). To get a conservation law, we must be able to write the right-hand side as a divergence of a tensor field T_{ik} in $\mathbb{R}^3$.

Let us examine the 1-component of the electric terms

$$\begin{aligned}
(\operatorname{div} \boldsymbol{E}) E_1 + (\operatorname{curl} \boldsymbol{E})_2 E_3 - (\operatorname{curl} \boldsymbol{E})_3 E_2 &= (E_{1,1} + E_{2,2} + E_{3,3}) E_1 \\
&\quad + (E_{1,3} - E_{3,1}) E_3 - (E_{2,1} - E_{1,2}) E_2 \\
&= \frac{\partial}{\partial x_1} E_1^2 + \frac{\partial}{\partial x_2} E_1 E_2 + \frac{\partial}{\partial x_3} E_1 E_3 - \frac{1}{2} \frac{\partial}{\partial x_1} \left(E_1^2 + E_2^2 + E_3^2 \right) \\
&= \sum_{k=1}^{3} \frac{\partial}{\partial x_k} \left(E_1 E_k - \frac{1}{2} \delta_{1k} \boldsymbol{E}^2 \right).
\end{aligned} \tag{2.6.15}$$

This is indeed a 3-dimensional divergence and the same follows for the other components by cyclic permutation. Thus

$$k_i + \varepsilon_0\mu_0 \frac{\partial}{\partial t} S_i = \sum_{k=1}^{3} \frac{\partial}{\partial x_k} \Big[\varepsilon_0 E_i E_k + \frac{1}{\mu_0} B_i B_k - \frac{1}{2}\delta_{ik} \Big(\varepsilon_0 \boldsymbol{E}^2 + \frac{1}{\mu_0} \boldsymbol{B}^2 \Big) \Big]. \tag{2.6.16}$$

The square bracket is **Maxwell's stress tensor**

$$T_{ik} = \varepsilon_0 E_i E_k + \frac{1}{\mu_0} B_i B_k - \frac{1}{2}\delta_{ik} \Big(\varepsilon_0 \boldsymbol{E}^2 + \frac{1}{\mu_0} \boldsymbol{B}^2 \Big), \tag{2.6.17}$$

which is a symmetrical tensor in $\mathbb{R}^3$. To interpret (2.6.16) as momentum conservation, we integrate again over a compact volume K_3 and use Gauss' theorem for a tensor field (0.1.40)

$$\int_{K_3} k_i \, d^3x + \frac{\partial}{\partial t} \int_{K_3} \frac{1}{c^2} S_i \, d^3x = \int_{\partial K_3} T_{ik} d\sigma_k. \tag{2.6.18}$$

The first term is the temporal change of momentum of the charge carriers. The second term must be the time derivative of the momentum of the electromagnetic field. Its momentum density is proportional to the energy-flow density $\boldsymbol{S}$ (2.6.13), where we have written

$$\varepsilon_0 \mu_0 = \frac{1}{c^2}. \tag{2.6.19}$$

For the moment let us take this only as a short notation, but in the following section we shall identify c as the velocity of light. The right-hand side of (2.6.18) must be the momentum flow of the fields through the boundary ∂K_3.

A momentum-flow density is a stress in elasticity theory. This explains Maxwell's notion stress tensor for T_{ik}. We want to follow this mechanical analogy a bit further. T_{ik} is the i-component of the force / m^2 on a surface with normal direction along the k-axis. The diagonal elements T_{ii} are the pressures, the non-diagonal elements $T_{ik}, k \neq i$ the shearing stresses. One decomposes T_{ik} into an isotropic and anisotropic part by separating a traceless tensor T^0_{ik} :

$$T_{ik} = T^0_{ik} - p\,\delta_{ik}, \tag{2.6.20}$$

$$\operatorname{Tr} T^0_{ik} = \sum_{i=1}^{3} T^0_{ii} = 0. \tag{2.6.21}$$

$$\operatorname{Tr} T_{ik} = -3p, \tag{2.6.22}$$

where p is the scalar presure. In case of the electromagnetic field it is called the radiation pressure

$$\begin{aligned} 3p &= -\varepsilon_0 \boldsymbol{E}^2 - \frac{1}{\mu_0} \boldsymbol{B}^2 + \frac{3}{2} \Big(\varepsilon_0 \boldsymbol{E}^2 + \frac{1}{\mu_0} \boldsymbol{B}^2 \Big) \\ &= \frac{1}{2} \Big(\varepsilon_0 \boldsymbol{E}^2 + \frac{1}{\mu_0} \boldsymbol{B}^2 \Big). \end{aligned} \tag{2.6.23}$$

Comparing this with (2.6.10), we realize that the pressure is proportional to the energy density

$$p = \frac{1}{3}u. \tag{2.6.24}$$

This relation is important in the theory of black-body radiation. The radiation pressure is small (Problem 11), but it can be seen in the turning of the comet's tail away from the sun.

We finally discuss **angular momentum conservation**. We start from momentum conservation (2.6.16)

$$k_i + \frac{1}{c^2}\frac{\partial}{\partial t}S_i = \frac{\partial}{\partial x_k}T_{ik} \tag{2.6.25}$$

and take the vector product with $\boldsymbol{x}$. For this purpose it is convenient to use the 3-dimensional ε-tensor (0.1.4): we multiply by $\varepsilon_{jki}x_k$ and sum over i and k

$$(\boldsymbol{x} \wedge \boldsymbol{k})_j + \frac{\partial}{\partial t}\frac{1}{c^2}(\boldsymbol{x} \wedge \boldsymbol{S})_j = \varepsilon_{jki}x_k\frac{\partial}{\partial x_l}T_{il}. \tag{2.6.26}$$

It must be possible to write the right-hand side as a divergence. In fact, it is equal to

$$\frac{\partial}{\partial x_l}\Big(\varepsilon_{jki}x_k T_{il}\Big) - \varepsilon_{jki}\delta_{lk}T_{il}.$$

Here the last term $\varepsilon_{jki}T_{ik}$ vanishes because T_{ik} is symmetric, but ε_{jki} antisymmetric. Integrating over a compact volume K_3, we obtain the following conservation equation

$$\int\limits_{K_3} (\boldsymbol{x} \wedge \boldsymbol{k})_j d^3x + \frac{\partial}{\partial t}\int\limits_{K_3} \frac{1}{c^2}\Big(\boldsymbol{x} \wedge \boldsymbol{S}\Big)_j d^3x = \int\limits_{\partial K_3} M_{jl}d\sigma_l, \tag{2.6.27}$$

where

$$M_{jl} = \varepsilon_{jki}x_k T_{il} \tag{2.6.28}$$

must be interpreted as the angular momentum-flow density. The first term in (2.6.27) is the moment of the Lorentz force on the charged particles. The second term is the temporal change of angular momentum

$$\boldsymbol{J} = \frac{1}{c^2}\int \boldsymbol{x} \wedge \boldsymbol{S}\, d^3x \tag{2.6.29}$$

of the electromagnetic field. We have found that the electromagnetic field has all conserved quantities (energy, momentum and angular momentum) like a material particle, although it has no mass. This mysterious fact will be elucidated in the following chapter, when we investigate electromagnetic waves.

2.7 Problems

1. How must Maxwell's equations be altered, if the magnetic field has a monopole density $\mu(x)$ as a source, which transforms as the zeroth component of a four-vector ? Calculate in all orders in the velocity v.
2. Calculate the electromagnetic field of a point charge e moving in the 1-direction with velocity v, by means of a Lorentz transformation. Show that the surfaces of constant potentials in the laboratory system are ellipsoids (Heaviside ellipsoids).
3. What are the transformation laws for the electromagnetic fields $\boldsymbol{E}$, $\boldsymbol{B}$ under a Lorentz boost with arbitrary velocity $\boldsymbol{v}$?
4. a) Show that the result of Problem 3 can be written as a complex rotation ($\in SO(3,\mathbb{C})$)

$$\boldsymbol{F}'_{\parallel} = \boldsymbol{F}_{\parallel}$$

$$\boldsymbol{F}'_{\perp} = \boldsymbol{F}_{\perp} \cos i\chi + \boldsymbol{n} \wedge \boldsymbol{F}_{\perp} \sin i\chi \tag{2.7.1}$$

   of a complex three-vector

$$\boldsymbol{F} = \frac{1}{c}\boldsymbol{E} - i\boldsymbol{B}, \tag{2.7.2}$$

   where

$$\tanh\chi = \frac{v}{c}, \quad \boldsymbol{n} = \frac{\boldsymbol{v}}{v}. \tag{2.7.3}$$

   b) From the invariance of $\boldsymbol{F}^2$, determine the invariants of the electromagnetic field under Lorentz transformations. Write them in terms of the field tensor $F^{\mu\nu}$. (Hint: Use the 4-dimensional antisymmetric tensor $\varepsilon^{\mu\nu\alpha\beta}$.)
5. Let constant uniform fields $\boldsymbol{E}$ and $\boldsymbol{B}$ be given. Try to find a reference frame so that
   a) the transformed fields $\boldsymbol{E}'$, $\boldsymbol{B}'$ are parallel (choose the relative velocity perpendicular to $\boldsymbol{E}$ and $\boldsymbol{B}$).
   b) a pure electric field ($\boldsymbol{B} = \boldsymbol{0}$) is changed into a pure magnetic field ($\boldsymbol{E}' = \boldsymbol{0}$).
6. Calculate the force between two point charges $+q$ and $-q$ of distance $2a$ by integrating Maxwell's stress tensor over the symmetric plane between the charges.
7. The energy flow due to the sun on the earth is 1.4 kW m^{-2} (solar constant). How big is the radiation pressure on a completely absorbing surface ? Calculate the total force on the earth and compare it with the gravitational force of the sun.

3. Electrodynamics in Vacuum

Maxwell's equations in vacuum together with the Lorentz force law have a huge variety of applications. In this chapter we discuss the most important of them. First of all we consider magnetic fields produced by stationary currents. This so-called magnetostatics has some similarity to electrostatics and the comparison between the two clarifies the different rôle which $\boldsymbol{E}$ and $\boldsymbol{B}$ play in electrodynamics. Then we investigate the motion of charged particles in electromagnetic fields. In Sect.3.3 we come to electromagnetic radiation which is certainly the most interesting and typical manifestation of the electromagnetic field. We first investigate the properties of electromagnetic waves and then, in Sect.3.4, analyse their production. Moving (accelerated) charged particles also emit radiation as studied in Sect.3.5. The chapter closes with a variational formulation of electrodynamics. This Lagrange formalism leads to a deeper understanding of relativistic field theory.

3.1 Stationary Magnetic Fields

We want to study the magnetic field

$$\boldsymbol{B}(\boldsymbol{x}) = \operatorname{curl} \boldsymbol{A}(\boldsymbol{x}) \tag{3.1.1}$$

generated by a stationary current density $\boldsymbol{j}(\boldsymbol{x})$

$$\boldsymbol{A}(\boldsymbol{x}) = \frac{\mu_0}{4\pi} \int d^3y \, \frac{\boldsymbol{j}(\boldsymbol{y})}{|\boldsymbol{x}-\boldsymbol{y}|}, \tag{3.1.2}$$

according to Ampère's law (2.5.14-16). As an example we consider a circular current in the xy-plane. Using polar coordinates (r, ϑ, φ) we take

$$\boldsymbol{j}(\boldsymbol{x}) = j(\boldsymbol{x})(-\boldsymbol{e}_1 \sin\varphi + \boldsymbol{e}_2 \cos\varphi), \tag{3.1.3}$$

where $\boldsymbol{e}_1$ and $\boldsymbol{e}_2$ are unit vectors in the x and y-directions. For a circular current J of radius a, flowing through an infinitely thin wire, we must choose

$$j(\boldsymbol{x}) = J\delta(\cos\vartheta)\frac{1}{a}\delta(r-a), \tag{3.1.4}$$

because

$$\int \boldsymbol{j}(\boldsymbol{x}) \cdot d\boldsymbol{\sigma} = J \int \delta(\cos\vartheta)\frac{1}{a}\delta(r-a)\, d\cos\vartheta\, r dr = J. \tag{3.1.5}$$

The resulting vector potential

$$\boldsymbol{A}(\boldsymbol{x}) = \frac{\mu_0}{4\pi}\frac{J}{a}\int dr'\, r'^2 d\cos\vartheta' d\varphi'\, \frac{\delta(\cos\vartheta')\delta(r'-a)}{|\boldsymbol{x}-\boldsymbol{y}|}\times$$
$$\times(-\boldsymbol{e}_1\sin\varphi' + \boldsymbol{e}_2\cos\varphi') \tag{3.1.6}$$

is axial symmetric. Therefore we choose $\boldsymbol{x} = (r, \vartheta, \varphi = 0)$ and $\boldsymbol{y} = (r', \vartheta', \varphi')$. Since

$$(\boldsymbol{x}-\boldsymbol{y})^2 = r^2 + r'^2 - 2rr'(\cos\vartheta\cos\vartheta' + \sin\vartheta\sin\vartheta'\cos(0-\varphi')),$$

only the last term in (3.1.6) $\sim \cos\varphi'$ contributes:

$$\boldsymbol{A}(\varphi = 0) = \boldsymbol{e}_2\frac{\mu_0}{4\pi}\frac{J}{a}\int dr'\, r'^2 d\cos\vartheta' d\varphi'\, \cos\varphi' \frac{\delta(\cos\vartheta')\delta(r'-a)}{|\boldsymbol{x}-\boldsymbol{y}|}. \tag{3.1.7}$$

Furthermore, $\boldsymbol{A}$ has only an azimuthal component (1.2.13)

$$A_\varphi = \frac{\mu_0}{4\pi}\frac{J}{a}\int dr' r'^2 d\cos\vartheta' d\varphi'\, \cos\varphi' \frac{\delta(\cos\vartheta')\delta(r'-a)}{|\boldsymbol{x}-\boldsymbol{y}|}. \tag{3.1.8}$$

Here we use the expansion (1.2.17) and the addition theorem (1.2.21)

$$\frac{1}{|\boldsymbol{x}-\boldsymbol{y}|} = \frac{4\pi}{r}\sum_{l=0}^{\infty}\sum_{m=-l}^{l}\left(\frac{r'}{r}\right)^l \frac{Y_l^m(\vartheta,\varphi)}{2l+1} Y_l^m(\vartheta',\varphi')^*. \tag{3.1.9}$$

This expression is real. Substituting $\cos\varphi' = \operatorname{Re}\exp i\varphi'$ into (3.1.8), we see that only $m = 1$ gives a non-vanishing φ'-integral:

$$\int dr' r'^2 \delta(r'-a)\int d\cos\vartheta' d\varphi' e^{i\varphi'}\,\delta(\cos\vartheta') r'^l Y_l^m(\vartheta',\varphi')^*$$
$$= \delta_{m1} a^{l+2} Y_l^1\left(\frac{\pi}{2}, 0\right), \tag{3.1.10}$$

hence

$$A_\varphi = \frac{\mu_0}{4\pi} J a \frac{8\pi^2}{r} \operatorname{Re}\sum_l \left(\frac{a}{r}\right)^l \frac{Y_l^1(\vartheta, 0)}{2l+1} Y_l^1\left(\frac{\pi}{2}, 0\right)^*. \tag{3.1.11}$$

Using the explicit expression

$$Y_l^1(\vartheta',\varphi')^* = \sqrt{\frac{2l+1}{4\pi}\frac{(l-1)!}{(l+1)!}} P_l^1(\cos\vartheta') e^{-i\varphi'}, \tag{3.1.12}$$

we arrive at

$$A_\varphi = \frac{\mu_0}{4\pi} Ja \frac{2\pi}{r} \sum_l \left(\frac{a}{r}\right)^l \frac{P_l^1(\cos\vartheta)}{l(l+1)} P_l^1(0). \tag{3.1.13}$$

This is a multipole expansion of the vector potential, in complete analogy with the corresponding expansion in electrostatics (1.2.23).

The radial component of the magnetic field is obtained from

$$B_r = (\operatorname{curl} \boldsymbol{A})_r = \frac{1}{r\sin\vartheta}\frac{\partial}{\partial\vartheta}\sin\vartheta A_\varphi = -\frac{1}{r}\frac{\partial}{\partial\cos\vartheta}\sin\vartheta A_\varphi. \tag{3.1.14}$$

Since

$$\frac{\partial}{\partial z}\sqrt{1-z^2}P_l^1(z) = l(l+1)P_l(z), \tag{3.1.15}$$

we get

$$B_r = -\frac{\mu_0}{4\pi} Ja \frac{2\pi}{r^2} \sum_l \left(\frac{a}{r}\right)^l P_l(\cos\vartheta) P_l^1(0). \tag{3.1.16}$$

The longitudinal component, parallel to the meridian, is equal to

$$B_\vartheta = (\operatorname{curl} \boldsymbol{A})_\vartheta = -\frac{1}{r}\frac{\partial}{\partial r} r A_\varphi \tag{3.1.17}$$

$$= \frac{\mu_0}{4\pi} Ja \frac{2\pi}{r^2} \sum_l \left(\frac{a}{r}\right)^l \frac{P_l^1(\cos\vartheta)}{l+1} P_l^1(0), \tag{3.1.18}$$

and the azimuthal component vanishes. Using the special values

$$P_0^1(0) = 0, \quad P_1^1(0) = -1, \quad P_2^1(0) = 0, \quad P_3^1(0) = \frac{15}{2},$$

we finally get for the leading contribution ($l = 1$)

$$B_r = \frac{\mu}{4\pi} J \frac{2\pi a^2}{r^3}\cos\vartheta = \frac{\mu_0}{4\pi} 2J\pi a^2 \frac{\cos\vartheta}{r^3}\left(1 + O\left(\frac{a^2}{r^2}\right)\right) \tag{3.1.19}$$

$$B_\vartheta = \frac{\mu_0}{4\pi} J\pi a^2 \frac{\sin\vartheta}{r^3}\left(1 + O\left(\frac{a^2}{r^2}\right)\right), \tag{3.1.20}$$

and $B_\varphi = 0$. Comparing this with (1.2.10) and (1.2.12), we recognize it as a dipole field with magnetic dipole moment

$$M = \mu_0 J\pi a^2 \tag{3.1.21}$$

proportional to the area of the closed loop. The quadrupole moment vanishes, but there are higher multipoles ($l = 3 \ldots$).

Let us now turn to the general multipole expansion in cartesian coordinates. In

$$A_k(\boldsymbol{x}) = \frac{\mu_0}{4\pi}\int d^3y\, \frac{j_k(\boldsymbol{y})}{|\boldsymbol{x}-\boldsymbol{y}|} \tag{3.1.22}$$

we expand (see (1.2.2))

$$\frac{1}{|\boldsymbol{x}-\boldsymbol{y}|} = \frac{1}{r} + \frac{\boldsymbol{x}\cdot\boldsymbol{y}}{r^3} + \ldots, \quad r = |\boldsymbol{x}|. \tag{3.1.23}$$

The first term gives no contribution because

$$\begin{aligned}\int d^3y\, j_k(\boldsymbol{y}) &= \int d^3y\, \boldsymbol{j}(\boldsymbol{y})\cdot \operatorname{grad} y_k \\ &= \int d^3y[\operatorname{div}(y_k\boldsymbol{j}) - y_k \operatorname{div}\boldsymbol{j}] = 0,\end{aligned} \tag{3.1.24}$$

since $\operatorname{div}\boldsymbol{j} = 0$ due to charge conservation and the first term is equal to a surface integral at infinity, that vanishes if the current distribution is concentrated in a compact region. Contrary to electrostatics there exists no magnetic monopole field. The second term of (3.1.23) leads to

$$A_k(\boldsymbol{x}) = \frac{\mu_0}{4\pi}\sum_{l=1}^{3}\frac{x_l}{r^3}\int d^3y\, y_l j_k(\boldsymbol{y})\Big(1+O\Big(\frac{a}{r}\Big)\Big), \tag{3.1.25}$$

where a is a measure of the spatial extension of the current distribution. We consider

$$\begin{aligned}\int d^3y\, y_l j_k &= \int d^3y\big[\operatorname{div}(y_k y_l \boldsymbol{j}) - y_k j_l - y_k y_l \operatorname{div}\boldsymbol{j}\big] \\ &= -\int d^3y\, y_k j_l,\end{aligned} \tag{3.1.26}$$

because the third term in the bracket vanishes by charge conservation and the first one is a surface integral. Consequently, (3.1.26) is antisymmetric in l, k. Then (3.1.25) can be written as follows

$$A_k(\boldsymbol{x}) = \frac{\mu_0}{4\pi}\frac{1}{2}\sum_l \frac{x_l}{r^3}\int d^3y\,(y_l j_k - y_k j_l). \tag{3.1.27}$$

An antisymmetric 3-dimensional tensor is always equivalent to a vector, indeed:

$$\begin{aligned}\boldsymbol{A}(\boldsymbol{x}) &= \frac{\mu_0}{8\pi}\int d^3y\Big(\frac{\boldsymbol{x}}{r^3}\cdot\boldsymbol{y}\boldsymbol{j} - \boldsymbol{y}\frac{\boldsymbol{x}}{r^3}\cdot\boldsymbol{j}\Big) \\ &= \Big(\frac{\mu_0}{8\pi}\int d^3y\,\boldsymbol{y}\wedge\boldsymbol{j}(\boldsymbol{y})\Big)\wedge\frac{\boldsymbol{x}}{r^3} \overset{\text{def}}{=} \frac{1}{4\pi}\boldsymbol{M}\wedge\frac{\boldsymbol{x}}{r^3}.\end{aligned} \tag{3.1.28}$$

Here

$$\boldsymbol{M} = \frac{\mu_0}{2}\int d^3y\,\boldsymbol{y}\wedge\boldsymbol{j}(\boldsymbol{y}) \tag{3.1.29}$$

is, by definition, the **magnetic moment of the current distribution**. It gives the dipole contribution in the magnetic multipole expansion (see (3.1.21)).

The corresponding magnetic field is readily obtained

$$
\begin{aligned}
\boldsymbol{B}(\boldsymbol{x}) &= \frac{1}{4\pi}\operatorname{curl}\Big(\boldsymbol{M}\wedge\frac{\boldsymbol{x}}{r^3}\Big)\\
&= \frac{1}{4\pi}\Big[-(\boldsymbol{M}\cdot\operatorname{grad})\frac{\boldsymbol{x}}{r^3}+\boldsymbol{M}\operatorname{div}\frac{\boldsymbol{x}}{r^3}\Big]\\
&= \frac{1}{4\pi}\Big[-\frac{\boldsymbol{M}}{r^3}+3\frac{\boldsymbol{M}\cdot\boldsymbol{x}}{r^5}\boldsymbol{x}+\boldsymbol{M}\Big(\frac{3}{r^3}+\boldsymbol{x}\cdot\operatorname{grad}\frac{1}{r^3}\Big)\Big]\\
&= \frac{1}{4\pi}\frac{3\boldsymbol{x}(\boldsymbol{M}\cdot\boldsymbol{x})-r^2\boldsymbol{M}}{r^5}.
\end{aligned}
\tag{3.1.30}
$$

Comparison with (1.2.8) shows that this is indeed a dipole field with the magnetic moment $\boldsymbol{M}$ replacing the electric dipole moment. From (3.1.29) we find for the magnetic moment of a plain current loop

$$
\boldsymbol{M} = \frac{\mu_0}{2}J\oint \boldsymbol{x}\wedge d\boldsymbol{s} = \frac{\mu_0}{2}J\cdot 2\boldsymbol{F} = \mu_0 J\boldsymbol{F}, \tag{3.1.31}
$$

where F is the area of the loop. This agrees, of course, with (3.1.21).

To calculate the field in more complicated geometry, the **integral form of Ampère's law** is often useful. Applying Stokes' theorem to $\operatorname{curl}\boldsymbol{B} = \mu_0\boldsymbol{j}$, we conclude

$$
\int_{K_2}\operatorname{curl}\boldsymbol{B}\cdot d\boldsymbol{\sigma} = \int_{\partial K_2}\boldsymbol{B}\cdot d\boldsymbol{s} = \mu_0 J \tag{3.1.32}
$$

for any two-dimensional surface K_2. Considering a ring solenoid, the field strength is constant $= B$ in the interior of the solenoid up to small corrections. Choosing the closed curve ∂K_2 completely inside the solenoid, we have

$$
\int_{\partial K_2}\boldsymbol{B}\cdot d\boldsymbol{s} = Bl = \mu_0 nJ, \tag{3.1.33}
$$

if the solenoid has length l and n windings. Thus

$$
B = \mu_0\frac{nJ}{l}, \tag{3.1.34}
$$

which holds approximately also for a long thin solenoid.

We now consider the **forces between stationary currents** (Ampère 1825). The force from the magnetic field $\boldsymbol{B}_2$ of the conductor Ω_2 on the current $\boldsymbol{j}_1(\boldsymbol{x}) = \rho_1(\boldsymbol{x})\boldsymbol{v}(\boldsymbol{x})$ flowing in Ω_1 is given by the Lorentz force law (2.3.2)

$$
\boldsymbol{K}_{12} = \int_{\Omega_1} d^3x\,\rho_1(\boldsymbol{x})\boldsymbol{v}(\boldsymbol{x})\wedge\boldsymbol{B}_2(\boldsymbol{x}). \tag{3.1.35}
$$

Using (2.5.9) we have

$$\boldsymbol{K}_{12} = \frac{\mu_0}{4\pi} \int\limits_{\Omega_1} d^3x \int\limits_{\Omega_2} d^3y \, \boldsymbol{j}_1(\boldsymbol{x}) \wedge \frac{\boldsymbol{j}_2(\boldsymbol{y}) \wedge (\boldsymbol{x}-\boldsymbol{y})}{|\boldsymbol{x}-\boldsymbol{y}|^3}. \tag{3.1.36}$$

This is analogous to Coulomb's law with the product of charge densities substituted by the vector product of current densities.

As an application we calculate the force between two thin wires. In this case we must insert

$$\int\limits_{\Omega_1'} \boldsymbol{j}_1(\boldsymbol{x}) d\sigma d\boldsymbol{s}_1 = J_1 d\boldsymbol{s}_1 \tag{3.1.37}$$

into (3.1.36) where Ω_1' is the cross section of the conductor 1 and similarly for the conductor 2. Thus

$$\boldsymbol{K}_{12} = \frac{\mu_0}{4\pi} J_1 J_2 \oint\limits_{C_1} \oint\limits_{C_2} \frac{d\boldsymbol{s}_1 \wedge (d\boldsymbol{s}_2 \wedge (\boldsymbol{x}-\boldsymbol{y}))}{|\boldsymbol{x}-\boldsymbol{y}|^3}, \tag{3.1.38}$$

where the integrals go along the two circuits. This can be simplified by means of

$$d\boldsymbol{s}_1 \wedge (d\boldsymbol{s}_2 \wedge (\boldsymbol{x}-\boldsymbol{y})) = -(d\boldsymbol{s}_1 \cdot d\boldsymbol{s}_2)(\boldsymbol{x}-\boldsymbol{y}) + d\boldsymbol{s}_2((\boldsymbol{x}-\boldsymbol{y}) \cdot d\boldsymbol{s}_1).$$

The last term gives no contribution because

$$\oint \frac{(\boldsymbol{x}-\boldsymbol{y}) \cdot d\boldsymbol{s}_1}{|\boldsymbol{x}-\boldsymbol{y}|^3} = -\oint \operatorname{grad}_x \frac{1}{|\boldsymbol{x}-\boldsymbol{y}|} \cdot d\boldsymbol{s}_1 = 0.$$

Hence

$$\boldsymbol{K}_{12} = -\frac{\mu_0}{4\pi} J_1 J_2 \oint \oint \frac{\boldsymbol{x}-\boldsymbol{y}}{|\boldsymbol{x}-\boldsymbol{y}|^3} d\boldsymbol{s}_1 \cdot d\boldsymbol{s}_2. \tag{3.1.39}$$

This result satisfies actio = reactio: $\boldsymbol{K}_{12} = -\boldsymbol{K}_{21}$.

For two parallel wires with distance a, the force is normal to the wires and attractive for parallel currents ($d\boldsymbol{s}_1 \cdot d\boldsymbol{s}_2 > 0$), repulsive for anti-parallel currents. The force per unit length is equal to

$$\frac{dK_{12}}{ds_1} = -\frac{\mu_0}{4\pi} J_1 J_2 \int\limits_{-\infty}^{+\infty} \frac{1}{a^2+y^2} \frac{a}{\sqrt{a^2+y^2}} dy. \tag{3.1.40}$$

The integral is elementary

$$\frac{dK_{12}}{ds_1} = -\frac{\mu_0}{4\pi} \frac{2J_1 J_2}{a} = -10^{-7} \frac{2J_1 J_2}{a} \mathrm{NA}^{-2}. \tag{3.1.41}$$

This force can be accurately measured and defines the unit Ampère: If the force in a distance $a = 1\mathrm{m}$ is $2 \cdot 10^{-7}$ N m^{-1}, then a current of 1 A is flowing.

3.2 Motion of Particles in Electromagnetic Fields

Let us consider N particles with charges q_j, masses m_j and velocities $\boldsymbol{v}_j$. They give rise to a current density

$$\boldsymbol{j}(\boldsymbol{x}) = \sum_{j=1}^{N} q_j \boldsymbol{v}_j \delta(\boldsymbol{x} - \boldsymbol{x}_j), \tag{3.2.1}$$

where $\boldsymbol{x}_j$ is the place of the particle j. The corresponding magnetic moment is given by (3.1.29)

$$\boldsymbol{M} = \frac{\mu_0}{2} \int \boldsymbol{x} \wedge \boldsymbol{j}(\boldsymbol{x})\, d^3x = \frac{\mu_0}{2} \sum_j q_j\, \boldsymbol{x}_j \wedge \boldsymbol{v}_j. \tag{3.2.2}$$

We wish to compare this with the total angular momentum of the particles

$$\boldsymbol{L} = \sum_j m_j\, \boldsymbol{x}_j \wedge \boldsymbol{v}_j. \tag{3.2.3}$$

If all particles are identical

$$\frac{q_j}{m_j} = \frac{e}{m}, \tag{3.2.4}$$

magnetic moment and angular momentum are proportional

$$\boldsymbol{M} = \frac{\mu_0}{2} \frac{e}{m} \boldsymbol{L}. \tag{3.2.5}$$

The ratio M/L is the so-called **gyromagnetic ratio**. It is a very interesting quantity because it gives information about the nature of magnetism in materials.

The gyromagnetic ratio can be measured by the Einstein-de Haas effect. It follows from (3.2.5) that a change of the magnetic moment $\boldsymbol{M}$ produces a change of the angular momentum $\boldsymbol{L}$. A change of angular momentum is equivalent to a torque. Consequently, if a probe hanging on a long thin wire is magnetized in the vertical direction, a rotation is observed (Fig.3). One expresses the result by the so-called g-factor of Landé

$$\frac{M}{L} = g \frac{\mu_0}{2} \frac{e}{m}. \tag{3.2.6}$$

From (3.2.5) we would expect $g = 1$, but in reality one finds $g \approx 2$ for ferromagnetic metals. This shows that ferromagnetism is not caused by circular motion of electrons, which was Ampère's hypothesis. Its origin is the permanent magnetic spin-moment of the electrons: An electron has an intrinsic angular momentum s of magnitude $s = \frac{1}{2}\hbar$ in its rest frame, where $\hbar$ is Planck's constant (divided by 2π). In addition it has a corresponding magnetic moment

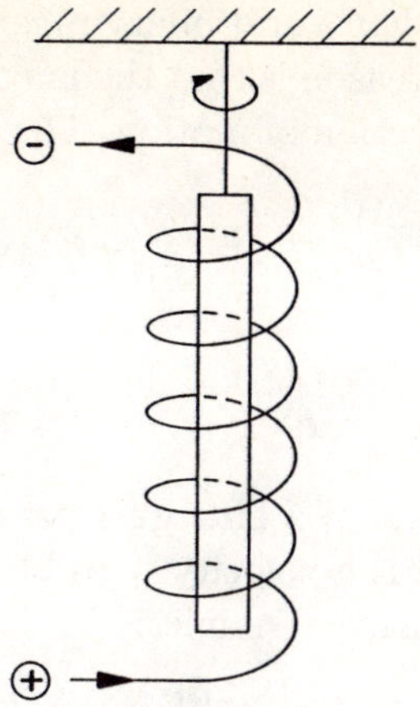

Fig. 3. Einstein-de Haas effect

$$m_s \approx \frac{\mu_0}{2} \frac{e}{m} \hbar, \tag{3.2.7}$$

which follows from relativistic quantum mechanics. The expression on the r.h.s. of (3.2.7) is called Bohr's magneton μ_B. This gives a gyromagnetic ratio

$$\frac{m_s}{s} \approx \mu_0 \frac{e}{m}, \tag{3.2.8}$$

that leads indeed to $g \approx 2$. Consequently, the Einstein-de Haas effect gives direct evidence for the spin of the electron.

The motion of point particles in electromagnetic fields is governed by the Lorentz force (2.3.2)

$$\boldsymbol{K}(\boldsymbol{x}) = q\boldsymbol{E}(\boldsymbol{x}) + q\boldsymbol{v} \wedge \boldsymbol{B}(\boldsymbol{x}). \tag{3.2.9}$$

It is the basis of particle optics with its typical applications in accelerators and the electron microscope, for example. We want first consider the mechanical work transferred to a particle in a stationary electromagnetic field:

$$-W = q \int_{\boldsymbol{x}_0}^{\boldsymbol{x}} \boldsymbol{E} \cdot d\boldsymbol{s} + q \int_{\boldsymbol{x}_0}^{\boldsymbol{x}} (\boldsymbol{v} \wedge \boldsymbol{B}) \cdot d\boldsymbol{s}.$$

Since the velocity is

$$\boldsymbol{v} = \frac{d\boldsymbol{s}}{dt}, \tag{3.2.10}$$

we may write this as follows

$$-W = q \int_{\boldsymbol{x}_0}^{\boldsymbol{x}} \boldsymbol{E} \cdot \boldsymbol{v}\, dt + q \int_{\boldsymbol{x}_0}^{\boldsymbol{x}} (\boldsymbol{v} \wedge \boldsymbol{B}) \cdot \boldsymbol{v}\, dt. \tag{3.2.11}$$

The last term vanishes, so that a stationary magnetic field does not transfer energy to the particle. The reason is that the magnetic force is perpendicular to $\boldsymbol{v}$, and, hence, to the direction of motion. The work per unit time is equal to

$$-\frac{dW}{dt} = q\boldsymbol{E} \cdot \boldsymbol{v} = \boldsymbol{K} \cdot \boldsymbol{v}, \tag{3.2.12}$$

and its density is given by

$$-\frac{dw}{dt} = \rho\boldsymbol{E} \cdot \boldsymbol{v} = \boldsymbol{j} \cdot \boldsymbol{E} = \boldsymbol{k} \cdot \boldsymbol{v}. \tag{3.2.13}$$

We now study the motion of a **charged particle in a constant magnetic field $\boldsymbol{B}$**. We assume the velocity $\boldsymbol{v}$ to be small compared with c, so that we can use nonrelativistic dynamics:

$$\boldsymbol{K} = \frac{d\boldsymbol{p}}{dt} = m\frac{d\boldsymbol{v}}{dt} = q\,\boldsymbol{v} \wedge \boldsymbol{B}. \tag{3.2.14}$$

Multiplying by $\boldsymbol{v}$,

$$m\boldsymbol{v} \cdot \frac{d\boldsymbol{v}}{dt} = 0 = \frac{d}{dt}\frac{m}{2}\boldsymbol{v}^2,$$

we see that the kinetic energy is indeed constant, as it must be because the magnetic field cannot give energy to the particle. Choosing $\boldsymbol{B}$ parallel to the 3-axis, Eq.(3.2.14) becomes

$$m\frac{dv_1}{dt} = qv_2B_3, \quad m\frac{dv_2}{dt} = -qv_1B_3 \tag{3.2.15}$$

$$m\frac{dv_3}{dt} = 0. \tag{3.2.16}$$

The last equation implies

$$v_3(t) = \text{const.} = v_3(0), \tag{3.2.17}$$

this is a uniform motion in the direction of the field. The first two equations (3.2.15) can be easily integrated:

$$\begin{aligned} v_1(t) &= v_1(0)\cos\omega t + v_2(0)\sin\omega t \\ v_2(t) &= -v_1(0)\sin\omega t + v_2(0)\cos\omega t. \end{aligned} \tag{3.2.18}$$

This is a precession in a left-screw (for positive q) around the direction of $\boldsymbol{B}$, with the so-called **cyclotron frequency**

$$\omega = \frac{q}{m}B_3. \tag{3.2.19}$$

One further integration yields the position of the particle

$$\begin{aligned} x_1(t) &= x_1(0) + \frac{v_1(0)}{\omega}\sin\omega t - \frac{v_2(0)}{\omega}\cos\omega t \\ x_2(t) &= x_2(0) + \frac{v_1(0)}{\omega}\cos\omega t + \frac{v_2(0)}{\omega}\sin\omega t \\ x_3(t) &= x_3(0) + v_3(0)t. \end{aligned} \tag{3.2.20}$$

This result describes a screw-motion around $\boldsymbol{B}$. We have a circular motion in the 1,2-plane with radius

$$\rho = \frac{\sqrt{v_1^2 + v_2^2}}{\omega}. \tag{3.2.21}$$

This radius is measured in a magnetic spectrometer. It allows the determination of the velocity $v_\perp$ perpendicular to $\boldsymbol{B}$, or, if this is known, the measurement of the mass through ω (3.2.19) (mass spectrometer).

In addition to the direct force on charged particles, the Lorentz force gives rise to a torque $\boldsymbol{N}$ on a current distribution:

$$\begin{aligned} \boldsymbol{N} &= \int d^3x\, \boldsymbol{x} \wedge \boldsymbol{k}(\boldsymbol{x}) \\ &= \int d^3x\, \boldsymbol{x} \wedge (\boldsymbol{j}(\boldsymbol{x}) \wedge \boldsymbol{B}(\boldsymbol{x})) \\ &= \int d^3x \Big(\boldsymbol{j}(\boldsymbol{x})(\boldsymbol{x} \cdot \boldsymbol{B}(\boldsymbol{x})) - \boldsymbol{B}(\boldsymbol{x})(\boldsymbol{x} \cdot \boldsymbol{j}(\boldsymbol{x}))\Big). \end{aligned} \tag{3.2.22}$$

In case of a constant magnetic field $\boldsymbol{B}(\boldsymbol{x}) = \boldsymbol{B} = \text{const.}$, the last term vanishes by charge conservation:

$$\int d^3x\, x_k j_k = \tfrac{1}{2} \int d^3x\, \operatorname{div}(x_k x_k \boldsymbol{j}) = 0,$$

hence

$$N_k = \int d^3x\, j_k x_l B_l = \tfrac{1}{2} \int d^3x\, (j_k x_l - j_l x_k) B_l, \tag{3.2.23}$$

where we have used the antisymmetry (3.1.26) again. Here the magnetic moment $\boldsymbol{M}$ (3.1.29) appears

$$\begin{aligned} \boldsymbol{N} &= \tfrac{1}{2} \int d^3x \Big(\boldsymbol{j}(\boldsymbol{x} \cdot \boldsymbol{B}) - \boldsymbol{x}(\boldsymbol{j} \cdot \boldsymbol{B})\Big) \\ &\Big(\tfrac{1}{2} \int d^3x\, \boldsymbol{x} \wedge \boldsymbol{j}(\boldsymbol{x})\Big) \wedge \boldsymbol{B} = \frac{1}{\mu_0} \boldsymbol{M} \wedge \boldsymbol{B}. \end{aligned} \tag{3.2.24}$$

This simple result for the torque remains obviously true if the magnetic field depends on time.

An important application is **magnetic resonance**. In this case $\boldsymbol{M}$ comes from the spin moment

$$\boldsymbol{M} = \mu_N \boldsymbol{S} \tag{3.2.25}$$

of some nucleus instead of the electron spin. μ_N is the so-called nuclear magneton. The equation of motion for the spin then reads

$$\boldsymbol{N} = \frac{d\boldsymbol{S}(t)}{dt} = \gamma \boldsymbol{S}(t) \wedge \boldsymbol{B}, \tag{3.2.26}$$

where $\gamma = \mu_N/\mu_B$ is the ratio of the nuclear to Bohr's magneton (see (3.2.7)). This equation agrees precisely with (3.2.14), so that we can immediately use the previous solution: The spin rotates around $\boldsymbol{B}$ with the so-called Larmor frequency

$$\omega = \gamma\,|\boldsymbol{B}|. \tag{3.2.27}$$

To observe this precession of the spins, one uses the following resonance method.

One applies an additional magnetic field in the xy-plane that varies in time:

$$\boldsymbol{B}(t) = (B_1 \cos\omega_1 t, -B_1 \sin\omega_1 t, B_3). \tag{3.2.28}$$

This is called a circular polarized field (see (3.3.69)). The equation of motion for the spin now reads as follows

$$\begin{aligned}\frac{dS_1}{dt} &= \gamma S_2 B_3 + \gamma S_3 B_1 \sin\omega_1 t\\ \frac{dS_2}{dt} &= \gamma S_3 B_1 \cos\omega_1 t - \gamma S_1 B_3\\ \frac{dS_3}{dt} &= -\gamma S_1 B_1 \sin\omega_1 t - \gamma S_2 B_1 \cos\omega_1 t.\end{aligned} \tag{3.2.29}$$

It can be solved by transforming to a rotating frame by means of the following ansatz

$$\begin{aligned}S_1 &= s_1(t)\cos\omega_1 t + s_2(t)\sin\omega_1 t\\ S_2 &= -s_1(t)\sin\omega_1 t + s_2(t)\cos\omega_1 t.\end{aligned} \tag{3.2.30}$$

One gets now a system of differential equations with time-independent coefficients

$$\begin{aligned}s_1' &= (\gamma B_3 - \omega_1)s_2\\ s_2' &= -(\gamma B_3 - \omega_1)s_1 + \gamma B_1 S_3\\ S_3' &= -\gamma B_1 s_2,\end{aligned} \tag{3.2.31}$$

where the prime denotes the time derivative. This system of linear differential equations with constant coefficients can be easily solved. We are particularly interested in the resonant case with

$$\omega_1 = \gamma B_3. \tag{3.2.32}$$

Then it follows $s_1 = \text{const}$, and

$$s_2' = \gamma B_1 S_3, \quad S_3' = -\gamma B_1 s_2. \tag{3.2.33}$$

This is a precession around the 1-axis.

To observe nuclear magnetic resonance, one can use the following method: In thermal equilibrium with only the static field B_3 present, there are more spins parallel to B_3 then antiparallel, because this lowers the energy.

We therefore have an excess magnetization in 3-direction. We now switch on the time-dependent field in the xy-plane, so that the excess spins precess around the 1-axis. After a 90°-rotation, that brings the spins into the xy-plane, the field B_1 is switched off. The spins now precess around the 3-axis with the Larmor frequency $\omega = \gamma|B_3|$. The induced circular polarized field can be accurately measured. This technique is called excitation by a 90°-pulse (Fig.4).

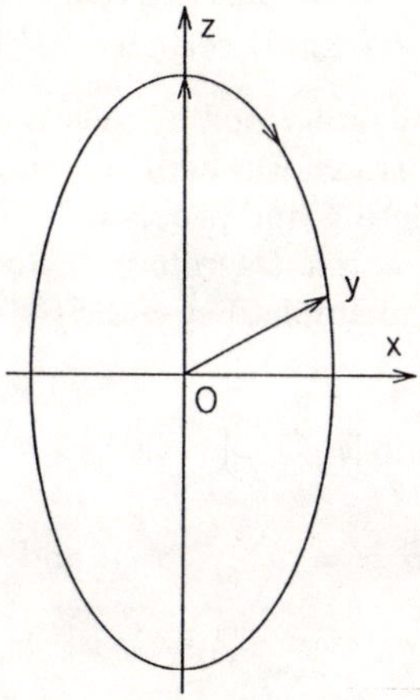

Fig. 4. 90°-pulse in nuclear magnetic resonance

3.3 Electromagnetic Radiation

We consider Maxwell's equations far away from the sources where $\rho = 0$, $\boldsymbol{j} = \boldsymbol{0}$:

$$\varepsilon_0 \mu_0 \boldsymbol{E}' = \operatorname{curl} \boldsymbol{B} \tag{3.3.1}$$

$$\boldsymbol{B}' = -\operatorname{curl} \boldsymbol{E} \tag{3.3.2}$$

$$\operatorname{div} \boldsymbol{E} = 0 \tag{3.3.3}$$

$$\operatorname{div} \boldsymbol{B} = 0. \tag{3.3.4}$$

Differentiating (3.3.1) with respect to t and substituting (3.3.2) on the right-hand side of (3.3.1), we get

$$\frac{1}{c^2} \boldsymbol{E}'' = \operatorname{curl} \boldsymbol{B}' = -\operatorname{curl} \operatorname{curl} \boldsymbol{E} = -\operatorname{grad} \operatorname{div} \boldsymbol{E} + \triangle \boldsymbol{E}, \tag{3.3.5}$$

where we have used (2.6.19) and the identity (0.1.26) which holds in cartesian coordinates. Since $\operatorname{div} \boldsymbol{E} = 0$, we arrive at the **wave equation**

$$\Big(\frac{1}{c^2}\frac{\partial^2}{\partial t^2}-\frac{\partial^2}{\partial x_1^2}-\frac{\partial^2}{\partial x_2^2}-\frac{\partial^2}{\partial x_3^2}\Big)\boldsymbol{E}=\Box\boldsymbol{E}=\boldsymbol{0}. \tag{3.3.6}$$

This equation describes the propagation of waves with velocity c, as we will discuss in detail in the following. This is the reason for the identification (2.6.19) of the light velocity. In the same way we get the wave equation for the magnetic field

$$\Box\boldsymbol{B}=\boldsymbol{0}. \tag{3.3.7}$$

However, by forming second time derivatives, some information of Maxwell's equations (3.3.1-4) is lost, we shall return to this point below.

The wave equation is best integrated by means of Fourier transformation. But we need distributive Fourier transform here. We make an interlude to present the necessary results. For more details and proofs we refer to *L.Hörmander, The Analysis of Linear Partial Differential Operators I, Second Edition, Springer-Verlag Berlin, Heidelberg, 1990*, for example. Let $\varphi\in S(\mathbb{R}^n)$ be a tempered test function, that means $\varphi\in C^\infty$ and

$$\sup_x |x^r D^s\varphi| = ||\varphi||_{r,s}<\infty \tag{3.3.8}$$

for all multi-indices r,s. Here $x=(x_1,\dots x_n)$, and

$$x^r = x_1^{r_1}\cdot\ldots\cdot x_n^{r_n},$$

$$D^s=\frac{\partial^{s_1}}{\partial x_1^{s_1}}\cdots\frac{\partial^{s_n}}{\partial x_n^{s_n}}. \tag{3.3.9}$$

Then the Fourier transform is simply defined by

$$F[\varphi](k)=(2\pi)^{-n/2}\int e^{-ik\cdot x}\varphi(x)\,d^nx=\hat{\varphi}(k). \tag{3.3.10}$$

We list the most important properties of this transformation:

1) Multiplication in x-space corresponds to differentiation of the Fourier transform:

$$\frac{\partial^r}{\partial k^r}\hat{\varphi}(k)=(2\pi)^{-n/2}\int(-ix)^r e^{-ikx}\varphi(x)\,d^nx=F_x\Big[(-ix)^r\varphi\Big]. \tag{3.3.11}$$

2) Differentiation is x-space corresponds to multiplication of the Fourier transform by powers. In fact, we find by partial integration

$$F_x\Big[\frac{\partial^r}{\partial x^r}\varphi\Big]=(2\pi)^{-n/2}\int e^{-ikx}\frac{\partial^r}{\partial x^r}\varphi(x)\,d^nx=(ik)^rF_x[\varphi]. \tag{3.3.12}$$

3) Translations are mapped as follows:

$$F[\varphi(x-x_0)]=e^{-ikx_0}F[\varphi] \tag{3.3.13}$$

$$F[\varphi](k+k_0)=F\Big[e^{-ikx_0}\varphi\Big](k). \tag{3.3.14}$$

4) The mapping F is continuous on $S(\mathbb{R}^n)$:

$$\left|k^r \frac{\partial^s}{\partial k^s}\hat{\varphi}\right| \leq (2\pi)^{-n/2} \int \frac{(1+|x|)^{n+1}}{(1+|x|)^{n+1}} \left|\frac{\partial^r}{\partial x^r} x^s \varphi(x)\right| d^n x$$

$$\leq \sup_x (1+|x|)^{n+1} \left|\frac{\partial^r}{\partial x^r} x^s \varphi(x)\right| (2\pi)^{-n/2} \int \frac{d^n x}{(1+|x|)^{n+1}}. \tag{3.3.15}$$

Since the last integral is finite, one finds

$$||\hat{\varphi}||_{r,s} \leq \text{linear combination of} \, (||\varphi||_{s-r,0}, \dots ||\varphi||_{s+n+1,r}). \tag{3.3.16}$$

This implies the continuity of F.

5) The inverse Fourier transform

$$F^{-1}[\varphi](x) = (2\pi)^{-n/2} \int e^{ikx} \varphi(k) \, d^n k$$

$$= F[\varphi(-k)] \tag{3.3.17}$$

has the same properties 1)–4) and it satisfies the

6) Inversion theorem on $S(\mathbb{R}^n)$:

$$F^{-1}[F(\varphi)] = \varphi = F[F^{-1}(\varphi)] \tag{3.3.18}$$

for all $\varphi \in S(\mathbb{R}^n)$.

7) Properties 5) and 6) imply

$$F^2[\varphi](x) = F\Big[F^{-1}[\varphi](-x)\Big] = \varphi(-x). \tag{3.3.19}$$

8) For $\varphi, \psi \in S(\mathbb{R}^n)$ we have Parseval's equation

$$\int \psi(k)\hat{\varphi}(k) \, d^n k = \int \hat{\psi}(x)\varphi(x) \, d^n x. \tag{3.3.20}$$

This last property (3.3.20), which may be written as a distributive equation

$$\langle \psi, \hat{\varphi} \rangle = \langle \hat{\psi}, \varphi \rangle, \tag{3.3.21}$$

is the starting point for introducing the Fourier transformation of tempered distributions:

Definition. *Let f be a tempered distribution, i.e. a continuous linear functional on $S(\mathbb{R}^n)$*

$$f: \quad \varphi \in S(\mathbb{R}^n) \longmapsto \langle f, \varphi \rangle \in \mathbb{R}, \quad f \in S'(\mathbb{R}^n). \tag{3.3.22}$$

Then

$$\langle F[f], \varphi \rangle \stackrel{\text{def}}{=} \langle f, \hat{\varphi} \rangle \tag{3.3.23}$$

also defines a tempered distribution $F[f]$, that is called the distributive Fourier transform of f.

From now on we reserve the symbol F for the distributive Fourier transform, whereas the ordinary Fourier transformation of test functions is denoted by a hat. $F[f]$ is a (weakly) continuous mapping of S' onto S'. In particular, if $f_j \rightharpoonup f$ in S' and $f_j \in L^1(\mathbb{R}^n)$, such that the ordinary Fourier transforms $\hat{f}_j$ exist, then

$$F[f] = \mathrm{w-}\lim_j \hat{f}_j, \quad \text{i.e.} \quad \langle f, \varphi\rangle = \lim_j \langle \hat{f}_j, \varphi\rangle \tag{3.3.24}$$

for all $\varphi \in S(\mathbb{R}^n)$. The properties 1)–3) and the inversion theorem go over to $S'(\mathbb{R}^n)$. For later use we give the following examples of distributive Fourier transforms:

Example 1. The Fourier transform of the δ-distribution follows from

$$\langle F[\delta], \varphi\rangle = \langle \delta, F[\varphi]\rangle = \hat{\varphi}(0) = (2\pi)^{-n/2} \int \varphi(x)\, d^n x,$$

$$F[\delta] = (2\pi)^{-n/2}. \tag{3.3.25}$$

In the same way

$$F^{-1}[\delta] = (2\pi)^{-n/2},$$

applying F on both sides, we get

$$F[1] = (2\pi)^{n/2}\delta(k). \tag{3.3.26}$$

Example 2. We want to compute the Fourier transform of

$$f(x) = \frac{\sin a|x|}{|x|}, \quad a > 0, \tag{3.3.27}$$

which is not in $L^1(\mathbb{R}^n)$. For $n = 1$ we have

$$F[f](k) = (2\pi)^{-1/2} \int_{-\infty}^{+\infty} dx\, \cos kx \frac{\sin ax}{x}$$

$$= \begin{cases} \frac{1}{2}\sqrt{2\pi}, & \text{if} \quad |k| < a \\ 0, & \text{if} \quad |k| > a \end{cases} = \frac{\sqrt{2\pi}}{2}\Theta(a^2 - k^2). \tag{3.3.28}$$

This Fourier integral is easily calculated by contour integration, or one may check the result by inverse Fourier transformation. In (3.3.28) we have introduced the step function (Heaviside function) $\Theta(x)$ which we will often use from now on. Its distributive derivative is the δ-distribution

$$\frac{d\Theta(x)}{dx} = \delta(x). \tag{3.3.29}$$

Example 3. The Fourier transform of (3.3.27) for $n = 2$

$$F[f](k) = (2\pi)^{-1}\mathrm{w-}\lim_{R\to\infty} \int_{|x|\le R} d^2x\, e^{-ik\cdot x}\frac{\sin ar}{r} \tag{3.3.30}$$

is calculated in polar coordinates $x_1 = r\cos\varphi$, $x_2 = r\sin\varphi$ with the 1-axis parallel to k:

$$= (2\pi)^{-1} \lim_{R\to\infty} \int_0^R dr\, r \int_0^{2\pi} d\varphi\, e^{-ikr\cos\varphi} \frac{\sin ar}{r}.$$

The φ-integral leads to a Bessel function

$$J_0(z) = \frac{1}{\pi} \int_0^{\pi} d\varphi\, e^{-iz\cos\varphi}, \tag{3.3.31}$$

thus

$$= \int_0^{\infty} dr\, J_0(kr) \sin ar = \begin{cases} 0, & a < |k| \\ \frac{1}{\sqrt{a^2-k^2}}, & a > |k| \end{cases} = \frac{\Theta(a^2-k^2)}{\sqrt{a^2-k^2}}. \tag{3.3.32}$$

We notice that the support of this distribution is the full circle $|k| \le a$.

Example 4. Similarly we compute the 3-dimensional Fourier transform using spherical coordinates with polar axis parallel to $\boldsymbol{k}$:

$$F[f](\boldsymbol{k}) = (2\pi)^{-3/2} \text{w-} \lim_{R\to\infty} \int_0^R dr\, r^2 \int_{-1}^{1} d\cos\vartheta \int_0^{2\pi} d\varphi\, e^{-ikr\cos\vartheta} \frac{\sin ar}{r}$$

$$= (2\pi)^{-1/2}\, \text{w-} \lim_{R\to\infty} \int_0^R dr\, r \sin ar \frac{e^{-ikr} - e^{ikr}}{-ikr}$$

$$= \sqrt{\frac{2}{\pi}} \frac{1}{k}\, \text{w-} \lim \int_0^R dr\, \sin ar \sin kr$$

$$= \sqrt{\frac{2}{\pi}} \frac{1}{k}\, \text{w-} \lim_{R\to\infty} \frac{1}{2} \int_0^R dr\, \Big(\cos(a-k)r - \cos(a+k)r\Big).$$

These weak limits are nothing but distributive Fourier transforms of 1 (3.3.26), hence

$$= \sqrt{\frac{2}{\pi}} \frac{\pi}{2k} \Big[\delta(a-k) - \delta(a+k)\Big]$$

$$= \sqrt{\frac{\pi}{2}} \frac{1}{a} \Big[\delta(a-k) + \delta(a+k)\Big] = \sqrt{2\pi}\,\delta(a^2-k^2). \tag{3.3.33}$$

This distribution has its support on the surface of the sphere $|\boldsymbol{k}| = a$.

We are now ready to determine the **Green's function** $G(t,x)$ **of the wave equation**. It is the distributive solution of

$$c^2\Box G = \frac{\partial^2 G}{\partial t^2} - c^2 \sum_{j=1}^{n} \frac{\partial^2 G}{\partial x_j^2} = \delta(t, x), \tag{3.3.34}$$

where $\delta(t, x) = \delta(t)\delta(x)$ is the $(n+1)$-dimensional δ-distribution. By n-dimensional Fourier transformation in x we obtain

$$\frac{\partial^2}{\partial t^2} F[G] + c^2|k|^2 F[G] = (2\pi)^{-n/2}\delta(t). \tag{3.3.35}$$

For fixed k this is an ordinary linear differential equation in t. If $t \neq 0$, the solutions are $\sin \omega t$ and $\cos \omega t$, with

$$\omega = c|k|. \tag{3.3.36}$$

To specify a unique solution of (3.3.35), we require

$$G(t, x) = 0, \quad \text{for} \quad t < 0. \tag{3.3.37}$$

This defines the so-called retarded Green's function.

The solution of (3.3.35) must have a discontinuous first t-derivative at $t = 0$. Then the second derivative is proportional to $\delta(t)$. That is achieved by switching the solution with a step function

$$F[G](t, k) = (2\pi)^{-n/2}\Theta(t)\, \frac{\sin \omega t}{\omega}. \tag{3.3.38}$$

In fact,

$$\frac{\partial}{\partial t} F[G] = (2\pi)^{-n/2}\delta(t)\, \frac{\sin \omega t}{\omega} + (2\pi)^{-n/2}\Theta(t)\, \cos \omega t,$$

since the first term vanishes we find the desired result

$$\frac{\partial^2}{\partial t^2} F[G] = (2\pi)^{-n/2}\delta(t) \cos \omega t - (2\pi)^{-n/2}\omega\Theta(t)\, \sin \omega t. \tag{3.3.39}$$

$G(t, x)$ is now given by inverse Fourier transformation

$$\begin{aligned} G(t, x) &= (2\pi)^{-n/2}\Theta(t)\, F^{-1}\left[\frac{\sin c|k|t}{c|k|}\right] \\ &= (2\pi)^{-n/2}\Theta(t)\, \frac{1}{c}\, F\left[\frac{\sin ct|k|}{|k|}\right], \end{aligned} \tag{3.3.40}$$

where (3.3.17) has been used. This Fourier transform has been computed in the above examples 2)–4). For $n = 1$ we have

$$G_1(t, x) = \frac{1}{2c}\, \Theta(t)\Theta(c^2t^2 - x^2) = \frac{1}{2c}\Theta(ct - |x|). \tag{3.3.41}$$

The results for $n = 2, 3$ are equal to

$$G_2(t,x) = \frac{1}{2\pi c}\frac{\Theta(c^2t^2-|x|^2)}{\sqrt{c^2t^2-|x|^2}} \tag{3.3.42}$$

$$G_3(t,\boldsymbol{x}) = \frac{1}{2\pi c}\Theta(t)\,\delta(c^2t^2-|\boldsymbol{x}|^2)$$
$$= \frac{1}{4\pi c^2 t}\Theta(t)\,\delta(ct-|\boldsymbol{x}|). \tag{3.3.43}$$

Let us now turn to the solution of the inhomogeneous wave equation

$$c^2\Box\, u(t,x) = f(t,x) \tag{3.3.44}$$

with an arbitrary (even distributive) source f that vanishes

$$f(t,x) = 0 \quad \text{for} \quad t<0. \tag{3.3.45}$$

We seek the solution $u(t,x)$ which also vanishes for negative t. Remembering the trivial representation of f

$$f(t,x) = \int \delta(t-t', x-x')\, f(t',x')\, dt'\, dx'$$

by means of δ-sources, we can immediately write down the solution

$$u(t,x) = \int\limits_{t>t'>0} G(t-t',x-x') f(t',x')\, dt'\, dx'$$
$$\stackrel{\text{def}}{=} (G*f)(t,x). \tag{3.3.46}$$

This is the convolution of the source f with the Green's function $G(t,x)$. Since the support of $G(t,x)$ is contained in the light cone

$$|x-x'| \le c(t-t'), \tag{3.3.47}$$

a disturbance at t', x' propagates at most with speed c. This is causality which is a consequence of the hyperbolic character of the wave equation, i.e. of the signs $+,-,-,-$ in the differential operator (3.3.6). The latter is a manifestation of Lorentz invariance of the wave equation. This is the important connection between the relativity principles and causality.

We want now to solve the **initial value problem for the homogeneous wave equation**

$$c^2\Box\, u(t,x) = 0 \tag{3.3.48}$$

with initial conditions

$$u(0,x) = u_0(x), \qquad \frac{\partial u}{\partial t}(0,x) = u_1(x). \tag{3.3.49}$$

We seek the solution for $t>0$, so that we may trivially set

$$u(t,x) = 0, \quad \text{for} \quad t<0. \tag{3.3.50}$$

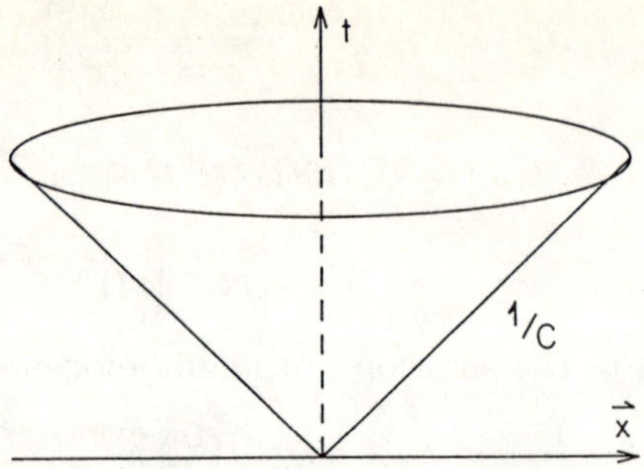

Fig. 5. The light cone

Then the (distributive) solution u must have discontinuous derivatives at $t = 0$, indeed for any $\varphi \in S(\mathbb{R}^{n+1})$ we find

$$\langle c^2 \Box u, \varphi \rangle \overset{\text{def}}{=} c^2 \int_0^\infty \int d^n x \, u \Box \varphi$$

$$= \lim_{\varepsilon \to 0} \int_\varepsilon^\infty \int d^n x \, u \Big(\frac{\partial^2 \varphi}{\partial t^2} - c^2 \bigtriangleup \varphi \Big)$$

$$= \lim_{\varepsilon \to 0} \Big[\int_\varepsilon^\infty dt \int d^n x \left(\frac{\partial^2 u}{\partial t^2} - c^2 \bigtriangleup u \right) \varphi$$

$$- \int d^n x \, u(\varepsilon, x) \frac{\partial \varphi}{\partial t}(\varepsilon, x) + \int d^n x \, \frac{\partial u}{\partial t}(\varepsilon, x) \varphi(\varepsilon, x) \Big]$$

by partial integration. Since u satisfies the homogeneous wave equation for $t > 0$ we conclude

$$= - \int d^n x \, u_0(x) \frac{\partial \varphi}{\partial t}(0, x) + \int d^n x \, u_1(x) \varphi(0, x),$$

or

$$c^2 \Box u = u_0(x) \delta'(t) + u_1(x) \delta(t). \tag{3.3.51}$$

We have here incorporated the initial conditions (3.3.49) into a singular distributive source at $t = 0$. The previous solution (3.3.46) can be used for such a source, too, thus

$$u(t, x) = \int G(t - t', x - x') \Big(u_0(x') \delta'(t') + u_1(x') \delta(t') \Big) \, dt' \, d^n x'$$

$$= \frac{\partial}{\partial t} \int G(t, x - x') u_0(x') \, d^n x' + \int G(t, x - x') u_1(x') \, d^n x'. \tag{3.3.52}$$

These are convolutions in x-space only.

We discuss the solutions in the physically relevant space dimensions $n = 1, 2, 3$. For $n = 1$ we get from (3.3.41)

$$G_1(t, x) = \frac{1}{2c}\Big[\Theta(x + ct) - \Theta(x - ct)\Big] \tag{3.3.53}$$

$$\frac{\partial G_1}{\partial t} = \frac{1}{2}\Big[\delta(x + ct) + \delta(x - ct)\Big], \tag{3.3.54}$$

thus

$$u(t, x) = \frac{1}{2}[u_0(x + ct) + u_0(x - ct)] + \frac{1}{2c}\int\limits_{x-ct}^{x+ct} dx'\, u_1(x') \tag{3.3.55}$$

$$= v_1(x + ct) + v_2(x - ct). \tag{3.3.56}$$

This last form is d'Alembert's solution. It has a simple interpretation for v_1, v_2 with compact disjoint supports: $v_1(x + ct)$ is a wave packet moving to the left and $v_2(x - ct)$ is moving to the right without changing their shapes.

For $n = 2$ we get from (3.3.42)

$$u(t, x) = \frac{1}{2\pi c}\frac{\partial}{\partial t}\int\limits_{|x'-x|\leq ct} d^2x' \frac{u_0(x')}{\sqrt{c^2t^2 - |x' - x|^2}}$$

$$+ \frac{1}{2\pi c}\int\limits_{|x'-x|\leq ct} d^2x' \frac{u_1(x')}{\sqrt{c^2t^2 - |x' - x|^2}}. \tag{3.3.57}$$

The result for $n = 3$ follows from (3.3.43)

$$u(t, \boldsymbol{x}) = \frac{1}{4\pi c^2}\frac{\partial}{\partial t}\frac{1}{t}\int\limits_{|\boldsymbol{x}'-\boldsymbol{x}|=ct} d^2x'\, u_0(\boldsymbol{x}') + \frac{1}{4\pi c^2 t}\int\limits_{|\boldsymbol{x}'-\boldsymbol{x}|=ct} d^2x'\, u_1(\boldsymbol{x}'). \tag{3.3.58}$$

The wave propagation in two and three dimensions is very different. In (3.3.57) the whole interior of the light cone contributes, while in (3.3.58) only the values on the surface of the cone enter. This has the consequence that an initial wave in a bounded region spreads out in 3-dimensional space in all directions, and after it has passed a fixed place, there is silence again (Fig.6). This is not so in two dimensions (3.3.57), where we find a slowly decaying echo. The latter would have catastrophic consequences for acoustical and electrical communication, if the mathematics were the other way around!

Finally let us return to Maxwell's equations. We consider **plane wave solutions** propagating in the positive $x_1 = x$ direction. According to (3.3.56) the electric and magnetic fields have the following form

$$\boldsymbol{E}(t, \boldsymbol{x}) = \boldsymbol{f}(x - ct), \quad \boldsymbol{B}(t, \boldsymbol{x}) = \boldsymbol{g}(x - ct). \tag{3.3.59}$$

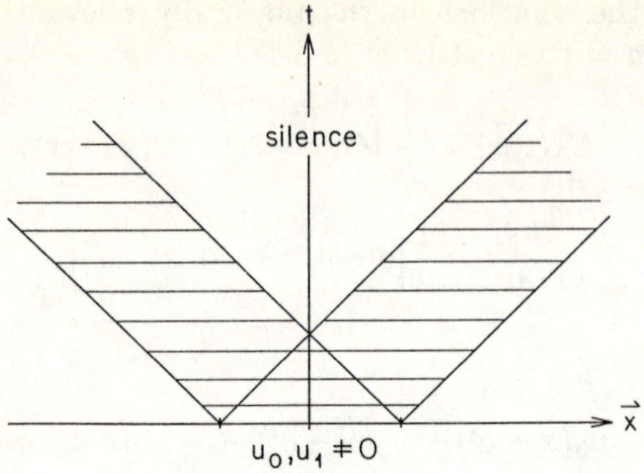

Fig. 6. Wave propagation in three dimensions

However, Maxwell's equations (3.3.2-4) give further restrictions on $\boldsymbol{E}$ and $\boldsymbol{B}$. From

$$\operatorname{div} \boldsymbol{E} = \frac{\partial f_1}{\partial x} = 0$$

for all x we get $f_1 = \text{const}$. We set this constant $= 0$, otherwise a static field would be superimposed. Then $\boldsymbol{E}$ (in the y, z-plane) is transversal to the propagation direction. The same is true for $\boldsymbol{B}$ due to (3.3.4). The induction law (3.3.2) gives a connection between $\boldsymbol{E}$ and $\boldsymbol{B}$. Substituting

$$\boldsymbol{E} = (0, f_2, f_3)(x - ct) \tag{3.3.60}$$

into $\boldsymbol{B}' = -\operatorname{curl} \boldsymbol{E}$, we find

$$-c\boldsymbol{g}' = -(0, -f_3, f_2)(x - ct),$$

and, choosing the constants of integration equal to 0 again, we finally get

$$\boldsymbol{B} = \frac{1}{c}(0, -f_3, f_2)(x - ct). \tag{3.3.61}$$

Therefore $\boldsymbol{B}$ is perpendicular to $\boldsymbol{E}$ and, together with the propagation direction $\boldsymbol{k}$ we have a positively oriented orthogonal triple (Fig.7)

$$\boldsymbol{B} = \frac{1}{c}\frac{\boldsymbol{k}}{|\boldsymbol{k}|} \wedge \boldsymbol{E}. \tag{3.3.62}$$

It is an interesting property of **electromagnetic waves** that they **transport energy, momentum and angular momentum through empty space**. To determine the energy flow, we calculate the Poynting vector for the plane wave

$$\boldsymbol{S} = \frac{1}{\mu_0}\boldsymbol{E} \wedge \boldsymbol{B} = \frac{1}{c\mu_0}\boldsymbol{E} \wedge \left(\frac{\boldsymbol{k}}{|\boldsymbol{k}|} \wedge \boldsymbol{E}\right) = \frac{\boldsymbol{E}^2}{c\mu_0}\frac{\boldsymbol{k}}{|\boldsymbol{k}|}. \tag{3.3.63}$$

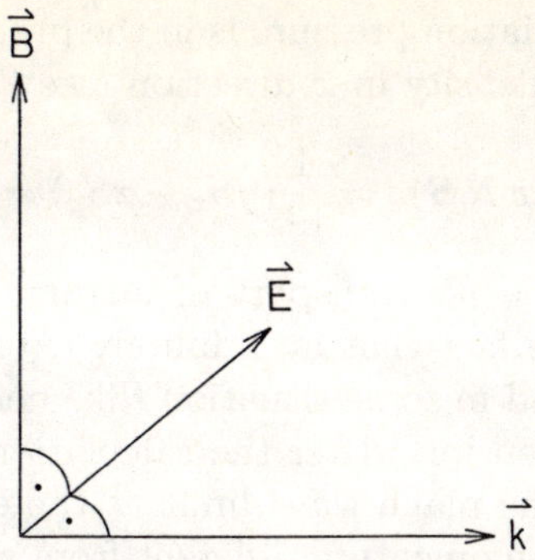

Fig. 7. Propagation of a plane wave

This flow is indeed in the propagation direction. The energy density is given by

$$u = \frac{1}{2}\Big(\varepsilon_0 \boldsymbol{E}^2 + \frac{1}{\mu_0}\boldsymbol{B}^2\Big) = \Big(\frac{\varepsilon_0}{2} + \frac{1}{2\mu_0 c^2}\Big)\boldsymbol{E}^2 = \frac{1}{c^2\mu_0}\boldsymbol{E}^2. \tag{3.3.64}$$

Since

$$\boldsymbol{S} = cu\,\frac{\boldsymbol{k}}{|\boldsymbol{k}|}, \tag{3.3.65a}$$

the total energy is transported with the speed of light c, nothing remains in a fixed finite region.

The momentum density of the plane wave is equal to

$$\boldsymbol{p} = \varepsilon_0\mu_0\boldsymbol{S} = \frac{u}{c}\frac{\boldsymbol{k}}{|\boldsymbol{k}|}, \tag{3.3.65b}$$

thus

$$u = c|\boldsymbol{p}|. \tag{3.3.66}$$

Comparing this with the energy-momentum relation of a particle of mass m

$$E = \sqrt{c^2\boldsymbol{p}^2 + m^2c^4},$$

we may attribute a particle with mass zero (photon) to electromagnetic radiation. This gets a precise meaning in quantum theory (see Epilogue). The momentum flow density

$$-T_{ik} = -\varepsilon_0 E_i E_k - \frac{1}{\mu_0}B_i B_k + \delta_{ik} u \tag{3.3.67}$$

has only one non-vanishing component

$$-T_{xx} = u.$$

Without surprise the radiation pressure is in the propagation direction. But the angular momentum density in x-direction (see (2.6.27))

$$\frac{1}{c^2}(\boldsymbol{x} \wedge \boldsymbol{S})_x = \frac{1}{c^2}(yS_z - zS_y) = 0$$

vanishes. There would be no transport of angular momentum, that is a paradox. It is due to the fact that an infinitely extended plane wave is an unphysical object. Instead in some situation (like here), we must consider a wave packet of finite extension. After the calculation of the required quantities we may perform the plane wave limit at the end. In this way we do get transport of angular momentum different from zero (Problem 3).

Finally we want to describe a **monochromatic plane wave**

$$\boldsymbol{E} = (0, f_2, f_3)(x - ct), \quad \boldsymbol{B} = \frac{1}{c}(0, -f_3, f_2). \tag{3.3.68}$$

In this case the functions f_2, f_3 are periodic with frequency $\omega = ck$. In particular, if these amplitudes are equal,

$$f_2 = E\cos(kx - \omega t), \quad f_3 = -E\sin(kx - \omega t) \tag{3.3.69}$$

we have a so-called left-handed circular polarized wave, or one with positive helicity. A right circularly polarized wave is given by

$$f_2 = E\cos(kx - \omega t), \quad f_3 = E\sin(kx - \omega t). \tag{3.3.70}$$

If the two amplitudes are unequal,

$$f_2 = E_2\cos(kx - \omega t), \quad f_3 = E_3\sin(kx - \omega t), \tag{3.3.71}$$

this is an elliptically polarized wave. For $E_3 = 0$ it is linearly polarized with $\boldsymbol{B}$ along the 3-axis. The direction od $\boldsymbol{B}$ is called the polarization direction in optics (see after (5.1.27)).

3.4 Production of Radiation, Electromagnetic Potentials

In the last section we have considered the propagation of electromagnetic waves in empty space, assuming that those waves have been produced somewhere. We now turn to the question how the waves are actually produced. **The sources of radiation can only be the charge and current densities in the inhomogeneous Maxwell's equations**. We therefore start from the full equations

$$\varepsilon_0\frac{\partial \boldsymbol{E}}{\partial t} = \frac{1}{\mu_0}\operatorname{curl}\boldsymbol{B} - \boldsymbol{j} \tag{3.4.1}$$

$$\frac{\partial \boldsymbol{B}}{\partial t} = -\operatorname{curl} \boldsymbol{E} \tag{3.4.2}$$

$$\varepsilon_0 \operatorname{div} \boldsymbol{E} = \rho \tag{3.4.3}$$

$$\operatorname{div} \boldsymbol{B} = 0. \tag{3.4.4}$$

The last equation implies by Theorem 4 (0.2.37) that there exists a **vector potential** $\boldsymbol{A}(t, \boldsymbol{x})$ such that

$$\boldsymbol{B}(t, \boldsymbol{x}) = \operatorname{curl} \boldsymbol{A}(t, \boldsymbol{x}), \tag{3.4.5}$$

where $\boldsymbol{A}(t, \boldsymbol{x})$ is now also depending on time. Using this representation of $\boldsymbol{B}$ in (3.4.2) we get

$$\operatorname{curl}(\boldsymbol{A}' + \boldsymbol{E}) = \boldsymbol{0}, \tag{3.4.6}$$

denoting the time derivative by a prime again. Thus the vector field $\boldsymbol{A}' + \boldsymbol{E}$ is vortex-free and, hence, there exists a **scalar potential** (0.1.45)

$$\boldsymbol{A}' + \boldsymbol{E} = -\operatorname{grad} V(t, \boldsymbol{x}), \quad \text{or}$$

$$\boldsymbol{E}(t, \boldsymbol{x}) = -\operatorname{grad} V - \frac{\partial \boldsymbol{A}}{\partial t}. \tag{3.4.7}$$

This is a generalization of electrostatics (1.1.18) to the time-dependent situation.

We have expressed the six components of the fields $\boldsymbol{E}$ and $\boldsymbol{B}$ by the four components of the potentials V and $\boldsymbol{A}$. In this way the homogeneous Maxwell's equations (3.4.2, 4) are fulfilled. This representation by the electromagnetic potentials must have a relativistic formulation. It is very plausible that V and $\boldsymbol{A}$ together form a **four-vector potential**

$$A^\mu = \left(\frac{1}{c} V, \boldsymbol{A}\right). \tag{3.4.8}$$

Differentiating it we get a second rank tensor (2.2.26). Since $\boldsymbol{E}$ and $\boldsymbol{B}$ appear in the antisymmetric field tensor $F^{\mu\nu}$ (2.3.20), we calculate the antisymmetric combination

$$F^{\mu\nu} = \partial^\mu A^\nu - \partial^\nu A^\mu, \tag{3.4.9}$$

and it is easy to verify that this is identical with (3.4.5) and (3.4.7).

However **the potentials are not unique, different potentials may give the same electromagnetic fields**. We know already from (0.2.33) that the vector potential can be changed by a gradient

$$\boldsymbol{A} \longrightarrow \boldsymbol{A} + \operatorname{grad} \chi, \tag{3.4.10}$$

without changing $\boldsymbol{B}$. $\chi = \chi(t, \boldsymbol{x})$ can be any differentiable function. But the transformation (3.4.10) alone would change the electric field (3.4.7). The additional term $-\operatorname{grad} \chi'$ is compensated by the following transformation of the scalar potential

$$V \longrightarrow V - \frac{\partial \chi}{\partial t}. \tag{3.4.11}$$

Then **the simultaneous transformations** (3.4.10, 11) **do not change the electromagnetic fields. These are the so-called gauge transformations, V and $\boldsymbol{A}$ are gauge fields**. The latter are not the directly observable fields, but nevertheless very important, in particular in quantum theory (see Epilogue). (3.4.10) and (3.4.11) can be combined into the **relativistic gauge transformation**

$$A^\mu \longrightarrow A^\mu - \partial^\mu \chi. \tag{3.4.12}$$

To verify the correct signs, we must remember the difference between covariant and contravariant components

$$\partial_\mu = \left(\frac{1}{c}\frac{\partial}{\partial t}, \frac{\partial}{\partial \boldsymbol{x}}\right), \quad \partial^\mu = \left(\frac{1}{c}\frac{\partial}{\partial t}, -\frac{\partial}{\partial \boldsymbol{x}}\right). \tag{3.4.13}$$

In terms of potentials the inhomogeneous Maxwell's equations read

$$\varepsilon_0(-\operatorname{grad} V' - \boldsymbol{A}'') = \frac{1}{\mu_0}\operatorname{curl}\operatorname{curl}\boldsymbol{A} - \boldsymbol{j}, \quad \text{or}$$

$$\frac{1}{c^2}\boldsymbol{A}'' + \frac{1}{c^2}\operatorname{grad} V' + \operatorname{grad}\operatorname{div}\boldsymbol{A} - \triangle\boldsymbol{A} = \mu_0\boldsymbol{j}, \tag{3.4.14}$$

where (0.1.26) and (2.6.19) have been used. If we choose the potentials in such a way that

$$\frac{1}{c^2}\frac{\partial V}{\partial t} + \operatorname{div}\boldsymbol{A} = 0, \tag{3.4.15}$$

then we arrive at the **inhomogeneous wave equation**

$$\frac{1}{c^2}\frac{\partial^2 \boldsymbol{A}}{\partial t^2} - \triangle\boldsymbol{A} = \Box\boldsymbol{A} = \mu_0\boldsymbol{j}. \tag{3.4.16}$$

The condition (3.4.15) is the so-called **Lorentz gauge condition**; in relativistic notation it simply reads

$$\partial_\mu A^\mu = 0. \tag{3.4.17}$$

Before showing that the Lorentz condition can be satisfied by modifying the potentials by a suitable gauge transformation, we want to derive the equation for the scalar potential. Substituting (3.4.7) into (3.4.3) we shall obtain

$$-\varepsilon_0(\operatorname{div}\operatorname{grad} V + \operatorname{div}\boldsymbol{A}') = \rho. \tag{3.4.18}$$

Using the Lorentz condition (3.4.15) again, we get

$$\frac{1}{c^2}\frac{\partial^2 V}{\partial t^2} - \triangle V = \frac{1}{\varepsilon_0}\rho. \tag{3.4.19}$$

The two wave equations can be combined into one **inhomogeneous wave equation for the four-vector potential**

$$\Box A^\mu = \mu_0 j^\mu. \tag{3.4.20}$$

The source is given by the charge-current density $j^\mu = (c\rho,\, \boldsymbol{j})$ (2.4.13).

Although everything fits together relativistically, we must still show that the Lorentz condition can be satisfied. Assuming V and $\boldsymbol{A}$ to be given, we look for a gauge transformation by means of a gauge function $\chi(t, \boldsymbol{x})$ such that the transformed potentials (3.4.10, 11) fulfill the Lorentz condition, i.e.

$$\frac{1}{c^2}(V' - \chi'') + \operatorname{div}(\boldsymbol{A} + \operatorname{grad}\chi) = 0. \tag{3.4.21}$$

This can be viewed as an equation for χ

$$\frac{1}{c^2}\frac{\partial^2 \chi}{\partial t^2} - \Delta\chi = \frac{1}{c^2}V' + \operatorname{div}\boldsymbol{A}. \tag{3.4.22}$$

Under weak assumptions on the potentials V, $\boldsymbol{A}$, this inhomogeneous wave equation can be solved for χ by (3.3.46). The solution is only determined up to a solution of the homogeneous wave equation. That means that there is still some gauge freedom within the class of potentials selected by the Lorentz condition (3.4.15). This is an important fact for quantum electrodynamics.

We want now to solve the inhomogeneous wave equations (3.4.16, 19). To use our previous result (3.3.46), let us assume that $\boldsymbol{j}(t, \boldsymbol{x})$ and $\rho(t, \boldsymbol{x})$ vanish for $t < 0$. The vector potential for $t > 0$ is then given by a convolution

$$\boldsymbol{A}(t, \boldsymbol{x}) = c^2 \mu_0 (G * \boldsymbol{j})(t, \boldsymbol{x}), \tag{3.4.23}$$

where the retarded Green's function (3.3.43) is equal to

$$G(t, \boldsymbol{x}) = \frac{1}{4\pi c^3 t}\Theta(t)\delta\Big(t - \frac{|\boldsymbol{x}|}{c}\Big). \tag{3.4.24}$$

This leads to the following result:

$$\begin{aligned}\boldsymbol{A}(t, \boldsymbol{x}) &= \frac{\mu_0}{4\pi}\int_0^t dt' \int d^3x'\, \delta\Big(t - t' - \frac{|\boldsymbol{x} - \boldsymbol{x}'|}{c}\Big)\frac{\boldsymbol{j}(t', \boldsymbol{x}')}{t - t'} \\ &= \frac{\mu_0}{4\pi}\int d^3x' \frac{\boldsymbol{j}\Big(t - \frac{|\boldsymbol{x}-\boldsymbol{x}'|}{c}, \boldsymbol{x}'\Big)}{|\boldsymbol{x} - \boldsymbol{x}'|}. \end{aligned} \tag{3.4.25}$$

This result is quite similar to the corresponding expression (3.1.2) in magnetostatics. The only difference is that the current density is not taken at the actual time t, but at the retarded time

$$t_{\text{ret}} = t - \frac{|\boldsymbol{x} - \boldsymbol{x}'|}{c}. \tag{3.4.26}$$

The difference $t - t_{\text{ret}}$ is just the time which the waves need to propagate from $\boldsymbol{x}'$ to $\boldsymbol{x}$ with velocity c. This reflects causality: a source at $\boldsymbol{x}'$ influences the fields at $\boldsymbol{x}$ only after the time $|\boldsymbol{x} - \boldsymbol{x}'|/c$ which is necessary for light to propagate from the source to the point $\boldsymbol{x}$ of observation. The solution for the scalar potential is obtained in the same way

$$V(t, \boldsymbol{x}) = \frac{1}{4\pi\varepsilon_0} \int d^3x' \, \frac{\rho\left(t - \frac{|\boldsymbol{x}-\boldsymbol{x}'|}{c}, \boldsymbol{x}'\right)}{|\boldsymbol{x} - \boldsymbol{x}'|}. \tag{3.4.27}$$

This is similar to electrostatics (1.1.19), apart from the retarded time in the argument of the source ρ. One therefore calls (3.4.25, 27) the **retarded potentials**. The most general solutions are obtained by adding solutions of the homogeneous wave equations. This would be the superposition of additional free waves which have been radiated in from outside (apart from homogeneous solutions that can be transformed to zero by a gauge transformation, so-called "pure gauges"). We shall disregard this in the following, because we are interested in the radiation produced by our sources.

Sources which vary periodically in time are very important in applications. It is convenient to represent the time dependence by a complex exponential factor

$$\boldsymbol{j}(t', \boldsymbol{x}') = \operatorname{Re}\left(e^{-i\omega t'} \boldsymbol{j}_0(\boldsymbol{x}')\right) \tag{3.4.28}$$

$$\rho(t', \boldsymbol{x}') = \operatorname{Re}\left(e^{-i\omega t'} \rho_0(\boldsymbol{x}')\right), \tag{3.4.29}$$

where ω is the frequency. It has become fashionable to omit the real part in classical electrodynamical calculations. This is convenient but dangerous because in this way one sweeps certain problems under the rug; for example, if one computes quantities depending non-linearly on the fields like the Poynting vector, one must insert real fields. We will always write Re if we mean real part, and we emphasize that **only real solutions of the classical Maxwell's equations have a physical meaning**. The densities (3.4.28, 29) satisfy the continuity equation

$$\frac{\partial \rho}{\partial t} + \operatorname{div} \boldsymbol{j} = 0,$$

which leads to

$$\operatorname{Re}\left[(-i\omega\rho_0 + \operatorname{div} \boldsymbol{j}_0) e^{-i\omega t'}\right] = 0. \tag{3.4.30}$$

Since this holds for all t', we may conclude

$$\operatorname{div} \boldsymbol{j}_0 = i\omega\rho_0. \tag{3.4.31}$$

This shows in particular that $\rho_0(\boldsymbol{x}')$, $\boldsymbol{j}_0(\boldsymbol{x}')$ are complex. The spatial extension $\boldsymbol{j}_0(\boldsymbol{x}')$ of the source is assumed to be contained in a sphere of radius d

around the origin. We will discuss the situation where d is small compared to the wave length $\lambda = 2\pi c/\omega$. This is true for emission of light by atoms, because the magnitude of the atoms is $d \approx 10^{-7}$ cm, whereas $\lambda \approx 10^{-4}$ cm for optical frequencies, and also for radio waves with $d \approx 10$ m and $\lambda \approx 100$ m, for example. The region $|\boldsymbol{x}| \gg d$ is called the wave zone and $|\boldsymbol{x}| \ll \lambda$ is the near zone.

In the wave zone one makes a multipole expansion which is slightly different from electrostatics (Sect.1.2). In

$$\boldsymbol{j}(t_{\mathrm{ret}}, \boldsymbol{x}') = \mathrm{Re}\left(e^{-i\omega t} e^{i\omega|\boldsymbol{x}-\boldsymbol{x}'|/c} \boldsymbol{j}_0(\boldsymbol{x}')\right) \tag{3.4.32}$$

we expand

$$|\boldsymbol{x} - \boldsymbol{x}'| = \left(r - \frac{\boldsymbol{x}\cdot\boldsymbol{x}'}{r}\right)\left(1 + O(\frac{d}{r})\right), \quad r = |\boldsymbol{x}|. \tag{3.4.33}$$

We introduce the wave vector

$$\boldsymbol{k} = \frac{\omega}{c}\frac{\boldsymbol{x}}{r}, \quad k = |\boldsymbol{k}| = \frac{2\pi}{\lambda} = \frac{\omega}{c}. \tag{3.4.34}$$

Then (3.4.25) becomes

$$\boldsymbol{A}(t, \boldsymbol{x}) = \frac{\mu_0}{4\pi} \mathrm{Re} \frac{e^{i(kr-\omega t)}}{r} \int e^{-i\boldsymbol{k}\cdot\boldsymbol{x}'} \boldsymbol{j}_0(\boldsymbol{x}')\, d^3x' \left(1 + O(\frac{d}{r}\right). \tag{3.4.35}$$

The first factor after the real part is a spherical wave, the remaining integral gives the directional dependence through $\boldsymbol{k}$. The exponential under the integral can be expanded into the power series

$$e^{-i\boldsymbol{k}\cdot\boldsymbol{x}'} = 1 - i\boldsymbol{k}\cdot\boldsymbol{x}' - \frac{1}{2}(\boldsymbol{k}\cdot\boldsymbol{x}')^2 + \ldots \tag{3.4.36}$$

because $d \ll \lambda$. The 1 gives the **electric dipole radiation**, the next term the **magnetic dipole and electric quadrupole radiation**, the quadratic term the magnetic quadrupole and electric octupole radiation, etc. The reason for this terminology becomes clear in the sequel.

For electric dipole radiation we have to consider

$$\int j_{0n}(\boldsymbol{x}')\, d^3x' = \int [\mathrm{div}\,(x'_n \boldsymbol{j}_0) - x'_n \mathrm{div}\, \boldsymbol{j}_0]\, d^3x'. \tag{3.4.37}$$

Here the first integral is equal to a surface integral by Gauss' theorem, which vanishes because $\boldsymbol{j}_0$ is concentrated in a compact region. In the remaining term we use the continuity equation (3.4.31), so that we arrive at

$$\int \boldsymbol{j}_0(\boldsymbol{x}')\, d^3x' = -i\omega \int \boldsymbol{x}' \rho_0(\boldsymbol{x}')\, d^3x'. \tag{3.4.38}$$

Here the dipole moment

$$\boldsymbol{P} = \int \boldsymbol{x}' \rho_0(\boldsymbol{x}')\, d^3x' \tag{3.4.39}$$

appears. This is the first non-vanishing moment of ρ_0 because

$$\int \rho_0(\boldsymbol{x}')\, d^3x' = \frac{1}{\omega} \int \operatorname{div} \boldsymbol{j}_0(\boldsymbol{x}')\, d^3x' = 0.$$

Without loss of generality we may assume that $\boldsymbol{P}$ is real, because this can simply be achieved by changing the phase of ρ_0; this is merely a shift of the origin of time. Now we get from (3.4.35) the following leading contribution

$$\begin{aligned} \boldsymbol{A}(t, \boldsymbol{x}) &= -\frac{\mu_0}{4\pi} \omega \boldsymbol{P} \operatorname{Re} i \frac{e^{i(kr-\omega t)}}{r} \\ &= \frac{\mu_0}{4\pi} \omega \boldsymbol{P} \frac{\sin(kr - \omega t)}{r}. \end{aligned} \tag{3.4.40}$$

Taking the curl, we obtain the magnetic field

$$\boldsymbol{B}(t, \boldsymbol{x}) = \frac{\mu_0}{4\pi} \omega \Big(\operatorname{grad} \frac{\sin(kr - \omega t)}{r} \Big) \wedge \boldsymbol{P}. \tag{3.4.41}$$

Let us choose the z-axis parallel to $\boldsymbol{P}$. Since the gradient in (3.4.41) is parallel to $\boldsymbol{x}/r = \boldsymbol{k}$, $\boldsymbol{B}$ is normal to the plane through $\boldsymbol{k}$ and the z-axis, the so-called longitudinal plane. Consequently $\boldsymbol{B}$ is in the xy-plane. The electric field is best calculated from Ampère's law

$$\frac{\partial \boldsymbol{E}}{\partial t} = \frac{1}{\varepsilon_0 \mu_0} \operatorname{curl} \boldsymbol{B}. \tag{3.4.42}$$

Taking the curl of (3.4.41) we find

$$\operatorname{curl} \boldsymbol{B} = \frac{\mu_0}{4\pi} \omega \Big[(\boldsymbol{P} \cdot \operatorname{grad}) \operatorname{grad} \frac{\sin(kr - \omega t)}{r} - \boldsymbol{P} \operatorname{div} \operatorname{grad} \frac{\sin(kr - \omega t)}{r} \Big]. \tag{3.4.43}$$

Integrating in t we obtain the electric field

$$\boldsymbol{E} = \frac{\mu_0 c^2}{4\pi} \Big[(\boldsymbol{P} \cdot \operatorname{grad}) \operatorname{grad} \frac{\cos(kr - \omega t)}{r} - \boldsymbol{P} \,\triangle\, \frac{\cos(kr - \omega t)}{r} \Big]. \tag{3.4.44}$$

To get a detailed picture of the **temporal evolution of the radiation process**, we follow the original discussion by the discoverer of electromagnetic waves Heinrich Hertz (*Untersuchungen über die Ausbreitung der elektrischen Kraft, 2.Auflage, Leipzig 1894, p.156*). We use cylindrical coordinates (ρ, φ, z)

$$\rho^2 = x^2 + y^2, \quad r^2 = \rho^2 + z^2, \quad \frac{\partial r}{\partial \rho} = \frac{\rho}{r} = \sin \vartheta. \tag{3.4.45}$$

Using the Laplace operator in cylindrical coordinates in the last term in (3.4.44), we get for the z-component of the electric field

$$E_z = \frac{1}{4\pi\varepsilon_0}\Big[P\frac{\partial^2 f}{\partial z^2} - P\Big(\frac{1}{\rho}\frac{\partial}{\partial\rho}\rho\frac{\partial f}{\partial\rho} + \frac{\partial^2 f}{\partial z^2}\Big)\Big], \tag{3.4.46}$$

with

$$f = \frac{\cos(kr - \omega t)}{r}. \tag{3.4.47}$$

This can be written as follows

$$E_z = -\frac{1}{4\pi\varepsilon_0}\frac{P}{\rho}\frac{\partial}{\partial\rho}\rho\frac{\partial f}{\partial\rho} = \frac{1}{\rho}\frac{\partial R}{\partial\rho}, \tag{3.4.48}$$

where we have introduced the function

$$\begin{aligned} R(\rho, z) &\stackrel{\text{def}}{=} -\frac{1}{4\pi\varepsilon_0}P\rho\frac{\partial f}{\partial\rho} = -\frac{1}{4\pi\varepsilon_0}P\rho\frac{\partial f}{\partial r}\frac{\rho}{r} = -\frac{1}{4\pi\varepsilon_0}Pr\sin^2\vartheta\frac{\partial f}{\partial r} \\ &= \frac{P\rho^2}{4\pi\varepsilon_0 r}\Big(k\frac{\sin(kr-\omega t)}{r} + \frac{\cos(kr-\omega t)}{r^2}\Big). \end{aligned} \tag{3.4.49}$$

Since $\boldsymbol{P}$ has only a z-component, the last term in (3.4.44) gives no contribution to E_x and E_y:

$$E_x = \frac{1}{4\pi\varepsilon_0}P\frac{\partial}{\partial z}\Big(\frac{x}{r}\frac{\partial f}{\partial r}\Big) \tag{3.4.50}$$

$$E_y = \frac{1}{4\pi\varepsilon_0}P\frac{\partial}{\partial z}\Big(\frac{y}{r}\frac{\partial f}{\partial r}\Big). \tag{3.4.51}$$

Obviously, $\boldsymbol{E}$ lies in the longitudinal plane (that is the plane through a meridian). Since the problem has cylindrical symmetry around the z-axis, instead of E_x, E_y, we better use the radial component

$$\begin{aligned} E_\rho &\stackrel{\text{def}}{=} \frac{x}{\rho}E_x + \frac{y}{\rho}E_y = \frac{P}{4\pi\varepsilon_0}\frac{x^2+y^2}{\rho}\frac{\partial}{\partial z}\Big(\frac{1}{r}\frac{\partial f}{\partial r}\Big) \\ &= \frac{1}{\rho}\frac{\partial}{\partial z}\frac{P\rho^2}{4\pi\varepsilon_0 r}\frac{\partial f}{\partial r} = -\frac{1}{\rho}\frac{\partial R}{\partial z}. \end{aligned} \tag{3.4.52}$$

Now comes the point. Consider the curves $R(\rho, z) = \text{const.}$ in the longitudinal plane. The two-dimensional gradient $(\partial R/\partial\rho, \partial R/\partial z)$ gives the normal direction to this curves, and the tangent is given by $(-\partial R/\partial z, \partial R/\partial\rho)$. A glance to (3.4.52) and (3.4.48) shows that $\boldsymbol{E}$ is parallel to the tangent. Hence, **the curves** $R = \text{const.}$ **are the electrical force lines**: a charge at rest feels a force along these lines. In Fig.8 the drawing of H.Hertz is reproduced for four different values of the phase ωt. If $\omega t = -\pi/2$, we have

$$R = \frac{Pk\rho^2}{4\pi\varepsilon_0 r^2}\Big(\cos kr - \frac{\sin kr}{kr}\Big), \tag{3.4.53}$$

according to (3.4.49). This goes to 0 for $r \to 0$, consequently there are no force lines going out of the dipole. This is connected with the fact that at this moment there is no charge on the dipole

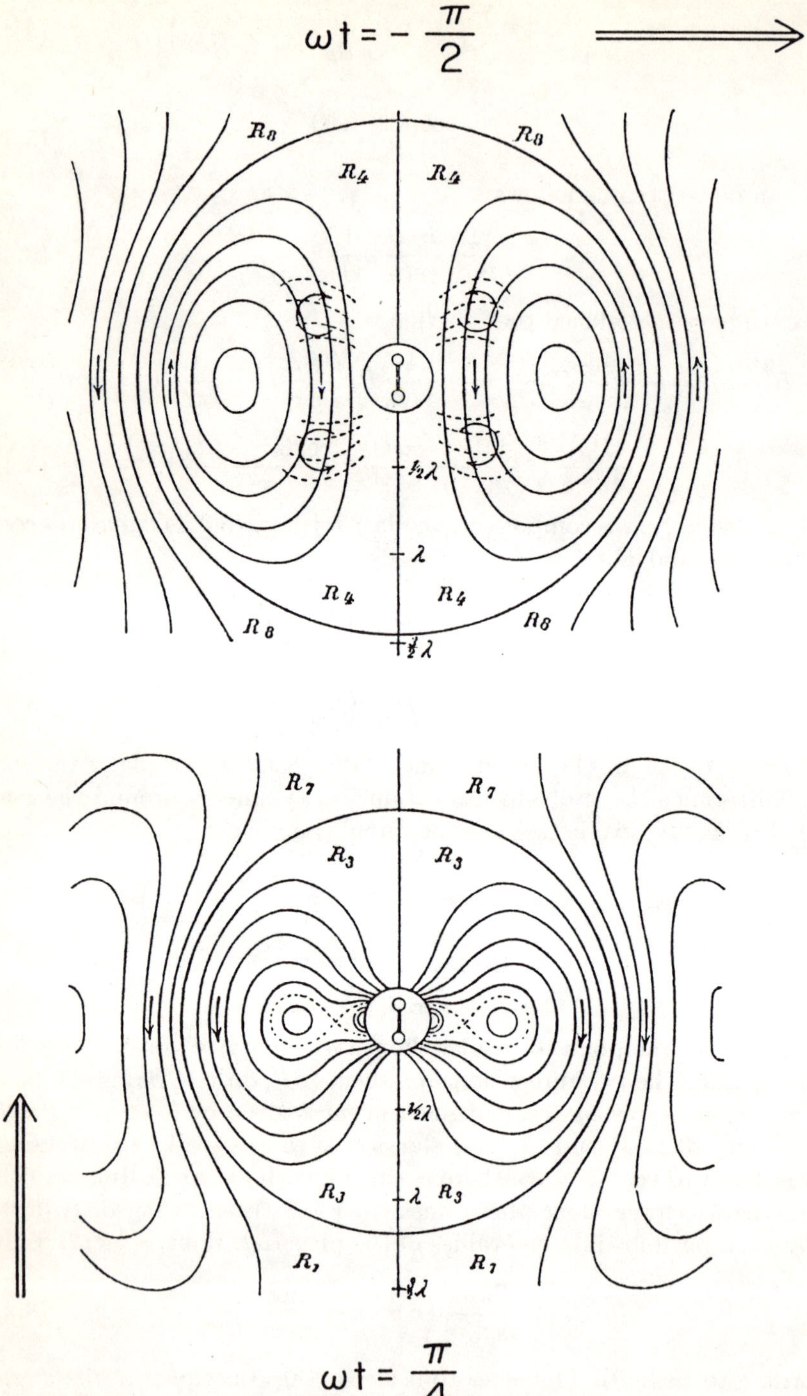

Fig. 8a. Production of dipole radiation after H.Hertz (1894)

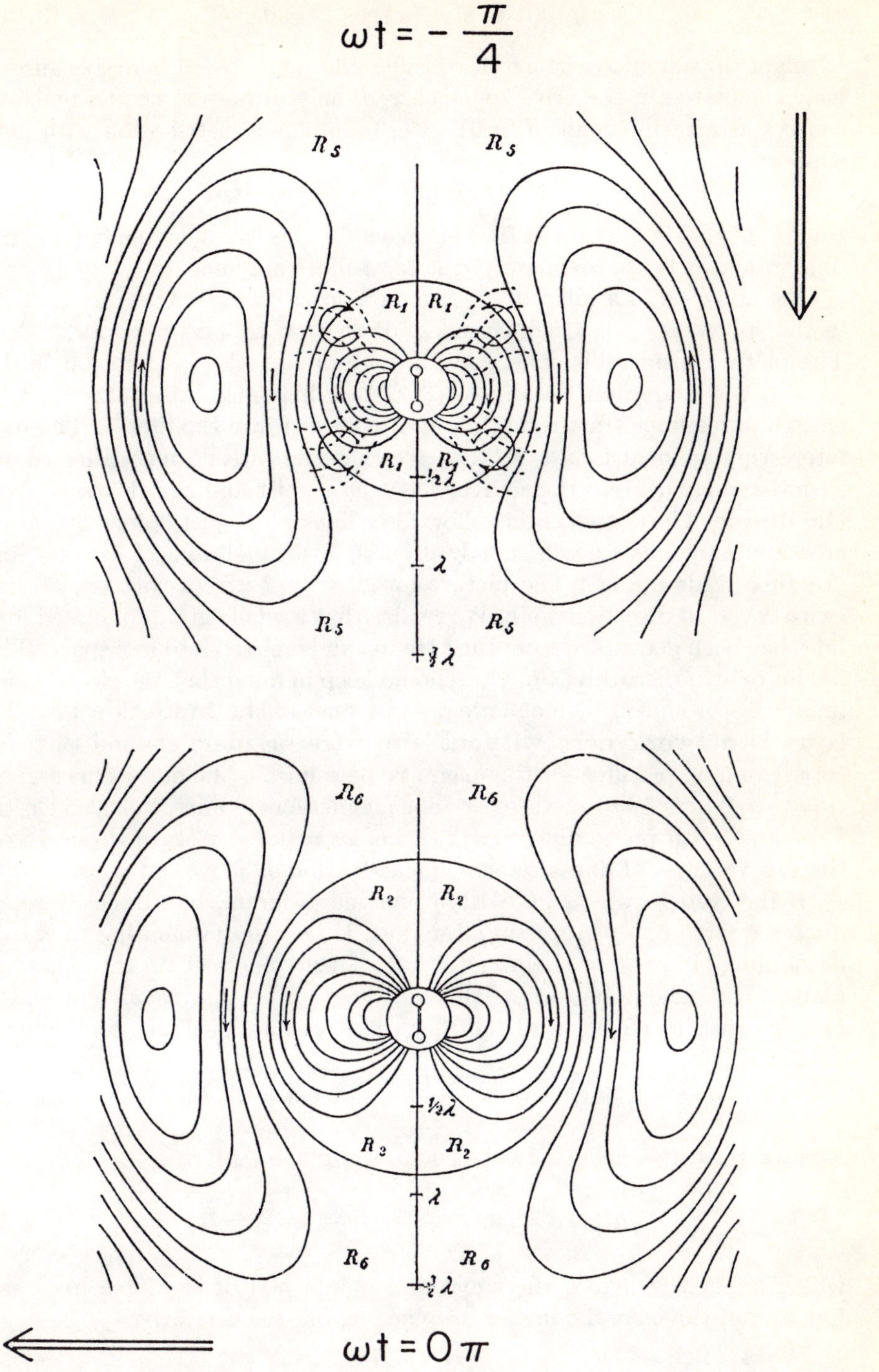

Fig. 8b. Continuation of Fig.8a, the phases $\omega t = -\pi/4$, 0, in between

$$\rho(t, \boldsymbol{x}) = \rho_0(\boldsymbol{x}) \cos \omega t = 0. \tag{3.4.54}$$

It might appear inconsistent to consider the limit $r \to 0$ here, because we have calculated in the wave zone $|\boldsymbol{x}| \gg d$, only. However, we can make the sources arbitrarily small ($d \to 0$), even becoming a distribution with point support

$$\rho_0(\boldsymbol{x}) = -\boldsymbol{P} \cdot \operatorname{grad} \delta(\boldsymbol{x}), \quad \boldsymbol{j}_0(\boldsymbol{x}) = -i\omega \boldsymbol{P} \delta(\boldsymbol{x})$$

(see (1.2.5)). The picture of Hertz is exact for this singular source (Hertzian dipole) and it is approximately true for a finite antenna.

For $\omega t = -\pi/4$ a full variety of force lines is formed. They are enclosed in the sphere $R = 0$, which corresponds to $r^2 = \rho^2 + z^2 = \text{const.}$, indeed. The picture inside this sphere looks more or less like a static dipole. For $\omega t = 0$ the charge on the dipole (3.4.54) is maximal. After this time the electric force lines start to shrink and to return into the dipole. The most interesting moment is $\omega t = \pi/4$: we see that **the outer force lines** cannot shrink and return into the source, they get a waist and **break away from the dipole**. The dotted curves show force lines which just "pinch off" at the crossing points, as we will shortly discuss. Two small almost circular lines are already decoupled. The picture for $\omega t = \pi/2$ agrees with $\omega t = -\pi/2$ (with reversed directions) etc. We realize that a whole group of closed force lines has been decoupled from the antenna and emitted into free space. **This is the origin of radiation.** One should keep in mind that the electric force lines are surrounded by magnetic ones because of **the induction law**. The latter **is at work here without any wire or magnet**, and there are corresponding pictures for the magnetic field lines. (The dotted lines in the $\omega t = -\pi/2$ and $-\pi/4$ pictures are not magnetic field lines, but also electric force lines with phase difference $\pi/2$, i.e. $\omega t = 0$ and $\pi/4$. In those regions the two varieties of lines are orthogonal, so that $\boldsymbol{E}$ moves on a circle.)

If the reader owns a PC with a plotting program, he can easily reproduce the force line pictures in all details. It is really fascinating to see the decoupling of force lines below $\omega t < \pi/2$. They "pinch off" in the equatorial plane $z = 0$. The points where this happens can be analytically calculated: Let us write the equation (3.4.49) for the force lines in dimensionless form

$$g \stackrel{\text{def}}{=} \frac{\bar{\rho}^2}{\bar{r}^2} \left(\sin(\bar{r} - \bar{t}) + \frac{\cos(\bar{r} - \bar{t})}{\bar{r}} \right) = \text{const.} \stackrel{\text{def}}{=} R, \tag{3.4.55}$$

where $\bar{t} = \omega t$, $\bar{r} = kr$, etc. In the equatorial plane we have

$$h(\bar{\rho}, \bar{t}) \stackrel{\text{def}}{=} \sin(\bar{\rho} - \bar{t}) + \frac{\cos(\bar{\rho} - \bar{t})}{\bar{\rho}} = R, \tag{3.4.56}$$

which implicitly defines the equatorial points $\bar{\rho}(\bar{t})$ of the force lines as a function of time. At the instant of pinching off, the derivative

$$\frac{d\bar{\rho}}{d\bar{t}} = -\frac{h_{,\bar{t}}}{h_{,\bar{\rho}}}$$

becomes infinite, that means

$$\frac{\partial h}{\partial \bar\rho} = \cos(\bar\rho - \bar t) - \frac{1}{\bar\rho}\Big(\sin(\bar\rho - \bar t) + \frac{\cos(\bar\rho - \bar t)}{\bar\rho}\Big)$$

$$= \cos(\bar\rho - \bar t) - \frac{R}{\bar\rho} = 0. \qquad (3.4.56a)$$

We denote partial derivatives by comma. From (3.4.56) and (3.4.56a) one finds the position of the pinching off points

$$\bar\rho^2 = \frac{2R}{R \pm \sqrt{4 - 3R^2}}.$$

The corresponding time then follows from (3.4.56a). We notice that only force lines with $|R| < 2/\sqrt{3}$ can decouple.

To investigate the force lines in the vicinity of a pinching off point, we return to (3.4.55) and consider the derivative

$$\frac{d\bar z}{d\bar\rho} = -\frac{g_{,\bar\rho}}{g_{,\bar z}}.$$

Since

$$\frac{\partial g}{\partial \bar z} = \frac{\partial g}{\partial \bar r}\frac{\bar z}{\bar r}, \qquad \frac{\partial g}{\partial \bar\rho} = 2\frac{g}{\bar\rho} + \frac{\partial g}{\partial \bar r}\frac{\bar\rho}{\bar r},$$

we get

$$\frac{d\bar z}{d\bar\rho} = -2\frac{g}{\bar\rho}\frac{\bar r}{\bar z}\Big(\frac{\partial g}{\partial \bar r}\Big)^{-1} - \frac{\bar\rho}{\bar z},$$

where the partial derivative of (3.4.55) is given by

$$\frac{\partial g}{\partial \bar r} = -2\frac{g}{\bar r} + \frac{\bar\rho^2}{\bar r^2}h_{,\bar r} = -2\frac{R}{\bar r} + \frac{\bar\rho^2}{\bar r^2}h_{,\bar r}\,.$$

Here $h_{,\bar r}$ can be taken from (3.4.56a) with $\bar\rho$ replaced by $\bar r$. We see that $h_{,\bar r}$ vanishes at the pinching off points, and it is a small quantity, say δ, in the vicinity. Then we find

$$\frac{d\bar z}{d\bar\rho} = -\frac{1}{\bar z}\Big(2\frac{R\bar r}{\bar\rho}\Big(-2\frac{R}{\bar r} + \frac{\bar\rho^2}{\bar r^2}\delta\Big)^{-1} + \bar\rho\Big)$$

$$= \frac{1}{\bar z\bar\rho}\Big(\bar z^2 + \frac{\bar\rho\bar r}{2R}\delta + O(\delta^2)\Big) = \frac{\bar z}{\bar\rho} + \frac{\bar r}{2R}\frac{\delta}{\bar z} + O(\delta^2).$$

For $\bar z \to 0$, the first term vanishes, but the second one has a finite limit α, which changes sign if the equator is approached from above or from below:

$$\frac{d\bar z}{d\bar\rho} \to \pm\frac{\bar r}{2R}\alpha.$$

Consequently, when a force line pinches off, its waist degenerates to a symmetrical edge. The two edges above and below the equator form a crossing point as shown by the dotted curve in Hertz' figure ($\omega t = \pi/4$). Then the outer and inner parts of the force line separate. The three stages are shown in Fig.9. Note that $\bar{\rho}(\bar{t}, \bar{z})$ is no longer differentiable at the pinching off point (3.4.56a). **Here is the true origin of classical radiation.**

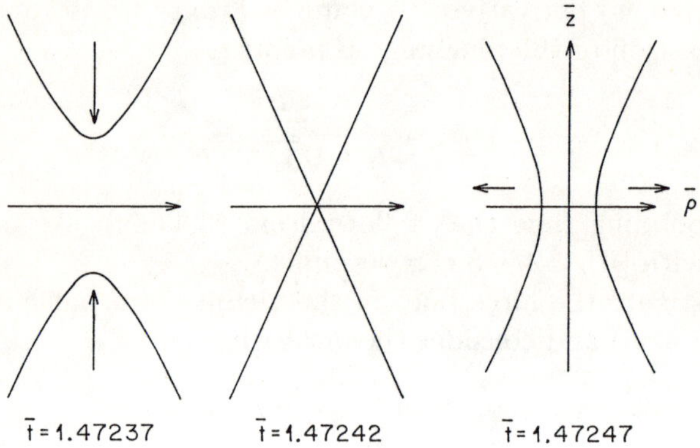

Fig. 9. A force line with $R = 0.3$ pinches off in the equatorial plane

The vectors $\boldsymbol{E}$, $\boldsymbol{B}$ and $\boldsymbol{k}$ form a positively oriented orthogonal system as in case of a plane wave. However, the relation $|\boldsymbol{E}| = c|\boldsymbol{B}|$ is only true in the wave zone $r \gg \lambda$. In the near zone $\boldsymbol{B}$ is smaller, as we will see below. To calculate the emitted radiation, we must only concentrate on the leading order in r/λ. Then we have only to differentiate the sin in (3.4.41)

$$\boldsymbol{B}(t, \boldsymbol{x}) = \frac{\mu_0}{4\pi} \omega \boldsymbol{k} \wedge \boldsymbol{P} \frac{\cos(kr - \omega t)}{r} \Big(1 + O\Big(\frac{\lambda}{r}\Big)\Big). \tag{3.4.56b}$$

As above (3.4.42), we shall obtain $\boldsymbol{E}$ from Ampère's law

$$\boldsymbol{E}(t, \boldsymbol{x}) = c\boldsymbol{B} \wedge \frac{\boldsymbol{k}}{|\boldsymbol{k}|} \Big(1 + O\Big(\frac{\lambda}{r}\Big)\Big). \tag{3.4.56c}$$

This is the plane wave result. The energy-flow density is asymptotically equal to

$$\boldsymbol{S} = \frac{1}{\mu_0} \boldsymbol{E} \wedge \boldsymbol{B} = \frac{c}{\mu_0} \Big(\boldsymbol{B} \wedge \frac{\boldsymbol{k}}{|\boldsymbol{k}|}\Big) \wedge \boldsymbol{B}$$

$$= \frac{c}{\mu_0} \boldsymbol{B}^2 \frac{\boldsymbol{k}}{|\boldsymbol{k}|} = \frac{\mu_0}{16\pi^2} \frac{\omega^4}{c} P^2 \frac{\cos^2(kr - \omega t)}{r^2} \sin^2 \vartheta \frac{\boldsymbol{k}}{|\boldsymbol{k}|}. \tag{3.4.57}$$

The polar angle ϑ is measured between $\boldsymbol{P}$ and the radiation direction $\boldsymbol{k}$. We notice that the maximal radiation is emitted perpendicular to the dipole ($\vartheta = \pi/2$), while the radiation in the direction of $\boldsymbol{P}$ is zero. This is intuitively clear from the decoupling of the force lines in the equatorial plane, discussed above. The radiation power grows with ω^4. Therefore, considerable radiation occurs only for high frequencies. Otherwise our 50 Hz electrical system would be impossible, because it would loose much energy by radiation.

Let us briefly consider the near zone $d, |\boldsymbol{x}| \ll \lambda$. In the original exact retarded potential (3.4.25)

$$\boldsymbol{A}(t,\boldsymbol{x}) = \frac{\mu_0}{4\pi} \mathrm{Re}\, e^{-i\omega t} \int d^3x'\, e^{ik|\boldsymbol{x}-\boldsymbol{x}'|} \frac{\boldsymbol{j}_0(\boldsymbol{x}')}{|\boldsymbol{x}-\boldsymbol{x}'|}$$

we may now neglect the exponential under the integral

$$= \frac{\mu_0}{4\pi} \mathrm{Re}\, e^{-i\omega t} \int d^3x' \frac{\boldsymbol{j}_0(\boldsymbol{x}')}{|\boldsymbol{x}-\boldsymbol{x}'|}. \tag{3.4.58}$$

This is an ordinary magnetostatic field that varies periodically in time. Let us further expand for $|\boldsymbol{x}| = r \gg d$:

$$\begin{aligned} \boldsymbol{A}(t,\boldsymbol{x}) &= \frac{\mu_0}{4\pi r} \mathrm{Re}\, e^{-i\omega t} \int d^3x'\, \boldsymbol{j}_0(\boldsymbol{x}') \\ &= \frac{\mu_0}{4\pi r} \mathrm{Re}\, e^{-i\omega t}(-i\omega)\boldsymbol{P} = -\frac{\mu_0}{4\pi r}\omega \boldsymbol{P} \sin\omega t, \end{aligned} \tag{3.4.59}$$

where (3.4.38, 39) have been used. Taking the curl, we get the magnetic field

$$\begin{aligned} \boldsymbol{B} &= -\frac{\mu_0}{4\pi}\omega \sin\omega t \,\mathrm{grad}\, \frac{1}{r} \wedge \boldsymbol{P} \\ &= \frac{\mu_0}{4\pi}\omega \sin\omega t\, \frac{\boldsymbol{x}}{r^3} \wedge \boldsymbol{P}. \end{aligned} \tag{3.4.60}$$

Another curl gives

$$\boldsymbol{E}' = \frac{\mu_0 c^2}{4\pi}\omega \sin\omega t \,\mathrm{curl} \left(\frac{\boldsymbol{x}}{r^3} \wedge \boldsymbol{P}\right). \tag{3.4.61}$$

Integrating in t and computing the curl on the r.h.s. by (0.1.23) we find

$$\boldsymbol{E} = \frac{\mu_0 c^2}{4\pi} \cos\omega t\, \frac{3\boldsymbol{x}(\boldsymbol{P}\cdot\boldsymbol{x}) - r^2 \boldsymbol{P}}{r^5}. \tag{3.4.62}$$

The space dependence is identical with a static dipole field (1.2.8). Comparing the factor in front with (3.4.60), we find a relative magnitude

$$\frac{c}{\omega r} = \frac{\lambda}{2\pi r} \gg 1.$$

This means that the electric field dominates over $\boldsymbol{B}$ in the near zone.

If the dipole moment $\boldsymbol{P}$ vanishes, the radiation emission is weaker and we must consider the next multipole in (3.4.35, 36)

$$\boldsymbol{A}(t,\boldsymbol{x}) = -\frac{\mu_0}{4\pi} \mathrm{Re}\, \frac{e^{i(kr-\omega t)}}{r} \int i\boldsymbol{k}\cdot\boldsymbol{x}'\boldsymbol{j}_0(\boldsymbol{x}')\, d^3x'. \tag{3.4.63}$$

In

$$A_l(t, \boldsymbol{x}) = -\frac{\mu_0}{4\pi}\mathrm{Re}\,\frac{e^{i(kr-\omega t)}}{r} i \int \sum_{n=1}^{3} k_n x'_n j_{0l}(\boldsymbol{x}')\, d^3x' \tag{3.4.64}$$

we decompose the integrand into a symmetric and antisymmetric part:

$$x'_n j_{0l} = \tfrac{1}{2}(x'_n j_{0l} + x'_l j_{0n}) + \tfrac{1}{2}(x'_n j_{0l} - x'_l j_{0n}).$$

The antisymmetric part gives

$$A^a_l = -\frac{\mu_0}{4\pi}\frac{1}{2}\mathrm{Re}\, i\frac{e^{i(kr-\omega t)}}{r} \int d^3x' \sum_n k_n (x'_n j_{0l} - x'_l j_{0n}). \tag{3.4.65}$$

A similar integral has appeared in magnetostatic (3.1.27)

$$\boldsymbol{A}^a = -\frac{\mu_0}{4\pi}\frac{1}{2}\mathrm{Re}\, i\frac{e^{i(kr-\omega t)}}{r} \int d^3x' \, (\boldsymbol{k}\cdot\boldsymbol{x}'\boldsymbol{j}_0(\boldsymbol{x}') - \boldsymbol{x}'\boldsymbol{k}\cdot\boldsymbol{j}_0(\boldsymbol{x}')). \tag{3.4.66}$$

Using the magnetic moment (3.1.29)

$$\boldsymbol{M} = \frac{\mu_0}{2}\int d^3x' \, \boldsymbol{x}' \wedge \boldsymbol{j}_0, \tag{3.4.67}$$

we arrive at

$$\begin{aligned} \boldsymbol{A}^a &= -\frac{1}{4\pi}\mathrm{Re}\, i\frac{e^{i(kr-\omega t)}}{r}\boldsymbol{M}\wedge\boldsymbol{k} \\ &= \frac{1}{4\pi}\frac{\sin(kr-\omega t)}{r}\boldsymbol{M}\wedge\boldsymbol{k}. \end{aligned} \tag{3.4.68}$$

We have assumed here that the magnetic moment (3.4.67) is real. This can always be achieved by choosing the phase of $\boldsymbol{j}_0$ suitably, which is again only a shift of the origin of time..

To calculate the fields from (3.4.68) to leading order in λ/r, we must again differentiate the trigonometric functions only

$$\boldsymbol{B}^a = \frac{1}{4\pi}\frac{\cos(kr-\omega t)}{r}\boldsymbol{k}\wedge(\boldsymbol{M}\wedge\boldsymbol{k}). \tag{3.4.69}$$

This looks like $\boldsymbol{E}$ (3.4.56c) for electrical dipole radiation. The result for $\boldsymbol{E}$ from Ampère's law

$$\boldsymbol{E}^a = \frac{\omega}{4\pi}\frac{\cos(kr-\omega t)}{r}\boldsymbol{M}\wedge\boldsymbol{k} \tag{3.4.70}$$

agrees with $\boldsymbol{B}$ (3.4.56b) of electric dipole radiation, up to a factor. For this reason one calls this antisymmetric part magnetic dipole radiation. A circular current (3.1.3) with sinusoidal time dependence does not emit electric, but only magnetic dipole radiation (Problem 10).

We briefly turn to the symmetric part in (3.4.64)

$$A_l^s = -\frac{\mu_0}{4\pi}\frac{1}{2}\mathrm{Re}\, i\frac{e^{i(kr-\omega t)}}{r}\sum_n k_n \int d^3x'\,(x'_n j_{0l} + x'_l j_{0n}). \tag{3.4.71}$$

Denoting the integral by Q'_{nl}, we transform it as follows

$$Q'_{nl} = \int d^3x'\,\boldsymbol{j}_0 \cdot \mathrm{grad}\,(x'_n x'_l) = \int d^3x' \Big(\mathrm{div}\,(x'_n x'_l \boldsymbol{j}_0) - x'_n x'_l \mathrm{div}\,\boldsymbol{j}_0\Big). \tag{3.4.72}$$

The divergence gives no contribution by Gauss' theorem. In the last term we use the continuity equation (3.4.37) and get essentially the electric quadrupole moment

$$Q'_{nl} = -i\omega \int d^3x'\, x'_n x'_l \rho_0(\boldsymbol{x}'). \tag{3.4.73}$$

For this reason, one calls this contribution (3.4.71) electric quadrupole radiation. In electrostatics (1.2.3) one defines the latter as follows

$$Q_{nl} = \int d^3x'\,(3x'_n x'_l - \boldsymbol{x}'^2\delta_{nl})\rho(\boldsymbol{x}'), \tag{3.4.74}$$

in order to have vanishing trace, $\sum Q_{nn} = 0$. We may safely add the isotropic part to (3.4.73), because it does not radiate:

$$A_l \propto \mathrm{Re}\,\frac{e^{i(kr-\omega t)}}{r} k_l$$

$$= \frac{\partial}{\partial x_l}\mathrm{Re}\,(-i)\frac{e^{i(kr-\omega t)}}{r}\Big(1 + O\Big(\frac{\lambda}{r}\Big)\Big).$$

Since this is a gradient, the magnetic field vanishes, $\mathrm{curl}\,\boldsymbol{A} = \boldsymbol{0}$.

Until now we have considered periodic time dependence of the sources. We want now to discuss dipole radiation without this restriction. We start from the retarded potential (3.4.25) and expand in the wave zone $|\boldsymbol{x}| = r \gg d$, where d is the extension of the current distribution. The leading order in λ/r is given by

$$\boldsymbol{A}(t,\boldsymbol{x}) = \frac{\mu_0}{4\pi r}\int d^3x'\,\boldsymbol{j}\Big(t - \frac{|\boldsymbol{x}-\boldsymbol{x}'|}{c}, \boldsymbol{x}'\Big). \tag{3.4.75}$$

We evaluate this for long waves with $d \ll \lambda = cT$, where $T = d/v$ is the time for a significant change of the current distribution, v is the mean velocity of the charges. This implies $v \ll c$, which means that the charges move slowly compared with the velocity of light. Then the retarded time argument in (3.4.75) becomes

$$t - \frac{|\boldsymbol{x}-\boldsymbol{x}'|}{c} = t - \frac{r}{c} + O\Big(\frac{d}{c}\Big). \tag{3.4.76}$$

Since $d/c = Tv/c$ is small compared with T, we may neglect the correction in (3.4.76). Consequently, the retardation is unimportant in the integral (3.4.75)

$$\boldsymbol{A}(t,\boldsymbol{x}) = \frac{\mu_0}{4\pi r}\int d^3x'\,\boldsymbol{j}(t-\frac{r}{c},\boldsymbol{x}'). \tag{3.4.77}$$

Let us assume that $\boldsymbol{j}$ comes from moving point charges

$$\boldsymbol{j}(t,\boldsymbol{x}) = \sum_j q_j \boldsymbol{v}_j(t)\delta(\boldsymbol{x}-\boldsymbol{x}_j(t)), \tag{3.4.78}$$

where

$$\boldsymbol{v}_j(t) = \boldsymbol{x}_j'(t) \tag{3.4.79}$$

is the velocity of the charge q_j. Then we obtain

$$\boldsymbol{A}(t,\boldsymbol{x}) = \frac{\mu_0}{4\pi r}\sum_j q_j \boldsymbol{v}_j(t-\frac{r}{c})$$

$$= \frac{\mu_0}{4\pi r}\frac{d}{dt}\sum_j q_j \boldsymbol{x}_j(t-\frac{r}{c}) \stackrel{\text{def}}{=} \frac{\mu_o}{4\pi r}\boldsymbol{P}'(t-\frac{r}{c}). \tag{3.4.80}$$

Here

$$\boldsymbol{P}(t) = \sum_j q_j \boldsymbol{x}_j(t) \tag{3.4.81}$$

is the dipole moment of the charge distribution. To calculate the fields for large r, we must only differentiate in the argument of $\boldsymbol{P}$

$$\boldsymbol{B}(t,\boldsymbol{x}) = \frac{\mu_0}{4\pi c r}\boldsymbol{P}''(t-\frac{r}{c})\wedge\frac{\boldsymbol{x}}{r}. \tag{3.4.82}$$

$\boldsymbol{E}$ follows again from Ampère's law, it is transverse

$$\boldsymbol{E}(t,\boldsymbol{x}) = c\boldsymbol{B}\wedge\frac{\boldsymbol{x}}{r} = \frac{\mu_0}{4\pi r}\Big[\boldsymbol{P}''\wedge\frac{\boldsymbol{x}}{r}\Big]\wedge\frac{\boldsymbol{x}}{r}. \tag{3.4.83}$$

This yields for the Poynting vector

$$\boldsymbol{S} = \frac{c}{\mu_0}\boldsymbol{B}^2\frac{\boldsymbol{x}}{r} = \frac{\mu_0}{(4\pi r)^2 c}|\boldsymbol{P}''|^2\sin^2\vartheta\,\frac{\boldsymbol{x}}{r}. \tag{3.4.84}$$

The corresponding power radiated per unit solid angle is given by

$$\frac{dN}{d\Omega} \stackrel{\text{def}}{=} r^2|\boldsymbol{S}| = \frac{\mu_0}{16\pi^2 c}|\boldsymbol{P}''|^2\sin^2\vartheta. \tag{3.4.85}$$

The total power or energy flux is obtained by integration over all directions

$$\int d\Omega \sin^2 \vartheta = 2\pi \int_{-1}^{+1} d\cos\vartheta \,(1 - \cos^2 \vartheta) = 2\pi \cdot \frac{4}{3},$$

$$N = \frac{\mu_0}{4\pi} \frac{2}{3} \frac{|\boldsymbol{P}''|^2}{c}. \tag{3.4.86}$$

If only one charge q is moving, $\boldsymbol{P} = q\boldsymbol{x}$, $\boldsymbol{P}'' = q\boldsymbol{x}''$, we get **Larmor's formula**

$$N = \frac{\mu_0}{4\pi} \frac{2}{3} \frac{q^2}{c} |\boldsymbol{x}''|^2 = \frac{1}{4\pi\varepsilon_0} \frac{2}{3} \frac{q^2}{c^3} |\boldsymbol{x}''(t)|^2. \tag{3.4.87}$$

The radiated power is proportional to the acceleration of the charge. A charge moving with constant velocity does not radiate, of course.

3.5 Electromagnetic Fields of Moving Particles

The naive picture of a moving charged particle is one where the Coulomb field of the particle moves with it along its trajectory. But we have observed at the end of the last section that an accelerated charge emits radiation. That means, the Coulomb field must be strongly disturbed in case of accelerated motion and part of it is radiated away, while the other part remains attached to the particle. It is our aim now to analyse this interesting phenomenon in detail.

We start from the retarded potentials (3.4.25, 27)

$$V(t, \boldsymbol{x}) = \frac{1}{4\pi\varepsilon_0} \int d^3x' \, \frac{\rho\left(t - \frac{|\boldsymbol{x}-\boldsymbol{x}'|}{c}, \boldsymbol{x}'\right)}{|\boldsymbol{x} - \boldsymbol{x}'|} \tag{3.5.1}$$

$$\boldsymbol{A}(t, \boldsymbol{x}) = \frac{\mu_0}{4\pi} \int d^3x' \, \frac{\boldsymbol{j}\left(t - \frac{|\boldsymbol{x}-\boldsymbol{x}'|}{c}, \boldsymbol{x}'\right)}{|\boldsymbol{x} - \boldsymbol{x}'|}. \tag{3.5.2}$$

We consider a point charge q that moves on a prescribed trajectory $\boldsymbol{x} = \boldsymbol{y}(t)$. This gives rise to a charge density

$$\rho(t, \boldsymbol{x}) = q\delta(\boldsymbol{x} - \boldsymbol{y}(t)) \tag{3.5.3}$$

and current density

$$\boldsymbol{j}(t, \boldsymbol{x}) = q\boldsymbol{v}(t)\delta(\boldsymbol{x} - \boldsymbol{y}(t)), \quad \boldsymbol{v}(t) = \boldsymbol{y}'(t). \tag{3.5.4}$$

In the scalar potential

$$V(t, \boldsymbol{x}) = \frac{q}{4\pi\varepsilon_0} \int d^3x' \, \frac{\delta\left(\boldsymbol{x}' - \boldsymbol{y}(t - \frac{|\boldsymbol{x}-\boldsymbol{x}'|}{c})\right)}{|\boldsymbol{x} - \boldsymbol{x}'|} \tag{3.5.5}$$

a δ-distribution at the retarded place $\boldsymbol{y}_{\text{ret}} = \boldsymbol{y}(t_{\text{ret}})$ appears. Here t_{ret} is the solution of the equation

$$t_{\text{ret}} = t - \frac{1}{c}|\boldsymbol{x} - \boldsymbol{y}(t_{\text{ret}})|, \tag{3.5.6}$$

which is unique for not too large t and always $< t$. It is conventionally called retarded time, although it is advanced with respect to the time t of observation. The 3-dimensional δ-distribution must be calculated by means of

$$\delta^3(\boldsymbol{f}(\boldsymbol{x}')) = \frac{\delta^3(\boldsymbol{x}' - \boldsymbol{y}_{\text{ret}})}{\left|\det \frac{\partial \boldsymbol{f}}{\partial \boldsymbol{x}'}\right|}, \tag{3.5.7}$$

where the Jacobian in the denominator has to be evaluated at the zero $\boldsymbol{y}_{\text{ret}}$ of $\boldsymbol{f}(\boldsymbol{x}')$.

We give some details of the calculation. From

$$\boldsymbol{f}(\boldsymbol{x}') = \boldsymbol{x}' - \boldsymbol{y}(t - |\boldsymbol{x} - \boldsymbol{x}'|/c) \tag{3.5.8}$$

we find

$$\left(\frac{\partial \boldsymbol{f}}{\partial \boldsymbol{x}'}\right)_{ik} = \frac{\partial f_i}{\partial x'_k} = \delta_{ik} + \frac{v_i}{c}\frac{x'_k - x_k}{|\boldsymbol{x} - \boldsymbol{x}'|}. \tag{3.5.9}$$

Introducing

$$\frac{\boldsymbol{v}}{c} = \boldsymbol{\beta}, \quad \frac{\boldsymbol{x} - \boldsymbol{y}_{\text{ret}}}{|\boldsymbol{x} - \boldsymbol{y}_{\text{ret}}|} = \boldsymbol{n}, \tag{3.5.10}$$

where $\boldsymbol{n}$ is the unit vector in the radiation (or rather observation) direction, we get for the Jacobian determinant

$$\begin{vmatrix} 1 - \beta_1 n_1 & -\beta_1 n_2 & -\beta_1 n_3 \\ -\beta_2 n_1 & 1 - \beta_2 n_2 & -\beta_2 n_3 \\ -\beta_3 n_1 & -\beta_3 n_2 & 1 - \beta_3 n_3 \end{vmatrix} = 1 - \boldsymbol{\beta} \cdot \boldsymbol{n} > 0. \tag{3.5.11}$$

This leads to

$$V(t, \boldsymbol{x}) = \frac{q}{4\pi\varepsilon_0}\frac{1}{1 - \boldsymbol{\beta} \cdot \boldsymbol{n}}\frac{1}{|\boldsymbol{x} - \boldsymbol{y}|}\bigg|_{\text{ret}}, \tag{3.5.12}$$

where the subscript "ret" means that we must take $t = t_{\text{ret}}(t, \boldsymbol{x})$ in $\boldsymbol{y}(t)$ and $\boldsymbol{v}(t)$. The vector potential is obtained in the same way from (3.5.2)

$$\boldsymbol{A}(t, \boldsymbol{x}) = \frac{q\mu_0}{4\pi}\frac{1}{1 - \boldsymbol{\beta} \cdot \boldsymbol{n}}\frac{\boldsymbol{v}}{|\boldsymbol{x} - \boldsymbol{y}|}\bigg|_{\text{ret}} = \frac{1}{c^2}V(t, \boldsymbol{x})\boldsymbol{v}(t_{\text{ret}}). \tag{3.5.13}$$

The potentials (3.5.12, 13) are the so-called **Liénard-Wiechert potentials**.

The computation of the electromagnetic fields

$$\boldsymbol{E} = -\text{grad}_{\,x} V - \frac{\partial \boldsymbol{A}}{\partial t} \tag{3.5.14}$$

$$\boldsymbol{B} = \operatorname{curl} \boldsymbol{A} \tag{3.5.15}$$

is complicated because everything depends on t_{ret}, which is implicitly defined by (3.5.6). We continue to calculate non-relativistically, at the end of this section we give the corresponding 4-dimensional calculation. Differentiating (3.5.6) with respect to t, we get

$$\begin{aligned} \frac{\partial t_{\text{ret}}}{\partial t} &= 1 - \frac{1}{c} \frac{\boldsymbol{x} - \boldsymbol{y}}{|\boldsymbol{x} - \boldsymbol{y}|} \cdot (-\boldsymbol{v}) \Big|_{\text{ret}} \frac{\partial t_{\text{ret}}}{\partial t} \\ &= 1 + \boldsymbol{\beta} \cdot \boldsymbol{n}|_{\text{ret}} \frac{\partial t_{\text{ret}}}{\partial t}. \end{aligned} \tag{3.5.16}$$

This leads to

$$\frac{\partial t_{\text{ret}}}{\partial t} = \frac{1}{1 - \boldsymbol{\beta} \cdot \boldsymbol{n}} \Big|_{\text{ret}}, \tag{3.5.17}$$

and similarly to

$$\operatorname{grad} t_{\text{ret}} = - \frac{\boldsymbol{n}}{c(1 - \boldsymbol{\beta} \cdot \boldsymbol{n})} \Big|_{\text{ret}}. \tag{3.5.18}$$

Now we are able to calculate the derivatives

$$\operatorname{grad}_x V = \frac{q}{4\pi\varepsilon_0} \left[\frac{\boldsymbol{\beta} - \boldsymbol{n}}{(1 - \boldsymbol{\beta} \cdot \boldsymbol{n})^2 |\boldsymbol{x} - \boldsymbol{y}|^2} + \boldsymbol{n} \frac{\beta^2 - \boldsymbol{\beta} \cdot \boldsymbol{n} - \boldsymbol{\beta}'(\boldsymbol{x} - \boldsymbol{y})/c}{(1 - \boldsymbol{\beta} \cdot \boldsymbol{n})^3 |\boldsymbol{x} - \boldsymbol{y}|^2} \right]_{\text{ret}}, \tag{3.5.19}$$

$$\frac{\partial \boldsymbol{A}}{\partial t} = \frac{q}{4\pi\varepsilon_0} \left[\frac{\boldsymbol{\beta}'/c}{(1 - \boldsymbol{\beta} \cdot \boldsymbol{n})^2 |\boldsymbol{x} - \boldsymbol{y}|} - \boldsymbol{\beta} \frac{\beta^2 - \boldsymbol{\beta} \cdot \boldsymbol{n} - \boldsymbol{\beta}' \cdot (\boldsymbol{x} - \boldsymbol{y})/c}{(1 - \boldsymbol{\beta} \cdot \boldsymbol{n})^3 |\boldsymbol{x} - \boldsymbol{y}|^2} \right]_{\text{ret}}. \tag{3.5.20}$$

This yields the electric field

$$\boldsymbol{E} = \frac{q}{4\pi\varepsilon_0} \left[(\boldsymbol{n} - \boldsymbol{\beta}) \frac{1 - \beta^2}{(1 - \boldsymbol{\beta} \cdot \boldsymbol{n})^3 |\boldsymbol{x} - \boldsymbol{y}|^2} \right. \tag{3.5.21a}$$

$$\left. + \frac{\boldsymbol{n} \wedge [(\boldsymbol{n} - \boldsymbol{\beta}) \wedge \boldsymbol{\beta}']}{c(1 - \boldsymbol{\beta} \cdot \boldsymbol{n})^3 |\boldsymbol{x} - \boldsymbol{y}|} \right]_{\text{ret}}. \tag{3.5.21b}$$

Here the first term (3.5.21a) falls off like r^{-2} as a static electric field. **This part, independent of the acceleration, remains attached to the moving particle**. The other term (3.5.21b), say $\boldsymbol{E}_a$, depending on the acceleration $\boldsymbol{\beta}'$, decreases like r^{-1} only. It transports energy to infinity, as we will shortly see, and represents radiation. We note already that $\boldsymbol{E}_a$ is perpendicular to $\boldsymbol{n}$.

For the magnetic field we still need

$$\begin{aligned} (\operatorname{curl} \boldsymbol{v}(t_{\text{ret}}))_j &= \frac{dv_l}{dt_{\text{ret}}} \frac{\partial t_{\text{ret}}}{\partial x_k} - \frac{dv_k}{dt_{\text{ret}}} \frac{\partial t_{\text{ret}}}{\partial x_l} \\ &= - \frac{1}{c(1 - \boldsymbol{\beta} \cdot \boldsymbol{n})} (v_l' n_k - v_k' n_l)_{\text{ret}}, \end{aligned} \tag{3.5.22}$$

where j, k, l is a cyclic permutation of 1,2,3. This is equal to

$$\operatorname{curl} \boldsymbol{v}(t_{\text{ret}}) = \frac{\boldsymbol{v}' \wedge \boldsymbol{n}}{c(1 - \boldsymbol{\beta} \cdot \boldsymbol{n})}\bigg|_{\text{ret}} = \frac{\boldsymbol{\beta}' \wedge \boldsymbol{n}}{1 - \boldsymbol{\beta} \cdot \boldsymbol{n}}\bigg|_{\text{ret}}. \tag{3.5.23}$$

Now we are ready to calculate the magnetic field

$$\boldsymbol{B} = \operatorname{curl} \frac{1}{c^2}\Big(V \boldsymbol{v}(t_{\text{ret}})\Big)$$

$$= \frac{1}{c^2} \operatorname{grad} V \wedge \boldsymbol{v}(t_{\text{ret}}) + \frac{V}{c^2} \operatorname{curl} \boldsymbol{v}(t_{\text{ret}})$$

$$= \frac{q}{4\pi\varepsilon_0 c^2}\left[\boldsymbol{v} \wedge \boldsymbol{n} \frac{1 - \beta^2}{(1 - \boldsymbol{\beta} \cdot \boldsymbol{n})^3 |\boldsymbol{x} - \boldsymbol{y}|^2}\right. \tag{3.5.24a}$$

$$\left. + \frac{(\boldsymbol{\beta}' \cdot \boldsymbol{n})\boldsymbol{\beta} \wedge \boldsymbol{n} + (1 - \boldsymbol{\beta} \cdot \boldsymbol{n})\boldsymbol{\beta}' \wedge \boldsymbol{n}}{(1 - \boldsymbol{\beta} \cdot \boldsymbol{n})^3 |\boldsymbol{x} - \boldsymbol{y}|}\right]_{\text{ret}}. \tag{3.5.24b}$$

We denote the second term, depending on the acceleration, by $\boldsymbol{B}_a$. It decreases like r^{-1} in contrast to (3.5.24a). A comparison with (3.5.21) shows that

$$\boldsymbol{B} = \frac{1}{c}(\boldsymbol{n} \wedge \boldsymbol{E}), \tag{3.5.25}$$

which was typical for electromagnetic radiation. But $\boldsymbol{E}$ is not orthogonal to $\boldsymbol{n}$, so that the whole field is not transverse. However, the acceleration part $\boldsymbol{E}_a$ is transverse to $\boldsymbol{n}$ and

$$\boldsymbol{B}_a = \frac{1}{c}\boldsymbol{n} \wedge \boldsymbol{E}_a, \tag{3.5.26}$$

as for an electromagnetic plane wave.

The true **identification of the radiation follows by considering the energy flow to infinity**. To get a non-vanishing flow to infinity, the Poynting vector

$$\boldsymbol{S} = \frac{1}{\mu_0}\boldsymbol{E} \wedge \boldsymbol{B} = \frac{1}{\mu_0 c}\boldsymbol{E} \wedge (\boldsymbol{n} \wedge \boldsymbol{E})$$

must decrease like r^{-2} (or less), in order that the total power

$$N = \int_{|\boldsymbol{x}-\boldsymbol{y}|=r} \boldsymbol{S} \cdot d\boldsymbol{\sigma}$$

remains finite for $r \to \infty$. This is only true for the terms depending on the accceleration $\boldsymbol{E}_a$, $\boldsymbol{B}_a$. They give

$$\boldsymbol{S}_a = \frac{1}{\mu_0 c}(\boldsymbol{n}\boldsymbol{E}_a^2 - \boldsymbol{E}_a \boldsymbol{E}_a \cdot \boldsymbol{n}) = \frac{1}{\mu_0 c}\boldsymbol{E}_a^2 \boldsymbol{n}. \tag{3.5.27}$$

First let us evaluate this for small velocities $\beta \ll 1$,

$$\boldsymbol{E}_a = \frac{q}{4\pi\varepsilon_0} \frac{\boldsymbol{n} \wedge [\boldsymbol{n} \wedge \boldsymbol{\beta}']}{r}\bigg|_{\text{ret}}, \quad r = |\boldsymbol{x} - \boldsymbol{y}|. \tag{3.5.28}$$

This agrees with the formula (3.4.83) for dipole radiation. Thus we get the same radiation power

$$\frac{dN}{d\Omega} = r^2 |\boldsymbol{S}_a| = \frac{1}{4\pi\varepsilon_0} \frac{q^2}{4\pi c^3} |\boldsymbol{v}'|^2 \sin^2\vartheta, \quad \text{and} \tag{3.5.29}$$

$$N = \frac{1}{4\pi\varepsilon_0} \frac{2}{3} \frac{q^2}{c^3} |\boldsymbol{v}'|^2, \tag{3.5.30}$$

in agreement with the Larmor formula (3.4.87).

Let us now evaluate the results for particle velocities which are comparable with c. From (3.5.21b)

$$\boldsymbol{E}_a = \frac{q}{4\pi\varepsilon_0 c} \frac{\boldsymbol{n} \wedge [(\boldsymbol{n} - \boldsymbol{\beta}) \wedge \boldsymbol{\beta}']}{(1 - \boldsymbol{\beta}\cdot\boldsymbol{n})^3 r}\bigg|_{\text{ret}} \tag{3.5.31}$$

we obtain

$$\boldsymbol{S}_a = \frac{1}{\mu_0 c} \boldsymbol{E}_a^2 \boldsymbol{n}, \quad \text{and} \tag{3.5.32}$$

$$\frac{dN(t)}{d\Omega} = r^2 |\boldsymbol{S}_a(t, \boldsymbol{n})| \stackrel{\text{def}}{=} |\boldsymbol{F}(t)|^2. \tag{3.5.33}$$

This is the energy per unit time and unit solid angle that is radiated in direction $\boldsymbol{n}$ at time t. The function $\boldsymbol{F}(t)$ introduced here is equal to

$$\boldsymbol{F}(t) = \frac{r}{\sqrt{\mu_0 c}} \boldsymbol{E}_a = \frac{1}{\sqrt{\mu_0 c}} \frac{q}{4\pi\varepsilon_0 c} \frac{\boldsymbol{n} \wedge [(\boldsymbol{n} - \boldsymbol{\beta}) \wedge \boldsymbol{\beta}']}{(1 - \boldsymbol{\beta}\cdot\boldsymbol{n})^3}\bigg|_{\text{ret}}. \tag{3.5.34}$$

We first want to compute the **total radiation rate** $N(t)$ at a fixed time t. Simplifying the vector product in (3.5.34) and calculating the square (3.5.33), we find

$$\frac{dN(t)}{d\Omega} = \frac{1}{\mu_0 c} \frac{q^2}{(4\pi\varepsilon_0 c)^2} \left[(\beta^2 - 1) \frac{(\boldsymbol{\beta}' \cdot \boldsymbol{n})^2}{(1 - \boldsymbol{\beta}\cdot\boldsymbol{n})^6} + \frac{2\boldsymbol{\beta}\cdot\boldsymbol{\beta}' \boldsymbol{\beta}' \cdot \boldsymbol{n}}{(1 - \boldsymbol{\beta}\cdot\boldsymbol{n})^5} \right.$$
$$\left. + \frac{\boldsymbol{\beta}'^2}{(1 - \boldsymbol{\beta}\cdot\boldsymbol{n})^4} \right]_{\text{ret}}. \tag{3.5.35}$$

For the angular integrations $d\cos\vartheta \, d\varphi$ over the unit sphere, we choose the polar axis $\vartheta = 0$ along $\boldsymbol{\beta}$, so that $\boldsymbol{\beta}\cdot\boldsymbol{n} = \beta\cos\vartheta$ (the length of 3-vectors is denoted by normal letters). Let α be the angle between $\boldsymbol{\beta}$ and $\boldsymbol{\beta}'$, thus $\boldsymbol{\beta}\cdot\boldsymbol{\beta}' = \beta\beta'\cos\alpha$. The angle α' between $\boldsymbol{n}$ and $\boldsymbol{\beta}'$ is given by the cosine-theorem in the spherical triangle $(\boldsymbol{\beta}/\beta, \boldsymbol{n}, \boldsymbol{\beta}'/\beta')$:

$$\cos\alpha' = \cos\alpha\cos\vartheta + \sin\alpha\sin\vartheta\cos\varphi, \tag{3.5.36}$$

where the azimuthal angle φ is the inner angle at $\boldsymbol{\beta}/\beta$ in the triangle. The integrations over φ and $\cos\vartheta$ in (3.5.35) are elementary. We finally obtain

$$N(t) = \frac{q^2}{4\pi\varepsilon_0 c} \frac{\beta'^2}{(1-\beta^2)^4} \frac{2}{3}\left[1 + \frac{\beta^2}{5} - \beta^2 \sin^2\alpha\left(\frac{4}{5} + \frac{2}{5}\beta^2\right)\right]_{\text{ret}} . \tag{3.5.37}$$

This **radiation rate $N(t)$ is the energy flux per unit time across a big spherical surface, measured in the laboratory frame.** It has a clear physical meaning based on the energy conservation law (2.6.12) and is directly measurable. In addition, one considers the **radiation power $P(t_{\text{ret}})$ measured per unit time in the rest system of the charge.** The two differ by the factor $\partial t/\partial t_{\text{ret}} = 1 - \boldsymbol{\beta}\cdot\boldsymbol{n}$ (3.5.17)

$$\frac{dP(t_{\text{ret}})}{d\Omega} = \frac{dN(t)}{d\Omega}(1 - \boldsymbol{\beta}\cdot\boldsymbol{n})_{\text{ret}}, \tag{3.5.38}$$

so that the powers in the denominators in (3.5.35) are reduced by one. Performing the angular integrations as above, we now get

$$P(t_{\text{ret}}) = \frac{q^2}{4\pi\varepsilon_0 c} \frac{\beta'^2}{(1-\beta^2)^3} \frac{2}{3}\left(1 - \beta^2 \sin^2\alpha\right)_{\text{ret}} . \tag{3.5.39}$$

At present, the physical meaning of this quantity is unclear, it is hard to imagine an arrangement where it can be experimentally measured. We shall see at the end of the following section that P is the radiated power in the rest frame of the moving charge. This radiation loss is important for the dynamics of the particle.

Next we want to integrate (3.5.33) over time, to get the total radiated energy per unit solid angle

$$\frac{dW}{d\Omega} = \int_{-\infty}^{+\infty} dt\, |\boldsymbol{F}(t)|^2 = \int_{-\infty}^{+\infty} d\omega\, |\hat{\boldsymbol{F}}(\omega)|^2. \tag{3.5.40}$$

We have expressed this square integral by means of the Fourier transform

$$\hat{\boldsymbol{F}}(\omega) = \frac{1}{\sqrt{2\pi}} \int_{-\infty}^{+\infty} dt\, \boldsymbol{F}(t) e^{i\omega t}, \tag{3.5.41}$$

using Parseval's equation (3.3.20). The Fourier transform obeys

$$\hat{\boldsymbol{F}}(-\omega) = F(\omega)^*, \tag{3.5.42}$$

because $\boldsymbol{F}(t)$ is real. The total energy can be decomposed into its spectral distribution

$$\frac{dW}{d\Omega} = \int_0^{\infty} d\omega\, \frac{d^2 I(\omega, \boldsymbol{n})}{d\omega d\Omega}.$$

Comparing this with (3.5.40) we obtain for the spectral energy density

$$\frac{d^2 I(\omega, \boldsymbol{n})}{d\omega d\Omega} = |\hat{\boldsymbol{F}}(\omega)|^2 + |\hat{\boldsymbol{F}}(-\omega)|^2 = 2|\hat{\boldsymbol{F}}(\omega)|^2, \tag{3.5.43}$$

where (3.5.42) has been taken into account.

There remains the Fourier transform

$$\hat{\boldsymbol{F}}(\omega) = \frac{1}{\sqrt{2\pi\mu_0 c}} \frac{q}{4\pi\varepsilon_0 c} \int\limits_{-\infty}^{+\infty} dt\, e^{i\omega t} \frac{\boldsymbol{n} \wedge [(\boldsymbol{n} - \boldsymbol{\beta}) \wedge \boldsymbol{\beta}']}{(1 - \boldsymbol{\beta} \cdot \boldsymbol{n})^3}\Big|_{\text{ret}} \tag{3.5.44}$$

to be calculated. We remember the connection (3.5.6) between t and the retarded time

$$t = t_{\text{ret}} + \frac{1}{c} r(t_{\text{ret}}), \tag{3.5.45}$$

where

$$r = |\boldsymbol{x} - \boldsymbol{y}(t)| = x - \boldsymbol{n} \cdot \boldsymbol{y} + O\Big(\frac{y^2}{x}\Big). \tag{3.5.46}$$

$\boldsymbol{x}$ is the point of observation of the radiation, while $\boldsymbol{y}(t)$ is the particle position. Using the new variable of integration t_{ret} and taking the inverse of (3.5.17) into account, we shall obtain

$$\hat{\boldsymbol{F}}(\omega) = \frac{1}{\sqrt{2\pi\mu_0 c}} \frac{q}{4\pi\varepsilon_0 c} \int\limits_{-\infty}^{+\infty} dt_{\text{ret}}\, e^{i\omega(t_{\text{ret}} + r/c)} \frac{\boldsymbol{n} \wedge [(\boldsymbol{n} - \boldsymbol{\beta}) \wedge \boldsymbol{\beta}']}{(1 - \boldsymbol{\beta} \cdot \boldsymbol{n})^2}\Big|_{\text{ret}}. \tag{3.5.47}$$

Writing t instead of t_{ret} as integration variable and using the expansion (3.5.46) for $x \gg y$, we get

$$\hat{\boldsymbol{F}}(\omega) = \frac{1}{\sqrt{2\pi\mu_0 c}} \frac{q}{4\pi\varepsilon_0 c} e^{i\omega x/c} \boldsymbol{I}(\omega, \boldsymbol{n}), \tag{3.5.48}$$

with

$$\boldsymbol{I}(\omega, \boldsymbol{n}) = \int\limits_{-\infty}^{+\infty} dt\, e^{i\omega(t - \boldsymbol{n} \cdot \boldsymbol{y}/c)} \frac{\boldsymbol{n} \wedge [(\boldsymbol{n} - \boldsymbol{\beta}) \wedge \boldsymbol{\beta}']}{(1 - \boldsymbol{\beta} \cdot \boldsymbol{n})^2}. \tag{3.5.49}$$

We are now ready to calculate (3.5.38)

$$\frac{d^2 I(\omega, \boldsymbol{n})}{d\omega\, d\Omega} = \frac{2}{2\pi\mu_0 c} \frac{q^2}{(4\pi\varepsilon_0 c)^2} |\boldsymbol{I}|^2 = \frac{1}{4\pi\varepsilon_0} \frac{q^2}{4\pi^2 c} |\boldsymbol{I}|^2. \tag{3.5.50}$$

In the integral (3.5.49) we make a partial integration, using

$$\begin{aligned} \frac{d\boldsymbol{n}}{dt} &= \frac{d}{dt} \frac{\boldsymbol{x} - \boldsymbol{y}(t)}{|\boldsymbol{x} - \boldsymbol{y}(t)|} = -\frac{\boldsymbol{v}}{|\boldsymbol{x} - \boldsymbol{y}|} - \frac{\boldsymbol{x} - \boldsymbol{y}}{|\boldsymbol{x} - \boldsymbol{y}|^2}(-\boldsymbol{n} \cdot \boldsymbol{v}) \\ &= \frac{c}{r}\Big(\boldsymbol{n}(\boldsymbol{\beta} \cdot \boldsymbol{n}) - \boldsymbol{\beta}\Big) \end{aligned} \tag{3.5.51}$$

and

$$\frac{d}{dt}\frac{\boldsymbol{n}\wedge(\boldsymbol{n}\wedge\boldsymbol{\beta})}{1-\boldsymbol{\beta}\cdot\boldsymbol{n}} = \frac{(\boldsymbol{n}-\boldsymbol{\beta})\boldsymbol{\beta}'\cdot\boldsymbol{n}-\boldsymbol{\beta}'(1-\boldsymbol{\beta}\cdot\boldsymbol{n})}{(1-\boldsymbol{\beta}\cdot\boldsymbol{n})^2}$$

$$= \frac{\boldsymbol{n}\wedge[(\boldsymbol{n}-\boldsymbol{\beta})\wedge\boldsymbol{\beta}']}{(1-\boldsymbol{\beta}\cdot\boldsymbol{n})^2}. \tag{3.5.52}$$

This agrees with the last fraction in (3.5.49), hence

$$\boldsymbol{I}(\omega,\boldsymbol{n}) = \int_{-\infty}^{+\infty} dt\, e^{i\omega(t-\boldsymbol{n}\cdot\boldsymbol{y}/c)}\frac{d}{dt}\frac{\boldsymbol{n}\wedge(\boldsymbol{n}-\boldsymbol{\beta})}{1-\boldsymbol{\beta}\cdot n}$$

$$= -i\omega\int_{-\infty}^{+\infty} dt\, e^{i\omega(t-\boldsymbol{n}\cdot\boldsymbol{y}/c)}(1-\boldsymbol{\beta}\cdot\boldsymbol{n})\frac{\boldsymbol{n}\wedge(\boldsymbol{n}\wedge\boldsymbol{\beta})}{1-\boldsymbol{\beta}\cdot\boldsymbol{n}} \tag{3.5.53}$$

$$+ \quad \text{boundary terms.}$$

Here, in differentiating the exponent, we have neglected the derivative of $\boldsymbol{n}$, because (3.5.51) vanishes for $r \to \infty$.

We want to apply this result to circular motion in the xy-plane. This is **synchrotron radiation**. To fix the geometry we choose the circular trajectory of radius ρ tangential to the x-axis (Fig.10). Since the problem is axial symmetric, only the angle ϑ between $\boldsymbol{n}$ and the xy-plane is important, so that the cartesian components of $\boldsymbol{n}$ are given by

$$\boldsymbol{n} = (\cos\vartheta,\, 0,\, \sin\vartheta). \tag{3.5.54}$$

The particle trajectory is represented by

$$\boldsymbol{y}(t) = (0,\rho,0) + \rho\Big(\sin\frac{vt}{\rho},\, \cos\frac{vt}{\rho},\, 0\Big). \tag{3.5.55}$$

Differentiating this we get the velocity

$$\boldsymbol{v}(t) = v\Big(\cos\frac{vt}{\rho}, -\sin\frac{vt}{\rho},\, 0\Big), \tag{3.5.56}$$

and the auxiliary vector

$$\boldsymbol{a} \overset{\text{def}}{=} \boldsymbol{n}\wedge(\boldsymbol{n}\wedge\boldsymbol{\beta}) = \boldsymbol{n}\beta\cos\frac{vt}{\rho}\cos\vartheta - \boldsymbol{\beta}$$

$$= \beta\Big(\cos\frac{vt}{\rho}(\cos^2\vartheta-1), -\sin\frac{vt}{\rho},\, \cos\frac{vt}{\rho}\sin\vartheta\cos\vartheta\Big). \tag{3.5.57}$$

$\boldsymbol{a}$ is perpendicular to $\boldsymbol{n}$, we therefore decompose it with respect to the basis vectors $\boldsymbol{e}_y = \boldsymbol{e}_\parallel$ and $\boldsymbol{n}\wedge\boldsymbol{e}_y = \boldsymbol{e}_\perp$

$$\boldsymbol{a} = \beta\Big(\boldsymbol{e}_\perp\cos\frac{vt}{\rho}\sin\vartheta - \boldsymbol{e}_\parallel\sin\frac{vt}{\rho}\Big). \tag{3.5.58}$$

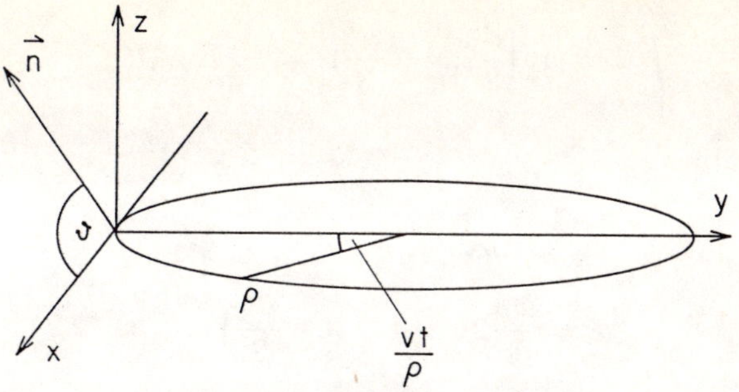

Fig. 10. Synchrotron radiation

We consider extremely relativistic particles with $v \approx c$, so that $1-\beta$ is small. Then, as we shall see, the radiation is emitted into a small cone around the instantaneous direction of velocity. We therefore expand for small ϑ and small t:

$$\boldsymbol{a} = \mathbf{e}_\perp \beta\vartheta - \mathbf{e}_\parallel \frac{vt}{\rho} + \ldots = \mathbf{e}_\perp \vartheta - \mathbf{e}_\parallel \frac{c}{\rho} t + \ldots \tag{3.5.59}$$

$$t - \frac{1}{c}\boldsymbol{n}\cdot\boldsymbol{y} = t - \frac{\rho}{c}\sin\frac{vt}{\rho}\cos\vartheta$$

$$= t(1-\beta+\frac{\vartheta^2}{2}) + \frac{t^3}{6}\frac{c^2}{\rho^2} + \ldots \tag{3.5.60}$$

This leads to

$$\boldsymbol{I} = \mathbf{e}_\perp I_\perp - \mathbf{e}_\parallel I_\parallel, \tag{3.5.61}$$

with

$$I_\perp(\omega,\vartheta) = -i\omega\vartheta \int\limits_{-\infty}^{+\infty} dt \, \exp i\omega\Big[t(1-\beta+\frac{\vartheta^2}{2}) + \frac{t^3}{6}\frac{c^2}{\rho^2}\Big] \tag{3.5.62}$$

$$I_\parallel(\omega,\vartheta) = -i\omega\frac{c}{\rho} \int\limits_{-\infty}^{+\infty} dt \, t \exp\Big[t(1-\beta+\frac{\vartheta^2}{2}) + \frac{t^3}{6}\frac{c^2}{\rho^2}\Big]. \tag{3.5.63}$$

We are interested in large frequencies ω, so that the exponentials are rapidly oscillating. Then, only the vicinity of $t=0$ (the actual place of the particle) contributes to the integrals. This is the reason for the above expansions and for the neglection of the boundary terms in (3.5.53).

The integrals (3.5.62, 63) lead to Airy functions

$$\int\limits_0^\infty dx \, \cos\Big[\frac{3}{2}\xi(x+\frac{x^3}{3})\Big] = \frac{1}{\sqrt{3}} K_{1/3}(\xi) \tag{3.5.64}$$

$$\int_0^\infty dx\, x \sin\Big[\frac{3}{2}\xi(x+\frac{x^3}{3})\Big] = \frac{1}{\sqrt{3}} K_{2/3}(\xi). \tag{3.5.65}$$

Using the variables

$$x = \frac{ct}{\rho\sqrt{2}(1-\beta+\frac{\vartheta^2}{2})^{1/2}} \tag{3.5.66}$$

$$\xi = \frac{\omega\rho}{3c} 2\sqrt{2}\Big(1-\beta+\frac{\vartheta^2}{2}\Big)^{3/2}, \tag{3.5.67}$$

we arrive at

$$\begin{aligned} I_\perp(\omega,\vartheta) &= -i\omega\frac{\rho}{c}\sqrt{2}\Big(1-\beta+\frac{\vartheta^2}{2}\Big)^{1/2} 2\int_0^\infty dx\, \cos\Big[\frac{3}{2}\xi\Big(x+\frac{x^3}{3}\Big)\Big] \\ &= -i\omega\frac{\rho}{c}2\sqrt{\frac{2}{3}}\Big(1-\beta+\frac{\vartheta^2}{2}\Big)^{1/2} K_{1/3}(\xi) \end{aligned} \tag{3.5.68}$$

$$I_\parallel(\omega,\vartheta) = \omega\frac{\rho}{c}\frac{4}{\sqrt{3}}\Big(1-\beta+\frac{\vartheta^2}{2}\Big) K_{2/3}(\xi). \tag{3.5.69}$$

This gives the following final result for the spectral energy distribution (3.5.43)

$$\begin{aligned} \frac{d^2 I(\omega,\vartheta)}{d\omega\, d\Omega} &= \frac{1}{4\pi\varepsilon_0}\frac{q^2}{4\pi^2 c}(|I_\perp|^2 + |I_\parallel|^2) \\ = \frac{1}{4\pi\varepsilon_0}\frac{4q^2\omega^2\rho^2}{3\pi^2 c^3}&\Big(1-\beta+\frac{\vartheta^2}{2}\Big)^2\Big(K_{2/3}(\xi)^2 + \frac{K_{1/3}(\xi)^2}{2(1-\beta+\frac{\vartheta^2}{2})}\Big). \end{aligned} \tag{3.5.70}$$

Since the Airy functions decrease exponentially for $\xi \gg 1$, it follows from (3.5.67) that ϑ must be small, thus the synchrotron radiation is mainly emitted in the plane of motion of the particle. We therefore put $\vartheta = 0$. Then, as a function of ω, there is also exponential decrease for $\omega \to \infty$. A broad maximum occurs around $\xi \approx 1$, i.e. around the characteristic frequency

$$\omega_c = \frac{3c}{2\sqrt{2}\rho}(1-\beta)^{-3/2}. \tag{3.5.71}$$

Expressing the velocity by the energy of the particle

$$E = \frac{mc^2}{\sqrt{1-\frac{v^2}{c^2}}},$$

we find

$$\omega_c = 3\Big(\frac{E}{mc^2}\Big)^3\frac{c}{\rho}.$$

This characteristic frequency ω_c gives the extension of the spectrum of synchrotron radiation. It is proportional to the turn-around frequency c/ρ of the particles. The highest frequencies are obtained with the lightest charged particles, i.e. electrons. Electron accelerators and storage rings are strong sources of synchrotron radiation. In fact, they are the strongest sources in the ultraviolet and X-ray region known today.

Next we want to discuss another application of (3.5.44) namely **bremsstrahlung**. It occurs if particles are stopped or deflected in collisions. If the particle is stopped or accelerated during a collision time τ, the radiation maximum occurs around a characteristic frequency $\omega_c \approx 1/\tau$. For high frequencies $\omega \gg 1/\tau$ the spectrum of bremsstrahlung must be calculated by quantum electrodynamics. But for small frequencies $\omega \ll 1/\tau$ classical electrodynamics is perfect. Let us assume that the acceleration $\boldsymbol{\beta}'$ in (3.5.44) is only different from 0 during the collision time $\tau = t_2 - t_1$. Since for small frequencies the exponential in (3.5.44) can be approximated by 1, we get

$$\hat{\boldsymbol{F}}(\omega) = \frac{1}{\sqrt{2\pi\mu_0 c}} \frac{q}{4\pi\varepsilon_0 c} \int_{t_1}^{t_2} dt \, \frac{\boldsymbol{n} \wedge [(\boldsymbol{n} - \boldsymbol{\beta}) \wedge \boldsymbol{\beta}']}{(1 - \boldsymbol{\beta} \cdot \boldsymbol{n})^2}, \tag{3.5.72}$$

where we have used the new variable of integration t_{ret} (3.5.17, 45) again. Here only the time dependence of the velocity $\boldsymbol{\beta}$ must be taken into account, because (3.5.51) vanishes for $r \to \infty$. A glance to (3.5.52) shows that

$$\hat{\boldsymbol{F}}(\omega) = \frac{1}{\sqrt{2\pi\mu_0 c}} \frac{q}{4\pi\varepsilon_0 c} \left(\frac{\boldsymbol{n} \wedge (\boldsymbol{n} \wedge \boldsymbol{\beta}_2)}{1 - \boldsymbol{\beta}_2 \cdot \boldsymbol{n}} - \frac{\boldsymbol{n} \wedge (\boldsymbol{n} \wedge \boldsymbol{\beta}_1)}{1 - \boldsymbol{\beta}_1 \cdot \boldsymbol{n}} \right). \tag{3.5.73}$$

Hence, the velocity components perpendicular to $\boldsymbol{n}$

$$\boldsymbol{\beta}^{\perp} = \boldsymbol{n} \wedge (\boldsymbol{\beta} \wedge \boldsymbol{n}) \tag{3.5.74}$$

are essential. $\boldsymbol{\beta}_1$ is the velocity before and $\boldsymbol{\beta}_2$ after the collision, respectively.

From (3.5.43) we finally obtain the low energy spectrum of bremsstrahlung

$$\frac{d^2 I(\omega, \boldsymbol{n})}{d\omega \, d\Omega} = \frac{1}{4\pi\varepsilon_0} \frac{q^2}{4\pi^2 c} \left(\frac{\boldsymbol{\beta}_2^{\perp}}{1 - \boldsymbol{\beta}_2 \cdot \boldsymbol{n}} - \frac{\boldsymbol{\beta}_1^{\perp}}{1 - \boldsymbol{\beta}_1 \cdot \boldsymbol{n}} \right)^2. \tag{3.5.75}$$

This has to be summed over all particles taking part in the collision. The point is that the right-hand side is independent of ω. Hence, **there is a constant radiation power of bremsstrahlung at small frequencies**. However, the energy $\hbar\omega$ of one photon goes to 0 for $\omega \to 0$. Consequently, the number of emitted photons must go to infinity. This is the so-called **infrared catastrophe** which causes considerable difficulties in quantum electrodynamics, while classical electrodynamics is perfect for small frequencies.

As a last application of (3.5.73) we consider **bremsstrahlung in small angle scattering** of relativistic particles. Let us assume that $\boldsymbol{\beta}_2 = \boldsymbol{\beta}_1 + \delta\boldsymbol{\beta}$, with $|\delta\boldsymbol{\beta}| \ll |\boldsymbol{\beta}_1|$. We may then expand

$$\frac{\boldsymbol{\beta}_2}{1-\boldsymbol{\beta}_2\cdot\boldsymbol{n}}-\frac{\boldsymbol{\beta}_1}{1-\boldsymbol{\beta}_1\cdot\boldsymbol{n}}=\frac{\boldsymbol{\beta}_1\delta\boldsymbol{\beta}\cdot\boldsymbol{n}+\delta\boldsymbol{\beta}(1-\boldsymbol{\beta}_1\cdot\boldsymbol{n})}{(1-\boldsymbol{\beta}_1\cdot\boldsymbol{n})^2}+O\Big((\delta\beta)^2\Big).$$

Taking the transversal component (3.5.74) of this expression

$$\boldsymbol{\beta}^{\perp}=\boldsymbol{n}(\boldsymbol{\beta}\cdot\boldsymbol{n})-\boldsymbol{\beta},$$

we shall obtain

$$\begin{aligned}\boldsymbol{D}&=\frac{(\boldsymbol{n}(\boldsymbol{\beta}_1\cdot\boldsymbol{n})-\boldsymbol{\beta}_1)\delta\boldsymbol{\beta}\cdot\boldsymbol{n}+(\boldsymbol{n}(\delta\boldsymbol{\beta}\cdot\boldsymbol{n})-\delta\boldsymbol{\beta})(1-\boldsymbol{\beta}_1\cdot\boldsymbol{n})}{(1-\boldsymbol{\beta}_1\cdot\boldsymbol{n})^2}\\&=\frac{(\boldsymbol{n}-\boldsymbol{\beta}_1)\delta\boldsymbol{\beta}\cdot\boldsymbol{n}-\delta\boldsymbol{\beta}(1-\boldsymbol{\beta}_1\cdot\boldsymbol{n})}{(1-\boldsymbol{\beta}_1\cdot\boldsymbol{n})^2}.\end{aligned}\tag{3.5.76}$$

We choose $\boldsymbol{\beta}_1=\beta(0,0,1)$ parallel to the z-axis. Since $|\boldsymbol{\beta}_2|\approx|\boldsymbol{\beta}_1|$, it follows that $\delta\boldsymbol{\beta}\perp\boldsymbol{\beta}_1$, thus, $\delta\boldsymbol{\beta}$ is in the xy-plane. Let $\boldsymbol{n}$ be in the xz-plane (Fig.11), thus

$$\delta\boldsymbol{\beta}=\delta\beta(\cos\varphi,\sin\varphi,0),\qquad\boldsymbol{n}=(\sin\vartheta,0,\cos\vartheta).\tag{3.5.77}$$

Substituting this into (3.5.76) and calculating the square, we get

$$\boldsymbol{D}^2=\frac{(\delta\beta)^2}{(1-\beta\cos\vartheta)^4}\Big[\cos^2\varphi(\beta-\cos\vartheta)^2+\sin^2\varphi(1-\beta\cos\vartheta)^2\Big].\tag{3.5.78}$$

This gives the following angular distribution of bremsstrahlung

$$\frac{d^2I}{d\omega\,d\Omega}=\frac{1}{4\pi\varepsilon_0}\frac{q^2(\delta\beta)^2}{4\pi^2c}\frac{\cos^2\varphi(\beta-\cos\vartheta)^2+\sin^2\varphi(1-\beta\cos\vartheta)^2}{(1-\beta\cos\vartheta)^4}.\tag{3.5.79}$$

If the scattered particle is not observed, we have to average over φ. Since

$$\overline{\cos^2\varphi}=\tfrac{1}{2}=\overline{\sin^2\varphi},\tag{3.5.80}$$

we shall obtain

$$\frac{d^2\bar{I}}{d\omega d\Omega}=\frac{1}{4\pi\varepsilon_0}\frac{q^2(\delta\beta)^2}{8\pi^2c}\frac{(\beta-\cos\vartheta)^2+(1-\beta\cos\vartheta)^2}{(1-\beta\cos\vartheta)^4}.\tag{3.5.81}$$

This has a maximum in the forward direction $\vartheta=0$. Expanding for small ϑ

$$1-\beta\cos\vartheta=1-\beta+\beta\frac{\vartheta^2}{2}=(1-\beta)\Big(1+\frac{\beta}{1-\beta}\frac{\vartheta^2}{2}\Big),$$

we find the width of this maximum

$$\delta\vartheta=\sqrt{2\frac{1-\beta}{\beta}}=\sqrt{2(\frac{c}{v}-1)}.\tag{3.5.82}$$

It goes to 0 for $v\to c$. Thus, for large v, bremsstrahlung is strongly peaked in the forward direction, similar to synchrotron radiation. The total spectral

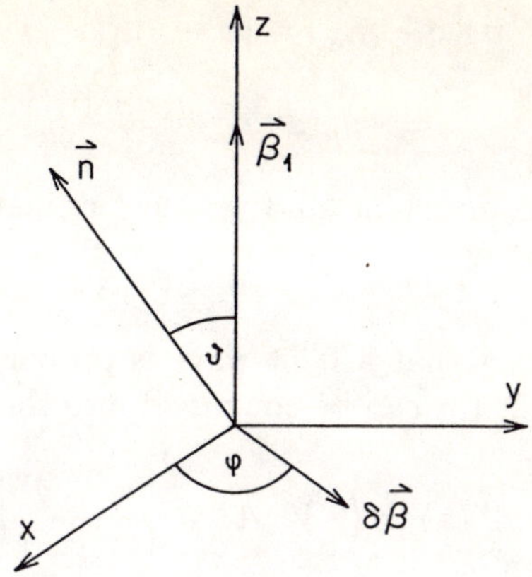

Fig. 11. Bremsstrahlung in small angle scattering

intensity is obtained by integrating (3.5.81) over $\cos\vartheta$ from -1 to 1 and trivially over φ. Since

$$\int_{-1}^{1} \frac{d\cos\vartheta}{(1-\beta\cos\vartheta)^2} = \frac{2}{1-\beta^2}$$

$$\int_{-1}^{1} d\cos\vartheta\, \frac{(\beta-\cos\vartheta)^2}{(1-\beta\cos\vartheta)^4} = \frac{2}{3}\frac{1}{1-\beta^2},$$

we get

$$\begin{aligned}\frac{d\bar{I}}{d\omega} &= \frac{1}{4\pi\varepsilon_0}\frac{q^2}{8\pi^2 c}2\pi\cdot\frac{8}{3}\frac{(\delta\beta)^2}{1-\beta^2}\\ &= \frac{1}{4\pi\varepsilon_0}\frac{q^2}{\pi c}\frac{2}{3}\frac{(\delta\beta)^2}{1-\beta^2}.\end{aligned} \tag{3.5.83}$$

For $\beta = v/c \to 1$, we observe stronger and stronger radiation.

The **Liénard-Wiechert potentials** (3.5.12, 13) can be written **in Lorentz-covariant form**. For this purpose we use the velocity four-vector (2.3.7)

$$u^\mu = \gamma\Big(1, \frac{\boldsymbol{v}}{c}\Big), \quad \gamma = 1/\sqrt{1-\frac{\boldsymbol{v}^2}{c^2}} \tag{3.5.84}$$

and the four-vector difference

$$x^\mu - y^\mu = (ct - ct_{\text{ret}}, \boldsymbol{x} - \boldsymbol{y}_{\text{ret}}) = (|\boldsymbol{x} - \boldsymbol{y}_{\text{ret}}|, \boldsymbol{x} - \boldsymbol{y}_{\text{ret}}), \tag{3.5.85}$$

where (3.5.6) has been taken into account. This leads to

$$(1-\boldsymbol{\beta}\cdot\boldsymbol{n})|\boldsymbol{x}-\boldsymbol{y}_{\text{ret}}| = |\boldsymbol{x}-\boldsymbol{y}_{\text{ret}}| - \boldsymbol{\beta}\cdot(\boldsymbol{x}-\boldsymbol{y}_{\text{ret}})$$
$$= \frac{1}{\gamma}u_\mu(x^\mu - y^\mu) \stackrel{\text{def}}{=} \frac{1}{\gamma}R, \tag{3.5.86}$$

where R is a Lorentz invariant, and we must remember

$$(x-y)^2 = 0, \quad x^0 > y^0. \tag{3.5.87}$$

This expresses the fact that a light wave is propagating from y to x. Now the potentials (3.5.12, 13) can be combined into the Liénard-Wiechert four-potential

$$A^\nu(x) = \left(\frac{1}{c}V,\, \boldsymbol{A}\right) = \frac{q}{4\pi\varepsilon_0 c}\left.\frac{u^\nu}{R}\right|_{\text{ret}}. \tag{3.5.88}$$

For the computation of the field tensor, we need various derivatives. The previous equations (3.5.17, 18) can be written as

$$\frac{\partial y^0}{\partial x^\mu} = \gamma\frac{x_\mu - y_\mu}{R}. \tag{3.5.89}$$

To complete this equation to get a covariant tensor equation, we notice that y^j is a function of $ct_{\text{ret}} = y^0$, given by the particle trajectory $\boldsymbol{y}(t)$. Hence,

$$\frac{\partial y^j}{\partial x^\mu} = \frac{\partial y^j}{\partial y^0}\frac{\partial y^0}{\partial x^\mu} = \frac{v^j}{c}\gamma\frac{x_\mu - y_\mu}{R}$$
$$\frac{\partial y^\nu}{\partial x^\mu} = u^\nu\frac{x_\mu - y_\mu}{R}. \tag{3.5.90}$$

We also need

$$\frac{\partial u^\nu}{\partial x^\mu} = \frac{\partial u^\nu}{\partial y^0}\frac{\partial y^0}{\partial x^\mu} = \gamma\frac{\partial u^\nu}{\partial y^0}\frac{x_\mu - y_\mu}{R},$$

where the first two factors must combine to a four-vector. In fact, remembering that $dy^0/\gamma = ds$ is the proper time element (2.3.5) of the particle,

$$a^\nu \stackrel{\text{def}}{=} \frac{\partial u^\nu}{\partial s} = \gamma\frac{\partial u^\nu}{\partial y^0} \tag{3.5.91}$$

is the four-acceleration, thus

$$\frac{\partial u^\nu}{\partial x^\mu} = a^\nu\frac{x_\mu - y_\mu}{R}. \tag{3.5.92}$$

In Minkowski space the acceleration is always orthogonal to the velocity:

$$u_\mu a^\mu = u_\mu\frac{\partial u^\mu}{\partial s} = \frac{1}{2}\frac{\partial}{\partial s}\left(u_\mu u^\mu\right) = 0. \tag{3.5.93}$$

We are now ready to calculate the field tensor. Using (3.5.90, 92), we find

$$\frac{\partial R}{\partial x^\mu} = \frac{w-1}{R}(x_\mu - y_\mu) - u_\mu, \tag{3.5.94}$$

where

$$w = a^\nu (x_\nu - y_\nu), \tag{3.5.95}$$

and

$$\frac{\partial A^\nu}{\partial x^\mu} = \frac{q}{4\pi\varepsilon_0 c}\left\{\frac{1}{R^2} a^\nu (x_\mu - y_\mu) - \frac{u^\nu}{R^2}\left[\frac{x_\mu - y_\mu}{R}(w-1) - u_\mu\right]\right\}_{\text{ret}} . \tag{3.5.96}$$

The field tensor $F^{\mu\nu}$ (3.4.9) is then equal to

$$\begin{aligned} F^{\mu\nu}(x) &= \partial^\mu A^\nu - \partial^\nu A^\mu \\ &= \frac{q}{4\pi\varepsilon_0 c R^2}\left\{ a^\nu (x^\mu - y^\mu) - a^\mu (x^\nu - y^u) \right. \\ &\quad \left. - \frac{w-1}{R}\left[u^\nu (x^\mu - y^\mu) - u^\mu (x^\nu - y^\nu)\right]\right\}_{\text{ret}} . \end{aligned} \tag{3.5.97}$$

As before, the first two terms depending on the acceleration a^ν decrease like R^{-1} for large R, whereas the last two terms are $\sim R^{-2}$. The former give rise to radiation. We will use this result in the following section to calculate the power of radiation in covariant form.

3.6 Lagrange Formalism of Electrodynamics

A very important tool in analytical mechanics is the possibility of deriving the equations of motion (Lagrange's equations or Hamilton's equations) from a variational principle. We want to do the same for electrodynamics, i.e. we wish to derive Maxwell's equations from a variational principle. Besides its own interest this is also important for the development of quantum electrodynamics.

The variational principle of mechanics which gives us the guiding principle, is Hamilton's principle of least action. It says that

$$\int_0^t L(q, q', t)\, dt = \text{stationary}, \tag{3.6.1}$$

if $q(t)$ and its time derivative $q'(t)$ describe a mechanical trajectory. Here L is the Lagrangian function of the mechanical system and the total integral (3.6.1) is called the action functional. In field theory the finitely many Lagrange coordinates $q(t)$, $q'(t)$ must be substituted by the basic fields $A(t, \boldsymbol{x})$, $A'(t, \boldsymbol{x})$. Indexed by $\boldsymbol{x}$, these are infinitely many degrees of freedom. But what are the fundamental fields? It is a highly important fact that the basic fields are not the electromagnetic fields $\boldsymbol{E}$, $\boldsymbol{B}$, but the potentials $A^\mu(x)$.

We directly refer to the four-vector potential (3.4.8) because the whole formalism is relativistic from the very beginning. The rôle of the potentials as Lagrange coordinates is the ultimate reason for their importance in quantum theory and for electrodynamics being a gauge theory (see *G.Scharf, Finite Quantum Electrodynamics, Springer Verlag 1989*).

Similarly to (3.6.1) we look for a variational principle

$$\int L(A, \partial_\mu A, t, \boldsymbol{x})\, d^3x\, dt = \text{stationary}, \tag{3.6.2}$$

where L is now the so-called Lagrangian density. The integration measures in (3.6.2) combine into a four-dimensional Lebesgue measure d^4x that is Lorentz invariant. Since the total action integral must also be Lorentz invariant, the Lagrangian density must be a Lorentz scalar. Since Maxwell's equations are linear in the electromagnetic fields, we look for a scalar that is quadratic in the field tensor $F^{\mu\nu}$. The only such scalar is

$$F^{\mu\nu} F_{\mu\nu} = 2\Big(\boldsymbol{B}^2 - \frac{1}{c^2}\boldsymbol{E}^2\Big). \tag{3.6.3}$$

(The second Lorentz invariant that we have found in Problem 4 of Chap.2 is a pseudo-scalar, that means it changes sign under spatial reflections, and, therefore, must not be considered here, because electrodynamics conserves parity, see (2.5.18).) To describe the interaction of the fields with the charge-current density $j^\mu(x)$, we must find a second scalar depending linearly on j^μ. The simplest such scalar must contain the potential: $j_\mu A^\mu$. This leads us to the following **Lagrangian density**

$$L = L_0 + L_{\text{int}} = -\frac{1}{4} F_{\mu\nu} F^{\mu\nu} - \mu_0 j_\mu A^\mu. \tag{3.6.4}$$

The first term L_0 is the Lagrangian density of the electromagnetic field alone, whereas the second term L_{int} describes the interaction with the sources. The factors have been chosen in such a manner that the correct Maxwell's equations come out. $L = L(x)$ contains the fields at the same space-time point x, only. This is therefore a local field theory with only local interactions.

To determine the fields A that make the action functional

$$S[A] = \int d^4x\, L(A^\mu, \partial_\nu A^\mu) \tag{3.6.5}$$

stationary, we must set the corresponding directional derivative (or Fréchet derivative) equal to zero. The latter is defined by

$$\begin{aligned}\Big(\frac{DS}{DA}\Big)(h) &\stackrel{\text{def}}{=} \lim_{\varepsilon \to 0} \frac{1}{\varepsilon}\Big(S[A + \varepsilon h] - S[A]\Big) \\ &= \frac{\partial}{\partial \varepsilon} S[A + \varepsilon h]\Big|_{\varepsilon = 0}.\end{aligned} \tag{3.6.6}$$

This is the derivative at the "point" A in the "direction" h. It should not be confounded with the variational derivative $\delta S/\delta A(x)$ which is explained below. To compute (3.6.6) for our functional (3.6.5), we use Taylor's formula

$$S[A+\varepsilon h]-S[A]=\int d^4x\left\{L(A^\mu+\varepsilon h^\mu,\,A^\mu{}_{,\nu}+\varepsilon h^\mu{}_{,\nu})\right.$$

$$\left.-L(A^\mu,\,A^\mu{}_{,\nu})\right\}$$

$$=\varepsilon\int d^4x\left(\frac{\partial L}{\partial A^\mu}h^\mu+\frac{\partial L}{\partial A^\mu{}_{,\nu}}\,\partial_\nu h^\mu\right)+O(\varepsilon^2). \tag{3.6.7}$$

Here we have used the comma notation to denote partial derivatives. Dividing by ε and taking the limit $\varepsilon\to 0$, we find

$$\left(\frac{DS}{DA}\right)(h)=\int d^4x\left(\frac{\partial L}{\partial A^\mu}h^\mu+\frac{\partial}{\partial x^\nu}\frac{\partial L}{\partial A^\mu{}_{,\nu}}\partial_\nu h^\mu\right). \tag{3.6.8}$$

We make a partial integration in the last term, assuming that $h^\mu(x)$ vanish at infinity. The condition for stationarity (3.6.5) then reads

$$\left(\frac{DS}{DA}\right)(h)=0=\int d^4x\left(\frac{\partial L}{\partial A^\mu}-\frac{\partial}{\partial x^\nu}\frac{\partial L}{\partial A^\mu{}_{,\nu}}\right)h^\mu(x) \tag{3.6.9}$$

for all $h(x)\in C^1(\mathbb{R}^4)$ vanishing at infinity. According to the fundamental lemma of variational calculus (lemma of Du Bois-Reymond) this implies

$$\frac{\partial L}{\partial A^\mu}-\frac{\partial}{\partial x^\nu}\frac{\partial L}{\partial A^\mu{}_{,\nu}}=0. \tag{3.6.10}$$

These are the well-known **Euler-Lagrange equations**. The left-hand side is identical with the variational derivative

$$\frac{\delta S}{\delta A(x)}\overset{\text{def}}{=}\frac{\partial L}{\partial A^\mu}-\frac{\partial}{\partial x^\nu}\frac{\partial L}{\partial A^\mu{}_{,\nu}}.$$

It is the integral kernel of the directional derivative. But we shall not use this notation anymore.

To prove that (3.6.4) is the correct Lagrange density for electrodynamics, we have to express the field tensor by the four-vector potential (3.4.9)

$$L=-\frac{1}{4}(A_\nu{}^{,\mu}-A^\mu{}_{,\nu})(A^\nu{}_{,\mu}-A_\mu{}^{,\nu})-\mu_0 j_\mu A^\mu \tag{3.6.11}$$

and calculate the derivatives in the Euler-Lagrange equations

$$\frac{\partial L}{\partial A^\mu}=-\mu_0 j_\mu \tag{3.6.12}$$

$$\frac{\partial L}{\partial A^\mu{}_{,\nu}}=-\frac{1}{2}(F^\nu{}_\mu-F_\mu{}^\nu)=F_\mu{}^\nu. \tag{3.6.13}$$

In doing so we use the fact that the double summation indices μ and ν can be arbitrarily substituted. Inserting this into (2.13.4), we obtain indeed the inhomogeneous Maxwell's equations (2.4.19)

$$-\mu_0 j_\mu - \frac{\partial}{\partial x^\nu} F_\mu{}^\nu = 0. \tag{3.6.14}$$

The homogeneous equations cannot be derived from the variational principle (3.6.5), because they are already contained in the existence of the potentials (see after (3.4.7)).

We now want to deduce the **conservation laws of electrodynamics from the Lagrange formalism.** This gives us a deeper relativistic understanding of these laws. Consider the Lagrangian density L_0 (3.6.4) of the electromagnetic field alone

$$L_0 = -\frac{1}{4} F^\mu{}_\nu F_\mu{}^\nu = L_0(A^\mu, \partial_\nu A^\mu), \tag{3.6.15}$$

assuming the four-vector current j^μ to be zero for a moment. L_0 does not explicitly depend on space-time x, only implicitly through the potentials $A^\mu(x)$. This expresses translation invariance of the free theory. This symmetry is disturbed if a given external source $j^\mu(x) \neq 0$ is present. As a consequence of translation invariance we get by means of the chain rule

$$\frac{\partial L_0}{\partial x^\nu} = \frac{\partial L_0}{\partial A^\lambda} \frac{\partial A^\lambda}{\partial x^\nu} + \frac{\partial L_0}{\partial (A^\lambda{}_{,\mu})} \frac{\partial A^\lambda{}_{,\mu}}{\partial x^\nu}. \tag{3.6.16}$$

Here we substitute the free Euler-Lagrange equation

$$\frac{\partial L_0}{\partial A^\lambda} = \frac{\partial}{\partial x^\mu} \frac{\partial L_0}{\partial A^\lambda{}_{,\mu}}, \tag{3.6.17}$$

and get

$$\begin{aligned} \frac{\partial L_0}{\partial x^\nu} &= \Big(\frac{\partial}{\partial x^\mu} \frac{\partial L_0}{\partial A^\lambda{}_{,\mu}} \Big) \frac{\partial A^\lambda}{\partial x^\nu} + \frac{\partial L_0}{\partial (A^\lambda{}_{,\mu})} A^\lambda{}_{,\mu\,,\nu} \\ &= \frac{\partial}{\partial x^\mu} \Big(\frac{\partial L_0}{\partial A^\lambda{}_{,\mu}} A^\lambda{}_{,\nu} \Big). \end{aligned} \tag{3.6.18}$$

The left-hand side can be trivially rewritten as a divergence

$$\frac{\partial L_0}{\partial x^\nu} = \frac{\partial}{\partial x^\mu} (\delta^\mu{}_\nu L_0). \tag{3.6.19}$$

Hence, the following tensor

$$\tilde{T}^\mu{}_\nu = \frac{\partial L_0}{\partial A^\lambda{}_{,\mu}} A^\lambda{}_{,\nu} - \delta^\mu{}_\nu L_0 \tag{3.6.20}$$

has vanishing divergence or is conserved

$$\partial_\mu \tilde{T}^\mu{}_\nu = 0 = \frac{\partial}{\partial t}\frac{1}{c}\tilde{T}^0{}_\nu + \frac{\partial}{\partial x^k}\tilde{T}^k{}_\nu. \tag{3.6.21}$$

In fact, the last equation is the continuity equation (2.4.11) which, by Gauss' theorem, always expresses conservation of the corresponding integrated quantity.

However, the tensor (3.6.20) has a serious physical defect. Using the previous derivative (3.6.13), we get

$$\tilde{T}^{\mu\nu} = F_\lambda{}^\mu A^{\lambda,\nu} + \frac{1}{4} g^{\mu\nu} F^{\alpha\beta} F_{\alpha\beta}, \tag{3.6.22}$$

where the second index μ has been raised by means of the metric tensor $g^{\mu\nu}$. The first term contains the four-vector potential that is gauge-dependent. Thus, $\tilde{T}^{\mu\nu}$ cannot have a direct physical meaning. This is a typical difficulty of gauge field theories. The method of correction is also typical: We try to add another conserved tensor in such a way that the result contains only the field tensor

$$F^{\lambda\nu} = \partial^\lambda A^\nu - \partial^\nu A^\lambda.$$

The term lacking in (3.6.22) is

$$-F_\lambda{}^\mu A^{\nu,\lambda} = -\partial^\lambda (F_\lambda{}^\mu A^\nu) + (\partial^\lambda F_\lambda{}^\mu) A^\nu. \tag{3.6.23}$$

This tensor has indeed vanishing divergence because

$$\partial_\mu \partial_\lambda (F^{\lambda\mu} A^\nu) = 0,$$

due to the antisymmetry of $F^{\lambda\mu}$ and the second term in (3.6.23) is anyway zero

$$\partial^\lambda F_\lambda{}^\mu = \mu_0 j^\mu = 0.$$

Hence the tensor

$$T^{\mu\nu} = F_\lambda{}^\mu (A^{\lambda,\nu} - A^{\nu,\lambda}) + \frac{1}{4} g^{\mu\nu} F^{\alpha\beta} F_{\alpha\beta}$$

$$= F^{\nu\lambda} F_\lambda{}^\mu - g^{\mu\nu} L_0 = -F^{\mu\lambda} F^\nu{}_\lambda - g^{\mu\nu} L_0 \tag{3.6.24}$$

is also conserved

$$\partial_\nu T^{\mu\nu} = 0. \tag{3.6.25}$$

This is the **energy-momentum tensor of the electromagnetic field.** It is obviously symmetric.

To understand the connection with the previous conservation laws in Sect.2.6, we compute the various components. For $\mu = 0$ we find

$$T^{00} = -F^{0\lambda} F^0{}_\lambda - L_0.$$

Using $F^0{}_\lambda = -F^{0\lambda}$ this gives

$$T^{00} = \frac{1}{c^2}\boldsymbol{E}^2 - \frac{1}{2c^2}\boldsymbol{E}^2 + \frac{1}{2}\boldsymbol{B}^2$$

$$= \frac{\mu_0}{2}(\varepsilon_0 \boldsymbol{E}^2 + \frac{1}{\mu_0}\boldsymbol{B}^2) = \mu_0 u, \tag{3.6.26}$$

where u is the energy density (2.6.10). The corresponding spatial components are

$$T^{0i} = \frac{1}{c}(\boldsymbol{E} \wedge \boldsymbol{B})^i = \frac{\mu_0}{c} S^i. \tag{3.6.27}$$

This is the Poynting vector, so that the $\mu = 0$ component of (3.6.25) is energy conservation in the free electromagnetic field.

We now compute the spatial components

$$\frac{1}{\mu_0}T^{ik} = -\frac{1}{\mu_0}F^{i\lambda}F^k{}_\lambda - \frac{1}{\mu_0}\delta^{ik}\frac{1}{2}\Big(\frac{1}{c^2}\boldsymbol{E}^2 - \boldsymbol{B}^2\Big)$$

$$= \varepsilon_0 E^i E^k + \frac{1}{\mu_0}B^i B^k - \frac{1}{2}\delta^{ik}(\varepsilon_0 \boldsymbol{E}^2 + \frac{1}{\mu_0}\boldsymbol{B}^2) = \tau^{ik}. \tag{3.6.28}$$

This agrees with Maxwell's stress tensor (2.6.17). Consequently, all quantities of Sect.2.6 are combined into the following symmetric Lorentz tensor

$$\frac{1}{\mu_0}T^{\mu\nu} = \begin{pmatrix} u & S^1/c & S^2/c & S^3/c \\ S^1/c & \tau^{11} & \tau^{12} & \tau^{13} \\ S^2/c & \tau^{21} & \tau^{22} & \tau^{23} \\ S^3/c & \tau^{31} & \tau^{32} & \tau^{33} \end{pmatrix}. \tag{3.6.29}$$

The spatial components $\mu = 1, 2, 3$ of (3.6.25) represent momentum conservation. From the symmetry of $T^{\mu\nu}$ it is immediately clear why the energy flow density $\boldsymbol{S}$ is essentially equal to the momentum density.

The four-momentum of the electromagnetic field at a time t can be obtained from $T^{\mu\nu}$ as the following 3-dimensional integral

$$P^\mu = \frac{1}{\mu_0} \int\limits_{x^0=ct} T^{\mu\nu} d\sigma_\nu. \tag{3.6.30}$$

The integral goes over the 3-dimensional hyperplane $x^0 = ct$, so that

$$d\sigma_\nu = (d^3x, \boldsymbol{0}). \tag{3.6.31}$$

That $P^\mu = \text{const}$ in time, follows directly from (3.6.25) by Gauss' theorem in four dimensions:

$$0 = \int\limits_{\Omega_4} \partial_\nu T^{\mu\nu} d^4x = \int\limits_{\partial\Omega_4} T^{\mu\nu} d\sigma_\nu. \tag{3.6.32}$$

Applying this to the region between the two hyperplanes $x^0 = ct_1$ and $x^0 = ct_2$, we get

$$0 = \int\limits_{x^0=ct_1} T^{\mu 0} d^3x - \int\limits_{x^0=ct_2} T^{\mu 0} d^3x, \tag{3.6.33}$$

where (3.6.31) has been taken into account. Since this is the difference of the four-momenta $P^\mu(t_1) - P^\mu(t_2)$, it proves

$$P^\mu(t) = \text{const.} \tag{3.6.34}$$

The symmetry of $T^{\mu\nu}$ is necessary for **angular momentum conservation**. The angular momentum tensor is defined by

$$\begin{aligned} M^{\mu\nu} &= \int (x^\mu dP^\nu - x^\nu dP^\mu) \\ &= \frac{1}{\mu_0} \int (x^\mu T^{\nu\lambda} - x^\nu T^{\mu\lambda}) d\sigma_\lambda. \end{aligned} \tag{3.6.35}$$

It is constant by the same argument (3.6.32, 33) as above, because

$$\partial_\lambda (x^\mu T^{\nu\lambda} - x^\nu T^{\mu\lambda}) = \delta^\mu_\lambda T^{\nu\lambda} - \delta^\nu_\lambda T^{\mu\lambda}, \tag{3.6.36}$$

by (3.6.25), and this vanishes if and only if $T^{\nu\mu} = T^{\mu\nu}$.

If the sources j^μ are different from zero, the divergence of $T^{\mu\nu}$ (3.6.24) no longer vanishes. Its value follows from Maxwell's equations. In

$$\partial_\nu T^{\mu\nu} = -F^{\mu\lambda}{}_{,\nu} F^\nu{}_\lambda - F^{\mu\lambda} F^\nu{}_{\lambda,\nu} + \frac{1}{2} F^{\alpha\beta,\mu} F_{\alpha\beta} \tag{3.6.37}$$

we insert the homogeneous

$$F^{\alpha\beta,\mu} = -F^{\beta\mu,\alpha} - F^{\mu\alpha,\beta} \tag{3.6.38}$$

and inhomogeneous Maxwell's equations

$$F^\nu{}_{\lambda,\nu} = -F_\lambda{}^\nu{}_{,\nu} = \mu_0 j_\lambda. \tag{3.6.39}$$

Then we shall obtain

$$\begin{aligned} \partial_\nu T^{\mu\nu} &= -F^{\mu\lambda,\nu} F_{\nu\lambda} - \mu_0 F^{\mu\lambda} j_\lambda \\ &\quad -\tfrac{1}{2} F^{\beta\mu,\alpha} F_{\alpha\beta} - \tfrac{1}{2} F^{\mu\alpha,\beta} F_{\alpha\beta}. \end{aligned} \tag{3.6.40}$$

By changing the summation indices we rewrite the first term as follows

$$\begin{aligned} -F^{\mu\lambda,\nu} F_{\nu\lambda} &= -\tfrac{1}{2}(F^{\mu\alpha,\beta} F_{\beta\alpha} + F^{\mu\beta,\alpha} F_{\alpha\beta}) \\ &= \tfrac{1}{2}(F^{\mu\alpha,\beta} F_{\alpha\beta} + F^{\beta\mu,\alpha} F_{\alpha\beta}). \end{aligned}$$

This result compensates the last two terms in (3.6.40), hence

$$\partial_\nu T^{\mu\nu} = -\mu_0 F^{\mu\lambda} j_\lambda = -\mu_0 k^\mu. \tag{3.6.41}$$

Here k^μ is the Lorentz force density (2.6.6) in relativistic form (2.3.18)

$$k^\mu = \left(\frac{1}{c}\boldsymbol{E}\cdot\boldsymbol{j},\, \boldsymbol{k}\right). \tag{3.6.42}$$

Consequently, the $\mu = 0$ component corresponds to energy conservation

$$\frac{1}{\mu_0}\partial_\nu T^{0\nu} = \frac{1}{c}\frac{\partial u}{\partial t} + \frac{1}{c}\operatorname{div}\boldsymbol{S} = -k^0 = -\frac{1}{c}\boldsymbol{E}\cdot\boldsymbol{j}, \tag{3.6.43}$$

including the external sources. The spatial components $\mu = i$ describe momentum conservation (2.6.25)

$$\frac{1}{c^2}\frac{\partial S^i}{\partial t} + \frac{\partial \tau^{ik}}{\partial x^k} = -k^i. \tag{3.6.44}$$

In the above equations $j^\mu(x)$ has been considered as a given external source. In a more general theory it should be described in more detail by additional field equations, for example by **hydrodynamic equations for a fluid of charged particles**. The conservation laws just studied must then remain valid. In fact, they can be used to find the correct field equations for the charged matter in interaction with the electromagnetic field.

The motion of a relativistic fluid is described by its four-velocity u^μ which is normalized (see (2.3.6))

$$u_\mu u^\mu = 1 \tag{3.6.45}$$

and by its mass density. The latter is assumed to be proportional to the charge density ρ_0, where the subscript 0 means that it is measured in the rest frame of the fluid. ρ_0 is a scalar in contrast to the charge density ρ in the laboratory system which is the zeroth component of the four-vector

$$j^\mu = c\rho_0 u^\mu. \tag{3.6.46}$$

The total energy-momentum tensor of the fluid plus electromagnetic field is simply additive

$$\bar{T}^{\mu\nu} = \frac{1}{\mu_0}T^{\mu\nu} + \frac{mc^2}{e}\rho_0 u^\mu u^\nu, \tag{3.6.47}$$

where $T^{\mu\nu}$ is the electromagnetic energy-momentum tensor (3.6.24).

From energy-momentum conservation of the total system, we shall obtain

$$0 = \partial_\mu \bar{T}^{\mu\nu} = F^{\lambda\nu} j_\lambda + \frac{mc^2}{e}\partial_\mu(\rho_0 u^\mu)u^\nu + \frac{mc^2}{e}\rho_0 u^\mu \partial_\mu u^\nu, \tag{3.6.48}$$

where (3.6.41) has been used. The second term vanishes by charge (or mass) conservation. Dividing by ρ_0, we end up with

$$u^\mu \partial_\mu u^\nu = \frac{e}{mc}F^{\nu\mu}u_\mu. \tag{3.6.49}$$

This is the **Euler equation of relativistic electro-hydrodynamics**. Together with Maxwell's equations this constitutes a mathematically perfect classical relativistic field theory of matter in interaction with the electromagnetic field. Its shortcoming is that it is not quite realistic because it neglects quantum phenomena. But it can be considered as a classical limit of quantum electrodynamics (see Epilogue). On the other hand, if one uses a mechanical description for the matter in terms of mass points and point charges, one gets serious mathematical and conceptual difficulties. An example is the problem of radiation reaction. We touch this very briefly in Problem 12 and 13. For detailed treatment we refer to the book by *A.D. Yaghjian, Relativistic Dynamics of a Charged Sphere, Springer Verlag 1992.* Obviously, nature does not like such a hybrid theory of mechanics and field theory. The field theoretical point of view is the more fundamental. We shall return to relativistic electro-hydrodynamics at the end of Sect.4.3, where we discuss an interesting application to superconductivity.

We close this chapter with a discussion of the **energy-momentum tensor** (3.6.24)

$$T^{\mu\nu} = F^{\mu\lambda} F_{\lambda}{}^{\nu} + \frac{1}{4} g^{\mu\nu} F^{\alpha\beta} F_{\alpha\beta} \tag{3.6.50}$$

for the Liénard-Wiechert field (3.5.97) of a moving point charge. In the calculation of this field we have introduced the time-like four-vector u^μ (3.5.84), the light vector $x^\mu - y^\mu$ (3.5.87) and the acceleration a^μ (3.5.91), which is space-like because it is orthogonal to u_μ (3.5.93). We have the following scalar products:

$$u_\mu u^\mu = 1, \quad u_\mu (x^\mu - y^\mu) = R, \quad u_\mu a^\mu = 0 \tag{3.6.51}$$

$$(x_\mu - y_\mu)(x^\mu - y^\mu) = 0, \quad (x_\mu - y_\mu) a^\mu = w. \tag{3.6.51a}$$

In addition, let us introduce the space-like unit vector

$$n^\mu = \frac{x^\mu - y^\mu}{R} - u^\mu, \tag{3.6.52}$$

with the scalar products

$$n_\mu n^\mu = -1, \quad n_\mu u^\mu = 0, \quad n_\mu (x^\mu - y^\mu) = -R. \tag{3.6.53}$$

All these four-vectors are part of the light cone with axis u^μ as shown in Fig.12.

Now we contract $T^{\mu\nu}$ with u_μ and n_ν to get a Lorentz invariant

$$T^{\mu\nu} u_\mu n_\nu = F^{\mu\lambda} u_\mu F_{\lambda}{}^{\nu} n_\nu. \tag{3.6.54}$$

This quantity has a simple interpretation in the rest frame of the particle where $u^\mu = (1, 0, 0, 0)$, $n^\mu = (0, \boldsymbol{n})$. Since $\boldsymbol{n}$ is normal to the sphere $|\boldsymbol{x} - \boldsymbol{y}| = R$ in 3-space, the surface integral of (3.6.54) over this sphere is equal to the energy flow per unit time across this surface. Due to (3.6.27), we find

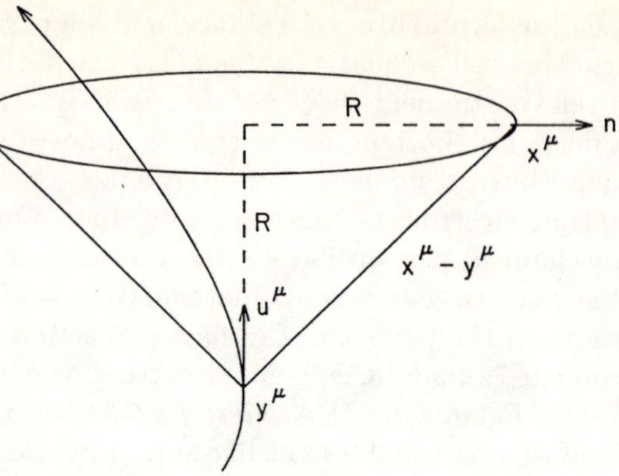

Fig. 12. Radiation of an accelerated particle in the rest frame

$$\int T^{0\nu} n_\nu R^2 \, d\Omega = -\frac{\mu_0}{c} \int_{|\boldsymbol{x}-\boldsymbol{y}|=R} \boldsymbol{S} \cdot d\boldsymbol{\sigma} = -\frac{\mu_0}{c} P. \tag{3.6.55}$$

This is the radiated power P in the rest frame (up to the factor μ_0/c), which we now want to evaluate.

Using the previous field tensor (3.5.97) for the Liénard-Wiechert field, we find

$$F^{\mu\nu} u_\mu = \frac{q}{4\pi\varepsilon_0 c R^2} \Big[(w-1) n^\lambda + R a^\lambda \Big] \tag{3.6.56}$$

$$F_{\lambda\nu} n^\nu = \frac{q}{4\pi\varepsilon_0 c R^2} (w \, n_\lambda + u_\lambda + R a_\lambda). \tag{3.6.57}$$

In the product (3.6.54)

$$T^{\mu\nu} u_\mu n_\nu = \frac{q^2}{(4\pi\varepsilon_0 c R^2)^2} \Big(w^2 + R^2 a^\lambda a_\lambda \Big), \tag{3.6.58}$$

only the terms depending on the acceleration survive. Here the second term has a trivial integral $\sim 4\pi$ over the unit sphere. In the first term we use

$$w^2 = R^2 (n_\lambda a^\lambda)^2 = R^2 n_j a^j n^k a_k$$

and

$$\int n_j n_k \, d\Omega = \delta_{jk} \int n_3^2 \, d\Omega = \delta_{jk} 2\pi \int_{-1}^{1} \cos^2\vartheta \, d\cos\vartheta$$

$$= \delta_{jk} \frac{4\pi}{3}.$$

Then (3.6.55) becomes independent of R:

$$P = -\frac{q^2 c}{(4\pi\varepsilon_0 c)^2 \mu_0}\Big(-\frac{4\pi}{3} + 4\pi\Big) a^\lambda a_\lambda$$

$$= -\frac{q^2 c}{4\pi\varepsilon_0}\frac{2}{3} a^\lambda a_\lambda. \tag{3.6.59}$$

Hence, the radiated power of the particle emitted at y is independent of the place x where the radiation is observed, as it has to be. This is a **covariant generalization of Larmor's formula** (3.5.30). But P must be measured in the rest frame of the charge, which is not quite realistic.

We now claim that this result is identical with (3.5.39). To verify this we express the four-acceleration by the velocity $\boldsymbol{\beta} = \boldsymbol{v}/c$

$$-a^\lambda a_\lambda = \Big(\frac{d\gamma\boldsymbol{\beta}}{ds}\Big)^2 - \Big(\frac{d\gamma}{ds}\Big)^2, \quad \gamma = (1 - \boldsymbol{\beta}^2)^{-1/2}. \tag{3.6.60}$$

Since

$$\frac{d\gamma}{ds} = \gamma^3 \boldsymbol{\beta} \cdot \frac{d\boldsymbol{\beta}}{ds},$$

we shall obtain

$$-a^\lambda a_\lambda = \Big(\frac{d\gamma}{ds}\Big)^2 (\boldsymbol{\beta}^2 - 1) + 2\gamma\frac{d\gamma}{ds}\boldsymbol{\beta}\cdot\frac{d\boldsymbol{\beta}}{ds} + \gamma^2\Big(\frac{d\boldsymbol{\beta}}{ds}\Big)^2$$

$$= \gamma^4\Big(\boldsymbol{\beta}\cdot\frac{d\boldsymbol{\beta}}{ds}\Big)^2 + \gamma^2\Big(\frac{d\boldsymbol{\beta}}{ds}\Big)^2. \tag{3.6.61}$$

Using the identity (0.1.9)

$$(\boldsymbol{\beta}\cdot\boldsymbol{\beta}')^2 = \boldsymbol{\beta}^2\boldsymbol{\beta}'^2 - (\boldsymbol{\beta}\wedge\boldsymbol{\beta}')^2,$$

we end up with

$$-a^\lambda a_\lambda = \gamma^4\Big[\Big(\frac{d\boldsymbol{\beta}}{ds}\Big)^2 - \Big(\boldsymbol{\beta}\wedge\frac{d\boldsymbol{\beta}}{ds}\Big)^2\Big]$$

$$= \frac{\gamma^6}{c^2}\Big[\Big(\frac{d\boldsymbol{\beta}}{dt}\Big)^2 - \Big(\boldsymbol{\beta}\wedge\frac{d\boldsymbol{\beta}}{dt}\Big)^2\Big]. \tag{3.6.62}$$

This enables us to express the radiated power in the rest frame P by quantities in the laboratory frame

$$P = \frac{q^2}{4\pi\varepsilon_0 c}\frac{2}{3}\gamma^6\Big[\Big(\frac{d\boldsymbol{\beta}}{dt}\Big)^2 - \Big(\boldsymbol{\beta}\wedge\frac{d\boldsymbol{\beta}}{dt}\Big)^2\Big]. \tag{3.6.63}$$

This is indeed identical with (3.5.39), because the absolute value of the vector product is equal to $\beta\beta'\sin\alpha$. However, P is not the radiation rate in the laboratory frame. As mentioned before, the latter is given by $N(t)$ (3.5.37), and only this quantity can be directly related to energy conservation.

3.7 Problems

1. Calculate the force between two parallel rectangular currents (currents J_1 and J_2, rectangle $a \times b$) in a distance c.
2. Write the force density $\boldsymbol{k} = \boldsymbol{j} \wedge \boldsymbol{B}$ in a magnetic field by means of Ampère's law as a divergence of a stress tensor T_{il}

$$k_i(\boldsymbol{x}) = \sum_{l=1}^{3} \frac{\partial}{\partial x_l} T_{il}(\boldsymbol{x}),$$

where T_{il} depends only on $\boldsymbol{B}$.

3. The equation of motion for a relativistic particle in a homogeneous electromagnetic field reads

$$\frac{d\boldsymbol{p}}{dt} = q(\boldsymbol{E} + \boldsymbol{v} \wedge \boldsymbol{B}), \tag{3.7.1}$$

where

$$\boldsymbol{p} = \frac{m\boldsymbol{v}}{\sqrt{1 - \frac{v^2}{c^2}}}. \tag{3.7.2}$$

Transform the equation by means of

$$p_0 = \sqrt{\boldsymbol{p}^2 + m^2c^2} \tag{3.7.3}$$

and the new variable s,

$$\frac{dt}{ds} = p_0 \tag{3.7.4}$$

into a four-dimensional linear system for the four-vector $(p_0, \boldsymbol{p})$. Determine the general solution $p(s)$ and, for $|\boldsymbol{p}| \ll mc$, the generalized cyclotron frequency. Show that for $\boldsymbol{E} \perp \boldsymbol{B}$, $E \ll cB$, you get a screw motion. Determine the axis of the screw.

4. Verify that the solution of the wave equation

$$u(t, \boldsymbol{x}) = \frac{1}{4\pi c^2} \frac{\partial}{\partial t} \frac{1}{t} \int\limits_{|\boldsymbol{x}'-\boldsymbol{x}|=ct} d^2x'\, u_0(\boldsymbol{x}') + \frac{1}{4\pi c^2 t} \int d^2x'\, u_1(\boldsymbol{x}') \tag{3.7.5}$$

satisfies for $t = 0$ the initial conditions $u(0, \boldsymbol{x}) = u_0(\boldsymbol{x})$, $(\partial_t u)(0, \boldsymbol{x}) = u_1(\boldsymbol{x})$.

5. Consider a monochromatic plane wave in x-direction of finite extension with a the following vector potential

$$\boldsymbol{A} = f(\varepsilon y, \varepsilon z)\Big(0, \sin(kx - \omega t), \cos(kx - \omega t)\Big), \tag{3.7.6}$$

where $f \in C_0^\infty(\mathbb{R}^2)$, $f(0) \neq 0$. In the limit $\varepsilon \to 0$ you get an infinitely extended plane wave.

a) Calculate from the Lorentz gauge condition the scalar potential and the fields $\boldsymbol{E}$ and $\boldsymbol{B}$. Show that this is a left-circular polarized wave. Show that Maxwell's equations are satisfied up to an error $O(\varepsilon^2)$.

b) Calculate the energy density and the Poynting vector, and the total angular momentum L_x in x-direction per unit length. Show that the wave transports angular momentum in the x-direction, also in the limit $\varepsilon \to 0$, which is proportional to the energy transport.

6. a) Write the energy W of the magnetic field generated by a stationary current distribution in terms of the current density $\boldsymbol{j}(\boldsymbol{x})$. Introducing a normalized current density $\boldsymbol{i}(\boldsymbol{x})$, so that $\boldsymbol{j}(\boldsymbol{x}) = I\boldsymbol{i}(\boldsymbol{x})$, one finds

$$W = \tfrac{1}{2}LI^2, \tag{3.7.7}$$

where L is the inductance of the system. How is L expressed by $\boldsymbol{i}(\boldsymbol{x})$?

b) Calculate the inductance of a long thin solenoid with winding number n, cross section Q and length l.

7. a) The magnetic energy W of a system of N separated conductors with currents I_n, $n = 1, \ldots N$, is equal to

$$W = \frac{1}{2}\sum_{n=1}^{N} L_n I_n^2 + \sum_{m>n} L_{mn} I_m I_n. \tag{3.7.8}$$

What are the general expressions for the inductance L_n and the mutual inductances L_{mn}?

b) Calculate L_{12} for two parallel, coaxial, circular currents with radii r_1, r_2 and distance a. Express the result by the complete elliptic integrals

$$K(k) = \int_0^{\pi/2} \frac{d\varphi}{\sqrt{1 - k^2 \sin^2\varphi}}, \quad E(k) = \int_0^{\pi/2} \sqrt{1 - k^2 \sin^2\varphi}\, d\varphi. \tag{3.7.9}$$

Compute the value of L_{12} for $a \gg r_1, r_2$.

8. a) Consider a current distribution that varies periodically in time. Calculate the temporal mean value of the energy per second that is radiated away by dipole radiation.

b) Apply the result a) to a dipole antenna of length l in z-direction where a current

$$I(t, z) = I_0\left(1 - 2\frac{|z|}{l}\right)\cos\omega(t + t_0)$$

is flowing.

9. A dipole is rotating with angular velocity ω in a plane. Calculate the radiation power per unit solid angle in the temporal mean and the total power. How does the vector $\boldsymbol{B}$ of the magnetic field (= direction of polarization) move?

10. In a circular wire of radius a flows an alternating current $I\cos\omega t$. Study the electric and magnetic dipole radiation and the electric quadrupole radiation. Calculate the radiated energy per second in the temporal mean.
11. Consider quadrupole radiation of frequency ω. Express the vector potential (3.4.71) in terms of the quadrupole moment

$$Q_{nl} = \int d^3x\,(3x_n x_l - x^2\delta_{nl})\rho_0(\boldsymbol{x}). \tag{3.7.10}$$

Determine the magnetic field and the Poynting vector in the wave zone. Calculate the energy flow per unit solid angle in the temporal mean. Specialize to the following diagonal quadrupole tensor

$$Q_{nl} = Q\begin{pmatrix} -\frac{1}{2} & 0 & 0 \\ 0 & -\frac{1}{2} & 0 \\ 0 & 0 & 1 \end{pmatrix}.$$

Draw a polar diagram and explain the notion quadrupole radiation.
12. Radiation reaction: Show that the damping force

$$\boldsymbol{K}_r(t) = \frac{1}{4\pi\varepsilon_0}\frac{2}{3}\frac{q^2}{c^3}\boldsymbol{x}'''(t) \tag{3.7.11}$$

acting on a periodically moving point charge q produces the same energy loss as the radiation according to the Larmor formula. Add K_r in Newton's equation of motion and discuss time reversal in the resulting equation.
13. Solve the equation of motion of Problem 12 for a harmonic oscillator with radiation damping

$$\boldsymbol{x}'' + \omega_0^2\boldsymbol{x} - \tau\boldsymbol{x}''' = \boldsymbol{0} \tag{3.7.12}$$

for $\omega_0\tau \ll 1$. Show that one fundamental solution is unphysical (runaway-solution). Calculate for the physical solution with

$$\boldsymbol{x}(0) = \boldsymbol{x}_0, \quad \boldsymbol{x}'(0) = \boldsymbol{0} \tag{3.7.13}$$

the total for $t \geq 0$ emitted energy

$$W = \int_0^\infty N(t)\,dt, \tag{3.7.14}$$

and its spectral distribution. Use always $\omega_0\tau \ll 1$.
14. Calculate the electric and magnetic field of a point charge q moving with constant velocity v in the positive 1-direction from the Liénard-Wiechert potentials. Express the fields by coordinates $(x_1', x_2' x_3')$ in the frame moving with the charge (rest frame). Compare the results $\boldsymbol{E}'(x')$,

$\boldsymbol{B}'(x')$ with the fields of a charge at rest and compare with Problem 2 of Chap.2.

15. In a classical picture of the β-decay of a nucleus one assumes that the emitted electron is instantaneously accelerated from $v = 0$ to its end velocity. Calculate the angular distribution of the radiation emitted by the electron and the total intensity per frequency interval. The recoil of the nucleus can be neglected, because it is small according to the mass ratio.
16. Determine the far field of a linearly accelerated charge q ($\boldsymbol{v} \parallel \boldsymbol{v}' \parallel z$-axis) and the corresponding Poynting vector. Discuss the angle ϑ where the radiation is maximal as a function of $\beta = v/c$. (Warning: Calculate in the laboratory system where the radiation is actually measured, not in the instantaneous rest frame of the charge.)
17. Hamilton formalism of electrodynamics:
    a) For a Lagrangian density $L(A^\mu, \partial_\nu A^\mu)$ one defines the canonical momenta Π_μ by
    $$\Pi_\mu = \frac{\partial L}{\partial A'^\mu}, \quad ' = \frac{\partial}{\partial x^0}. \tag{3.7.15}$$
    The Hamiltonian density is then obtained by a Legendre transformation
    $$H = \Pi^\mu A'_\mu - L. \tag{3.7.16}$$
    Calculate H for electrodynamics, using the Lagrangian (3.6.11). Transform the total
    $$\mathcal{H} = \int H \, d^3x \tag{3.7.17}$$
    by partial integration, using Gauss' law as a subsidiary condition. Show that (3.7.17) is essentially the energy.
    b) The equations of motion for A^i and Π^i can be written as canonical equations
    $$\frac{\partial \Pi^i}{\partial t} = \frac{\delta \tilde{\mathcal{H}}}{\delta A^i}, \quad \frac{\partial A^i}{\partial t} = -\frac{\delta \tilde{\mathcal{H}}}{\delta \Pi^i}, \tag{3.7.18}$$
    where
    $$\tilde{\mathcal{H}} = \int \tilde{H} \, d^3x, \tag{3.7.19}$$
    and
    $$\frac{\delta \tilde{\mathcal{H}}}{\delta A^i} = \frac{\partial \tilde{H}}{\partial A^i} - \frac{\partial}{\partial x^k} \frac{\partial \tilde{H}}{\partial A^i{}_{,k}}. \tag{3.7.20}$$
    In $\tilde{H}$ the subsidiary condition of Gauss' law is taken into account by means of a Lagrange multiplier.
18. Show that the current density
    $$\boldsymbol{j}(\boldsymbol{x}) = \operatorname{curl}(\boldsymbol{M}\delta(\boldsymbol{x})) \tag{3.7.21}$$
    is the source of a magnetic dipole field. If this source is moving with velocity v in the x-direction, it generates an electric dipole field in addition. Determine the dipole moment $\boldsymbol{P}$ of the latter.

4. Phenomenological Electrodynamics in Simple Matter

Maxwell's equations, as we have discussed them in the last chapter, are also valid in matter, in principle. The reason is that matter has an atomistic structure: it consists mainly of vacuum, containing very small elementary particles. The latter could in principle be described by charge and current densities $\rho(t, \boldsymbol{x})$, $\boldsymbol{j}(t, \boldsymbol{x})$, which, however, vary over microscopic distances, i.e. $\approx 10^{-8}$ cm. In addition there are variations in time by thermal oscillations of the atoms on a time scale of 10^{-13} sec. The resulting electric and magnetic fields have corresponding temporal and spatial variations. Macroscopic measurements, on the other hand, give mean values over a length scale

$$L \approx 10^{-6}\text{cm}. \tag{4.1}$$

The corresponding volume $L^3 = 10^{-18}$ cm^3 contains still 10^6 atoms, so that the atomistic structure is certainly averaged out. But the corresponding time $t = 10^{-6}$ cm/c $\approx 10^{-15}$ sec is short compared with the thermal oscillations. Consequently a temporal averaging is not justified. We shall, therefore, derive the macroscopic Maxwell's equations by only spatial averaging.

There are some weak points in this procedure. First, Lorentz covariance is destroyed. We will live with this because this chapter is mainly non-relativistic. This is not in conflict with the relativity principle, because the matter defines a distinct Lorentz frame. Secondly, strictly speaking the atoms must be described by quantum mechanics and, thirdly, the spatial averaging made below should be replaced by a (quantum) statistical ensemble averaging. This would be too much for a concise course. If the reader will come across such a more sophisticated quantum mechanical treatment, he will quickly realize that this runs quite similar to our treatment below: the spatial mean values are replaced by quantum ensemble averages, but otherwise, the procedure is essentially the same.

It is clear from these introductory remarks that phenomenological electrodynamics is much less fundamental than Maxwell's theory in vacuum. For conceptual reasons, any mixing of the two should be avoided. The fact that the mathematics is quite the same must not obscure the fact that the physics is totally different. The physical assumptions which constitute the basis of phenomenological electrodynamics, will be discussed at the beginning of Sect.4.2. There we are also concerned with the limited validity of the averaging procedure and of macroscopic electrodynamics on the whole. In the further applications of the theory (Sect.4.4), we concentrate again on radiation phenomena.

4.1 Derivation of the Phenomenological Maxwell's Equations

The basis for this derivation are the microscopic Maxwell's equations

$$\varepsilon_0 \frac{\partial \boldsymbol{E}}{\partial t} = \frac{1}{\mu_0} \operatorname{curl} \boldsymbol{B} - \boldsymbol{j} \tag{4.1.1}$$

$$\frac{\partial \boldsymbol{B}}{\partial t} = -\operatorname{curl} \boldsymbol{E} \tag{4.1.2}$$

$$\varepsilon_0 \operatorname{div} \boldsymbol{E} = \rho \tag{4.1.3}$$

$$\operatorname{div} \boldsymbol{B} = 0. \tag{4.1.4}$$

The spatial averaging will be carried out by convolution with a positive function $a(\boldsymbol{x})$; we denote it by angular brackets

$$\langle \boldsymbol{E}(t,\boldsymbol{x}) \rangle \stackrel{\text{def}}{=} \int a(\boldsymbol{x}-\boldsymbol{x}')\boldsymbol{E}(t,\boldsymbol{x}')\, d^3x'. \tag{4.1.5}$$

The integral over $a(\boldsymbol{x})$ must be equal to

$$\int a(\boldsymbol{x})\, d^3x = 1 \tag{4.1.6}$$

and otherwise we assume $a(\boldsymbol{x})$ to be infinitely differentiable with a compact support of an extension $L \approx 10^{-6}$ cm, large compared with the atomic dimensions.

Differentiating (4.1.5), we find

$$\begin{aligned} \frac{\partial}{\partial x_i} \langle \boldsymbol{E}(t,\boldsymbol{x}) \rangle &= \int \frac{\partial}{\partial x_i} a(\boldsymbol{x}-\boldsymbol{x}')\boldsymbol{E}(t,\boldsymbol{x}')\, d^3x' \\ &= \int a(\boldsymbol{x}-\boldsymbol{x}') \frac{\partial}{\partial x_i'} \boldsymbol{E}(t,\boldsymbol{x}')\, d^3x' = \left\langle \frac{\partial \boldsymbol{E}(t,\boldsymbol{x})}{\partial x_i} \right\rangle, \end{aligned} \tag{4.1.7}$$

where we have shifted the derivative from $\boldsymbol{x}$ to $\boldsymbol{x}'$ and performed a partial integration. This means that averaging and spatial derivatives can be interchanged. The same is trivially true for partial derivatives with respect to t. Consequently, the homogeneous Maxwell's equations (4.1.2, 4) are valid for the averaged fields, too,

$$\frac{\partial}{\partial t}\langle \boldsymbol{B}(t,\boldsymbol{x})\rangle = -\operatorname{curl}\langle \boldsymbol{E}(t,\boldsymbol{x})\rangle \tag{4.1.8}$$

$$\operatorname{div}\langle \boldsymbol{B}\rangle = 0. \tag{4.1.9}$$

This was very cheap, but the inhomogeneous equations require much more work. The reason is that the averaged densities $\langle\rho(t,\boldsymbol{x})\rangle$ and $\langle\boldsymbol{j}(t,\boldsymbol{x})\rangle$ can no longer be assumed as given as in microscopic electrodynamics. They usually depend strongly on the applied fields and, therefore, must be worked out in detail, taking the atomistic structure of the matter into account.

We assume that the microscopic charge density can be written as a sum

$$\rho(t,\boldsymbol{x}) = \sum_i \rho_i(t,\boldsymbol{x}), \tag{4.1.10}$$

of the charge densities of individual "atoms". Here "atom" stands also for free electrons, ions etc. This essential assumption (4.1.10) requires discussion. We postpone this important discussion to the beginning of the next section. The charge density ρ_i of atom i is concentrated at its position $\boldsymbol{y}_i(t)$. In the average

$$\langle\rho(t,\boldsymbol{x})\rangle = \sum_i \int a(\boldsymbol{x}-\boldsymbol{x}')\rho_i(t,\boldsymbol{x}')\,d^3x' \tag{4.1.11}$$

$a(\boldsymbol{x}-\boldsymbol{y}_i+\boldsymbol{y}_i-\boldsymbol{x}')$ is slowly varying over the microscopic support of ρ_i. We therefore use a Taylor expansion and neglect quadratic and higher contributions:

$$\begin{aligned}\langle\rho(t,\boldsymbol{x})\rangle &= \sum_i \int d^3x'\Big[a(\boldsymbol{x}-\boldsymbol{y}_i) + \frac{\partial a(\boldsymbol{x}-\boldsymbol{y}_i)}{\partial x_k}(y_{ik}-x'_k)\Big]\rho_i(t,\boldsymbol{x}') \\ &= \sum_i q_i a(\boldsymbol{x}-\boldsymbol{y}_i) - \sum_i \boldsymbol{p}_i\cdot\frac{\partial a(\boldsymbol{x}-\boldsymbol{y}_i)}{\partial \boldsymbol{x}}.\end{aligned} \tag{4.1.12}$$

Here

$$q_i(t) = \int d^3x'\,\rho_i(t,\boldsymbol{x}') \tag{4.1.13}$$

is the total charge of the atom i, and

$$\boldsymbol{p}_i(t) = \int d^3x'\,(\boldsymbol{x}'-\boldsymbol{y}_i(t))\rho_i(t,\boldsymbol{x}') \tag{4.1.14}$$

is its dipole moment with respect to the center $\boldsymbol{y}_i$ (see (1.2.3)). Writing the first term on the r.h.s. of (4.1.12) as follows

$$q_i a(\boldsymbol{x}-\boldsymbol{y}_i) = \langle q_i\delta(\boldsymbol{x}-\boldsymbol{y}_i)\rangle, \tag{4.1.15}$$

we see that it represents the atoms by point charges (monopoles) at the positions $\boldsymbol{y}_i(t)$ of the atoms. The sum

$$\sum_i q_i a(\boldsymbol{x} - \boldsymbol{y}_i) \stackrel{\text{def}}{=} \rho_c(t, \boldsymbol{x}) \tag{4.1.16}$$

is the conduction charge density. Here only charged atoms ($q_i \neq 0$) contribute, which explains the name "conduction charge". The individual charges (electrons and nuclei) of neutral atoms are not resolved, they add up to zero in the integral (4.1.13).

The dipole term in (4.1.12) can be written as follows

$$\boldsymbol{p}_i \cdot \frac{\partial a(\boldsymbol{x} - \boldsymbol{y}_i)}{\partial \boldsymbol{x}} = \operatorname{div}_x \boldsymbol{p}_i a(\boldsymbol{x} - \boldsymbol{y}_i) = \operatorname{div}_x \langle \boldsymbol{p}_i \delta(\boldsymbol{x} - \boldsymbol{y}_i) \rangle. \tag{4.1.17}$$

Comparing this with (1.2.5), we realize that this represents a dipole moment at the position of the atom. The atoms are represented by their lowest multipole moments. Summing over all atoms, we get the so-called **polarization charge density**

$$\sum_i \boldsymbol{p}_i \cdot \frac{\partial a(\boldsymbol{x} - \boldsymbol{y}_i)}{\partial \boldsymbol{x}} \stackrel{\text{def}}{=} \operatorname{div} \boldsymbol{P}(t, \boldsymbol{x}), \tag{4.1.18}$$

where

$$\boldsymbol{P} = \sum_i \langle \boldsymbol{p}_i \delta(\boldsymbol{x} - \boldsymbol{y}_i) \rangle \tag{4.1.19}$$

is the electric polarization. Omitting higher multipole contributions, we obtain for the averaged charge density

$$\langle \rho \rangle = \rho_c - \operatorname{div} \boldsymbol{P}. \tag{4.1.20}$$

Then averaging of Gauss' law (4.1.3) leads to

$$\varepsilon_0 \operatorname{div} \langle \boldsymbol{E} \rangle = \rho_c - \operatorname{div} \boldsymbol{P}. \tag{4.1.21}$$

Introducing the new phenomenological field

$$\boldsymbol{D} = \varepsilon_0 \langle \boldsymbol{E} \rangle + \boldsymbol{P}, \tag{4.1.22}$$

the **macroscopic Gauss' law** assumes the same form as the microscopic one

$$\operatorname{div} \boldsymbol{D} = \rho_c. \tag{4.1.23}$$

The field $\boldsymbol{D}$ is usually called the electric displacement, because the macroscopic electric field $\langle \boldsymbol{E} \rangle$ is displaced by the polarization $\boldsymbol{P}$ in (4.1.22). The source of $\boldsymbol{D}$ is only the conduction charge density ρ_c, not the total charge density (4.1.20).

In a similar manner we **consider the current density**

$$\boldsymbol{j}(t, \boldsymbol{x}) = \sum_i \boldsymbol{j}_i(t, \boldsymbol{x}), \tag{4.1.24}$$

where $\boldsymbol{j}_i$ is the current density produced by the atom i. Charge conservation for the atom i implies

$$\int d^3x'\,\partial_t\rho_i = -\int \operatorname{div}\boldsymbol{j}_i\,d^3x' = -\int \boldsymbol{j}_i\cdot d\boldsymbol{\sigma} = 0. \tag{4.1.25}$$

Similarly we treat the next moments

$$\int d^3x'\,\partial_t\rho_i x'_k = -\int d^3x'\,x'_k \operatorname{div}\boldsymbol{j}_i$$

$$= -\int d^3x'\,\partial'_l(x'_k j_{il}) + \int d^3x'\,j_{ik} = \int d^3x'\,j_{ik} \tag{4.1.26}$$

$$\int d^3x'\,\partial_t\rho_i x'_k x'_l = -\int d^3x'\,x'_k x'_l \operatorname{div}\boldsymbol{j}_i$$

$$= \int d^3x'\,x'_l j_{ik} + \int d^3x'\,x'_k j_{il}. \tag{4.1.27}$$

Now we are ready to expand the averaged current density

$$\langle\boldsymbol{j}_i\rangle = \int d^3x'\left[a(\boldsymbol{x}-\boldsymbol{y}_i) + \frac{\partial a(\boldsymbol{x}-\boldsymbol{y}_i)}{\partial\boldsymbol{x}}\cdot(\boldsymbol{y}_i-\boldsymbol{x}')\right]\boldsymbol{j}(t,\boldsymbol{x}') \tag{4.1.28}$$

$$\stackrel{\text{def}}{=} \boldsymbol{I}_i + \boldsymbol{II}_i,$$

By means of (4.1.26, 25), the first term $\boldsymbol{I}_i$ in (4.1.28) can be expressed by the dipole moment (4.1.14)

$$\boldsymbol{I}_i = a(\boldsymbol{x}-\boldsymbol{y}_i)\int d^3x'\,\partial_t\rho_i(t,\boldsymbol{x}')(\boldsymbol{x}'-\boldsymbol{y}_i(t))$$

$$= a(\boldsymbol{x}-\boldsymbol{y}_i)\partial_t\boldsymbol{p}_i + a(\boldsymbol{x}-\boldsymbol{y}_i)q_i\partial_t\boldsymbol{y}_i$$

$$= \langle(\partial_t\boldsymbol{p}_i)\delta(\boldsymbol{x}-\boldsymbol{y}_i)\rangle + \langle q_i(\partial_t\boldsymbol{y}_i)\delta(\boldsymbol{x}-\boldsymbol{y}_i)\rangle. \tag{4.1.29}$$

On the other hand, the time derivative of (4.1.19) yields

$$\partial_t\boldsymbol{P} = \sum_i\langle(\partial_t\boldsymbol{p}_i)\delta(\boldsymbol{x}-\boldsymbol{y}_i)\rangle + \sum_i\langle\boldsymbol{p}_i\partial_t\delta(\boldsymbol{x}-\boldsymbol{y}_i(t))\rangle$$

$$= \sum_i\langle(\partial_t\boldsymbol{p}_i)\delta(\boldsymbol{x}-\boldsymbol{y}_i)\rangle - \frac{\partial}{\partial x_k}\sum_i\langle\boldsymbol{p}_i\delta(\boldsymbol{x}-\boldsymbol{y}_i)\partial_t y_{ik}\rangle. \tag{4.1.30}$$

For the last term in (4.1.29) we introduce the **conduction current density**

$$\boldsymbol{j}_c(t,\boldsymbol{x}) \stackrel{\text{def}}{=} \sum_i\langle q_i(\partial_t\boldsymbol{y}_i)\delta(\boldsymbol{x}-\boldsymbol{y}_i)\rangle. \tag{4.1.31}$$

Then the sum of (4.1.29) over all atoms yields

$$\sum_i \boldsymbol{I}_i = \partial_t \boldsymbol{P} + \frac{\partial}{\partial x_k} \sum_i \langle \boldsymbol{p}_i \delta(\boldsymbol{x} - \boldsymbol{y}_i) \partial_t y_{ik} \rangle + \boldsymbol{j}_c. \tag{4.1.32}$$

To compute the second term $\boldsymbol{II}_i$ in (4.1.28), we decompose

$$\int d^3x' \, (y_{ik} - x'_k) j_{il}(t, \boldsymbol{x}')$$

$$= \tfrac{1}{2} \int d^3x' \left[(y_{ik} - x'_k) j_{il} - (y_{il} - x'_l) j_{ik} \right]$$

$$+ \tfrac{1}{2} \int d^3x' \left[(y_{ik} - x'_k) j_{il} + (y_{il} - x'_l) j_{ik} \right] \tag{4.1.33}$$

into an antisymmetric and symmetric part. Using (4.1.26, 27) we get

$$= \tfrac{1}{2} \Big(\int d^3x' \, (\boldsymbol{y}_i - \boldsymbol{x}') \wedge \boldsymbol{j}_i \Big)_m + \tfrac{1}{2} \Big[y_{ik} \int d^3x' \, \partial_t \rho_i x'_l$$

$$- \int d^3x' \, x'_l j_{il} + y_{il} \int d^3x' \, \partial_t \rho_i x'_k - \int d^3x' \, x'_l j_{ik} \Big]$$

$$= -\frac{1}{\mu_0} (\boldsymbol{m}_i)_m + \tfrac{1}{2} \Big[y_{ik} \int d^3x' \, \partial_t \rho_i x'_l - \int d^3x' \, \partial_t \rho_i x'_k x'_l + y_{il} \int d^3x' \, \partial_t \rho_i x'_k \Big]. \tag{4.1.34}$$

Here $\boldsymbol{m}_i$ is the magnetic moment of atom i (3.1.29) with respect to its center $\boldsymbol{y}_i$, and the index m is such that k, l, m is a cyclic permutation of 1,2,3. The last term in (4.1.34) leads to the quadrupole moment (3.4.74)

$$q_{kl}^{(i)} = \int d^3x' \, \rho_i (x'_k - y_{ik})(x'_l - y_{il}). \tag{4.1.35}$$

Then (4.1.34) yields

$$= -\frac{1}{\mu_0} (\boldsymbol{m})_m - \tfrac{1}{2} \int d^3x' \, (\partial_t \rho_i)(x'_k - y_{ik})(x'_l - y_{il})$$

$$= -\frac{1}{\mu_0} (\boldsymbol{m}_i)_m - \tfrac{1}{2} \Big[\partial_t q_{kl}^{(i)} + \int d^3x' \, \rho_i (\partial_t y_{ik})(x'_l - y_{il})$$

$$+ \int d^3x' \, \rho_i (x'_k - y_{ik}) \partial_t y_{il} \Big]$$

$$= -\frac{1}{\mu_0} (\boldsymbol{m}_i)_m - \tfrac{1}{2} \partial_t q_{kl}^{(i)} - \tfrac{1}{2} [p_{il} \partial_t y_{ik} + p_{ik} \partial_t y_{il}], \tag{4.1.36}$$

where again the dipole moment (4.1.14) appears. According to (4.1.28), this result must be multiplied by the gradient of $a(\boldsymbol{x} - \boldsymbol{y}_i)$

$$(\boldsymbol{II}_i)_l = -\frac{1}{\mu_o}\frac{\partial a}{\partial x_k}(\boldsymbol{m}_i)_m - \frac{1}{2}\frac{\partial a}{\partial x_k}(p_{ik}\partial_t y_{il} + p_{il}\partial_t y_{ik})$$

$$= \frac{1}{\mu_0}\Big(\operatorname{grad}_x a(\boldsymbol{x}-\boldsymbol{y}_i) \wedge \boldsymbol{m}_i\Big)_l - \frac{1}{2}\frac{\partial a}{\partial x_k}(p_{ik}\partial_t y_{il} + p_{il}\partial_t y_{ik}). \tag{4.1.37}$$

This gives the following result for the vector $\boldsymbol{II}_i$

$$\boldsymbol{II}_i = \frac{1}{\mu_0}\operatorname{curl}_x a(\boldsymbol{x}-\boldsymbol{y}_i)\boldsymbol{m}_i - \frac{1}{2}\frac{\partial a}{\partial x_k}(p_{ik}\partial_t \boldsymbol{y}_i + \boldsymbol{p}_i\partial_t y_{ik}). \tag{4.1.38}$$

Here we introduce the **macroscopic magnetic moment density, or magnetization**

$$\boldsymbol{M} \stackrel{\text{def}}{=} \sum_i a(\boldsymbol{x}-\boldsymbol{y}_i)\boldsymbol{m}_i = \sum_i \langle \boldsymbol{m}_i \delta(\boldsymbol{x}-\boldsymbol{y}_i)\rangle. \tag{4.1.39}$$

Then the total contribution of $\boldsymbol{II}_i$ is equal to

$$\sum_i \boldsymbol{II}_i = \frac{1}{\mu_0}\operatorname{curl}\boldsymbol{M} - \frac{1}{2}\frac{\partial}{\partial x_k}\sum_i\Big(\langle p_{ik}\delta(\boldsymbol{x}-\boldsymbol{y}_i)\partial_t \boldsymbol{y}_i\rangle$$

$$+ \langle \boldsymbol{p}_i\delta(\boldsymbol{x}-\boldsymbol{y}_i)\partial_t y_{ik}\rangle\Big).$$

The last term herein can be combined with the middle term in (4.1.32). We then obtain the following final result

$$\langle \boldsymbol{j}\rangle = \boldsymbol{j}_c + \frac{\partial \boldsymbol{P}}{\partial t} + \frac{1}{\mu_0}\operatorname{curl}\boldsymbol{M} + \boldsymbol{V}, \tag{4.1.40}$$

with

$$V_l = \frac{1}{2}\frac{\partial}{\partial x_k}\Big\langle \sum_i (p_{il}\partial_t y_{ik} - p_{ik}\partial_t y_{il})\delta(\boldsymbol{x}-\boldsymbol{y}_i)\Big\rangle. \tag{4.1.41}$$

We notice that the divergence vanishes

$$\operatorname{div}\boldsymbol{V} = \frac{\partial V_l}{\partial x_l} = \frac{1}{2}\frac{\partial^2}{\partial x_l \partial x_k}\Big\langle \dots \Big\rangle = 0,$$

because the sum in (4.1.41) is antisymmetric in k, l, whereas the second derivative is symmetric. Then the microscopic current conservation

$$\partial_t\langle\rho\rangle + \operatorname{div}\langle \boldsymbol{j}\rangle = 0$$

implies

$$\partial_t \rho_c + \operatorname{div}\boldsymbol{j}_c = 0 \tag{4.1.42}$$

by means of (4.1.20) and (4.1.40). Hence, the conduction current alone is also conserved. For small velocities or small dipole moments, $\boldsymbol{V}$ (4.1.41) can be neglected, as well as the higher order multipole contributions. Then the last phenomenological Maxwell's equation, the **macroscopic Ampère's law** assumes the following form

$$\varepsilon_0 \frac{\partial \langle \boldsymbol{E} \rangle}{\partial t} = \frac{1}{\mu_0} \operatorname{curl} \langle \boldsymbol{B} \rangle - \boldsymbol{j}_c - \frac{\partial \boldsymbol{P}}{\partial t} - \frac{1}{\mu_0} \operatorname{curl} \boldsymbol{M}. \tag{4.1.43}$$

Introducing the new field

$$\boldsymbol{H} = \frac{1}{\mu_0} \Big(\langle \boldsymbol{B} \rangle - \boldsymbol{M} \Big) \tag{4.1.44}$$

and using the electric displacement (4.1.22), we finally get

$$\frac{\partial \boldsymbol{D}}{\partial t} = \operatorname{curl} \boldsymbol{H} - \boldsymbol{j}_c. \tag{4.1.45}$$

Unfortunately, the field $\boldsymbol{H}$ is usually called "magnetic field" and $\boldsymbol{B}$ magnetic induction, because it appears in the induction law (4.1.8). Of course, $\boldsymbol{B}$ is the fundamental (microscopic) magnetic field and $\boldsymbol{H}$ is derived from it (4.1.44). To avoid notational confusion, we will simply say $\boldsymbol{B}$-field and $\boldsymbol{H}$-field in the following.

Looking again at this derivation, we notice that only the following properties of the averaging operation $\langle \ldots \rangle$ have been used: linearity and commutability with derivatives. Ensemble averaging in statistical mechanics usually has also these properties, so that the same derivation goes through in this case.

4.2 Phenomenological Maxwell's Equations and Constitutive Relations

In the phenomenological Maxwell's equations

$$\frac{\partial \boldsymbol{D}}{\partial t} = \operatorname{curl} \boldsymbol{H} - \boldsymbol{j}_c \tag{4.2.1}$$

$$\frac{\partial \boldsymbol{B}}{\partial t} = -\operatorname{curl} \boldsymbol{E} \tag{4.2.2}$$

$$\operatorname{div} \boldsymbol{D} = \rho_c \tag{4.2.3}$$

$$\operatorname{div} \boldsymbol{B} = 0 \tag{4.2.4}$$

we drop the angular brackets. From now on all fields will be macroscopic, averaged fields, if nothing else is explicitly said. The equations as they stand are rather void, because in the inhomogeneous equations (4.2.1, 3) two new macroscopic fields $\boldsymbol{D}$, $\boldsymbol{H}$ appear, so that we have no longer a closed system. This shows that the derivation in the last section, although mathematically straightforward, omits the physical problems which we now want to discuss.

The key assumption was the decomposition of the microscopic charge density into a sum over individual "atoms" (4.1.10)

$$\rho(t, \boldsymbol{x}) = \sum_i \rho_i(t, \boldsymbol{x}), \tag{4.2.5}$$

and the analogous equation (4.1.24) for the current density. Such a **decomposition is to a certain degree arbitrary**. For example, some electron could be attached to different ions (all particles are treated classically distinguishable), or it could be treated as a free electron in (4.2.5). In the latter case, this electron would contribute to the conduction charge density ρ_c. Consequently, the splitting (4.1.20) of the averaged charge density, $\langle\rho\rangle = \rho_c - \operatorname{div} \boldsymbol{P}$, is also ambiguous. **For a precise definition of the macroscopic quantities, we have to know the detailed microscopic structure of the substance**. This ultimately requires the solution of a many-body problem. A second important fact is that **the atomic charge densities ρ_i depend on the applied electromagnetic fields**. For example, an electric field polarizes the atoms, so that ρ_i and the dipole moment $\boldsymbol{p}_i$ (4.1.14) change with the field. Consequently, the macroscopic polarization $\boldsymbol{P}$ and $\boldsymbol{D}$ are functions of $\boldsymbol{E}$, and similarly for the magnetic quantities. These relations are the **constitutive relations** (material equations), they characterize the electromagnetic properties of the matter under consideration. If such relations are known, the phenomenological equations (4.2.1-4) become a closed system and macroscopic electrodynamics can be worked out.

We conclude from this discussion that the constitutive relations are the heart of the matter. They again can only be provided by a microscopic many-body theory. Furthermore, for the constitutive relations we must know the solution of the many-body problem in much detail. This was not necessary in the derivation of the phenomenological Maxwell's equations themselves. There only the existence of a reasonable decomposition (4.2.5) of the microscopic charge and current densities was required. Here we need the precise dependence of these quantities upon the applied electromagnetic fields. We will restrict ourselves in the following to so-called simple matter, where reasonable constitutive relations can be written down without entering into many-body theory. We will get those relations either from simple models or from phenomenological considerations or assume them given by experiment. For further discussion of constitutive relations we refer to *R.Balian, From Microphysics to Macrophysics, Springer-Verlag 1991, Sect.11.3.* Let us finally point out that there are situations where the whole concept of spatial averaging breaks down, so that phenomenological electrodynamics is no longer an adequate description. This obviously is the case if the wavelength λ of the fields becomes comparable with the interatomic distance, but also if λ is comparable with the mean free path of conduction electrons (as in the anomalous skin effect). In superconductors spatial averaging is also not appropriate. But here, phenomenological equations make still sense. They can be obtained, for example, from models like the charged fluid model, together with the constitutive relations. This will be discussed at the end of the next section (4.3.26). A complete description of superconductivity, however, requires a microscopic theory.

Let us now turn to a discussion of the simplest constitutive relations and some of their consequences. For many applications the fields can be considered to be small. Then, if the material is not ferroelectric or ferromagnetic, simple linear material equations can be used:

$$\boldsymbol{D} = \varepsilon_0 \boldsymbol{E} + \boldsymbol{P} = \varepsilon_0 (1 + \chi_e) \boldsymbol{E} \stackrel{\text{def}}{=} \varepsilon \boldsymbol{E}. \tag{4.2.6}$$

Here χ_e is called the electric susceptibility and ε is the **dielectric constant**. Similarly, for the magnetic fields one writes

$$\boldsymbol{B} = \mu_0 \boldsymbol{H} + \boldsymbol{M} = \mu_0 (1 + \chi_m) \boldsymbol{H} \stackrel{\text{def}}{=} \mu \boldsymbol{H},$$

where χ_m is the **magnetic susceptibility** and μ the permeability. For the moment, all these quantities are assumed to be constant. Then Maxwell's equations (4.2.1-4) are identical with the vacuum equations, where only ε_0 is replaced by ε and μ_0 by μ. Most of the discussion in Chap.3, therefore, remains valid. In particular (compare (3.3.1-6)), we can have electromagnetic waves in matter with a light velocity

$$c = \frac{1}{\sqrt{\varepsilon\mu}} \stackrel{\text{def}}{=} \frac{c_0}{n}. \tag{4.2.7}$$

Here c_0 is the light velocity in vacuum. At the boundary of the material there are discontinuities in ε, μ and c. In the next chapter we will see that this leads to the interpretation of n as the index of refraction. It can be expressed in terms of ε and μ. Introducing the relative quantities

$$\varepsilon_r = \frac{\varepsilon}{\varepsilon_0} = 1 + \chi_e, \quad \mu_r = \frac{\mu}{\mu_0} = 1 + \chi_m, \tag{4.2.8}$$

we get Maxwell's relation

$$n = \sqrt{\varepsilon_r \mu_r}. \tag{4.2.9}$$

It is roughly valid in reality; more cannot be expected because it is impossible to describe the electromagnetic properties of real substances by only two constants, since the constitutive relations are usually much more complicated. The light velocity c (4.2.7) is the velocity that appears in the wave equation. This is the so-called phase velocity. Consider a plane wave solution (3.3.59)

$$E(t, \boldsymbol{x}) = f(ct - \boldsymbol{e} \cdot \boldsymbol{x}) = \tilde{f}(c_0 t - n\boldsymbol{e} \cdot \boldsymbol{x}), \tag{4.2.10}$$

where $\boldsymbol{e}$ is the propagation direction, $|\boldsymbol{e}| = 1$. Then the argument of f is the phase and the points of constant phase move with velocity c.

There are additional constitutive equations for the sources in Maxwell's equations. The conduction current density $\boldsymbol{j}_c$ is no longer independent, but is related to the electric field strength by **Ohm's law**

$$\boldsymbol{j}_c = \sigma \boldsymbol{E}. \tag{4.2.11}$$

σ is the conductivity, it depends on the material, only, not on the geometry of the arrangement. Ohm's law is often stated for the total current J in the form $U/J = R$, where U is the voltage and R the resistance. However, R depends on the geometry of the arrangement. To see this, we consider a homogeneous current distribution in a parallelepiped of cross section F and length l. Then we have $J = j_c F$. Multiplying (4.2.11) by F and taking the line integral along the current flow between two points 1,2 of distance l, we find

$$J \cdot l = \sigma F \cdot U, \quad U = \int_1^2 \boldsymbol{E} \cdot d\boldsymbol{s}.$$

This implies

$$\frac{U}{J} = \frac{1}{\sigma} \frac{l}{F} \stackrel{\text{def}}{=} R, \tag{4.2.12}$$

and this depends indeed on geometric quantities l, F.

The reason for Ohm's law is the following simple kinetic argument: The conduction electrons in the material gain energy from the Lorentz force when they move in the electric field, and they lose this energy in collisions with impurities and lattice vibrations, so that it is finally converted into heat. These two processes must be in equilibrium. If the mean free time between two collisions is τ, an electron with charge e and velocity $\boldsymbol{v}$ gains the energy $\varepsilon = e\tau \boldsymbol{v} \cdot \boldsymbol{E}$ during this time. This energy gain ε corresponds to a drift velocity $\boldsymbol{w}$ in the direction of $\boldsymbol{E}$

$$e\tau \boldsymbol{v} \cdot \boldsymbol{E} = \frac{m}{2}(\boldsymbol{v} + \boldsymbol{w})^2 - \frac{m}{2}\boldsymbol{v}^2 = m\boldsymbol{v} \cdot \boldsymbol{w} + O(w^2),$$

thus

$$\boldsymbol{w} = \frac{e\tau}{m}\boldsymbol{E}.$$

If $\rho_c = n_e e$ is the charge density of the conduction electrons (n_e is the electron density), the corresponding current density is equal to

$$\boldsymbol{j}_c = \rho_c \boldsymbol{w} = \frac{n_e e^2 \tau}{m} \boldsymbol{E} \stackrel{\text{def}}{=} \sigma \boldsymbol{E}.$$

From the energy conservation law (2.6.9) we know the energy q_c per unit time and volume

$$q_c = j_c \cdot \boldsymbol{E} = \sigma \boldsymbol{E}^2, \tag{4.2.13}$$

that is converted into heat. This is Joule's heat which is quadratic in $\boldsymbol{E}$ and, therefore, in linear approximation is often neglected.

The material "constants" ε, μ, σ are in general not constant. In anisotropic substances (crystals) they depend on the spatial directions. Then the linear constitutive equations become (3-dimensional) tensor equations

$$D_i = \varepsilon_{ik} E_k, \quad B_i = \mu_{ik} H_k, \quad j_i = \sigma_{ik} E_k. \tag{4.2.14}$$

Moreover, these quantities are generally frequency dependent, in particular for high (optical) frequencies. This is called **dispersion**. If the fields are still weak (and the substance is not ferroelectric), we have a linear relation between the temporal Fourier transforms

$$\hat{D}_j(\omega, \boldsymbol{x}) = \sum_k \varepsilon_{jk}(\omega) \hat{E}_k(\omega, \boldsymbol{x}),$$

where

$$\boldsymbol{D}(t, \boldsymbol{x}) = \frac{1}{\sqrt{2\pi}} \int_{-\infty}^{+\infty} \hat{\boldsymbol{D}}(\omega, \boldsymbol{x}) e^{-i\omega t} d\omega. \tag{4.2.15}$$

This leads to a non-local relation in time between $\boldsymbol{E}$ and $\boldsymbol{D}$:

$$D_j(t, \boldsymbol{x}) = \frac{1}{\sqrt{2\pi}} \int_{-\infty}^{+\infty} \varepsilon_{jk}(\omega) \hat{E}_k(\omega, \boldsymbol{x}) e^{-i\omega t} d\omega$$

$$= \frac{1}{2\pi} \int d\omega\, \varepsilon_{jk}(\omega) e^{-i\omega t} \int dt'\, E_k(t', \boldsymbol{x}) e^{i\omega t'}. \tag{4.2.16}$$

The dielectric "constant" $\varepsilon(\omega)$ has only a distributive Fourier transform, because $\varepsilon(\omega) \to \varepsilon_0$ for $\omega \to \infty$. This is due to the fact that the polarization cannot follow very rapid oscillations of the electric field. To get better convergence of the Fourier integral, we therefore consider the difference

$$\frac{1}{2\pi} \int_{-\infty}^{+\infty} [\varepsilon_{jk}(\omega) - \varepsilon_0 \delta_{jk}] e^{-i\omega t} d\omega \stackrel{\text{def}}{=} R_{jk}(t). \tag{4.2.17}$$

We have omitted a space dependence, assuming a spatial homogeneous situation. The function $R_{jk}(t)$ is called the response function. Instead of (4.2.16) we now get the following non-local time dependence of the fields

$$D_j(t) = \varepsilon_0 E_j(t) + \int_{-\infty}^{+\infty} R_{jk}(t - t') E_k(t')\, dt'$$

$$= \varepsilon_0 E_j(t) + \int_{-\infty}^{+\infty} R_{jk}(\tau) E_k(t - \tau)\, d\tau. \tag{4.2.18}$$

There exists a very **important restriction on the response function $R_{jk}(t)$ which is a consequence of causality**: $D(t)$ can only depend on $E(t')$ of earlier times $t' < t$, thus

$$R_{jk}(\tau) = 0, \quad \text{for} \quad \tau < 0. \tag{4.2.19}$$

Using this in the inversion of (4.2.17), we find

$$\varepsilon_{jk}(\omega) = \varepsilon_0 \delta_{jk} + \int_0^\infty R_{jk}(\tau) e^{i\omega\tau} d\tau. \tag{4.2.20}$$

Since the lower limit of integration is 0, we can analytically continue this function to complex ω into the upper half-plane $\operatorname{Im}\omega > 0$. For real ω we get by complex conjugation of (4.2.20)

$$\varepsilon_{jk}(\omega)^* = \varepsilon_{jk}(-\omega), \quad \omega \text{ real}, \tag{4.2.21}$$

because $R_{jk}(\tau)$ is real.

We now want to express the causality contained in (4.2.19, 20) by relations involving $\varepsilon(\omega)$, only. For simplicity we omit the indices j, k. We start from Cauchy's theorem for a contour C in the upper half plane $\operatorname{Im} z > 0$ where $\varepsilon(z) - \varepsilon_0$ is regular

$$\varepsilon(z) = \varepsilon_0 + \frac{1}{2\pi i} \int_C \frac{\varepsilon(z') - \varepsilon_0}{z' - z} dz'. \tag{4.2.22}$$

Let us assume that

$$|\varepsilon(z') - \varepsilon_0| \leq \frac{\text{const}}{|z'|} \quad \text{for} \quad |z'| \to \infty, \tag{4.2.23}$$

(for more general conditions see Problem 1). For C we may then take the real axis, closed by a large semi-circle in the upper half-plane. By (4.2.23) the integral over the large semi-circle vanishes and the integral over the real axis converges for $|z'| \to \infty$. Hence,

$$\varepsilon(z) = \varepsilon_0 + \frac{1}{2\pi i} \int_{-\infty}^{+\infty} \frac{\varepsilon(\omega') - \varepsilon_0}{\omega' - z} d\omega'. \tag{4.2.24}$$

Here we consider the limit where $z = \omega + i\eta$ approaches the real axis $\eta \to 0$:

$$\varepsilon(\omega) = \varepsilon_0 + \frac{1}{2\pi i} \lim_{\eta \to 0} \int_{-\infty}^{+\infty} \frac{\varepsilon(\omega') - \varepsilon_0}{\omega' - \omega - i\eta} d\omega'. \tag{4.2.25}$$

We use the important distributive equation

$$\frac{1}{x - i0} = \mathrm{P}\frac{1}{x} + i\pi\delta(x), \tag{4.2.26}$$

where P means principle value integral. For the sake of completeness we give a short derivation of this relation: for any test function φ with support in the interval $[-a, +a]$ we have

$$\lim_{\varepsilon\to 0}\int_{-a}^{a}\frac{\varphi(x)-\varphi(0)+\varphi(0)}{x-i\varepsilon}\,dx = \lim_{\varepsilon\to 0}\int_{-a}^{a}\frac{\varphi(x)-\varphi(0)}{x-i\varepsilon}\,dx +$$

$$+\varphi(0)\lim_{\varepsilon\to 0}\int_{-a}^{a}\frac{dx}{x-i\varepsilon}.$$

Since the last integral tends to $\log(-1) = i\pi$ in the limit $\varepsilon \to 0$, we get

$$= \lim_{\delta\to 0}\left[\int_{-a}^{-\delta}+\int_{\delta}^{a}\right]\left(\frac{\varphi(x)}{x}-\frac{\varphi(0)}{x}\right)dx + i\pi\varphi(0) =$$

$$= \mathrm{P}\int_{-a}^{a}\frac{\varphi(x)}{x}\,dx + i\pi\varphi(0).$$

Using (4.2.26) in (4.2.25), we arrive at

$$\varepsilon(\omega) = \varepsilon_0 + \frac{1}{2\pi i}\,\mathrm{P}\int_{-\infty}^{+\infty}\frac{\varepsilon(\omega')-\varepsilon_0}{\omega'-\omega}\,d\omega' + \tfrac{1}{2}(\varepsilon(\omega)-\varepsilon_0),$$

or

$$\varepsilon(\omega) = \varepsilon_0 + \frac{1}{\pi i}\,\mathrm{P}\int_{-\infty}^{+\infty}\frac{\varepsilon(\omega')-\varepsilon_0}{\omega'-\omega}\,d\omega'. \tag{4.2.27}$$

We decompose (4.2.27) into real and imaginary parts:

$$\mathrm{Re}\,\varepsilon(\omega) = \varepsilon_0 + \frac{1}{\pi}\,\mathrm{P}\int_{-\infty}^{+\infty}\frac{\mathrm{Im}\,\varepsilon(\omega')}{\omega'-\omega}\,d\omega' \tag{4.2.28}$$

$$\mathrm{Im}\,\varepsilon(\omega) = -\frac{1}{\pi}\int_{-\infty}^{+\infty}\frac{\mathrm{Re}\,\varepsilon(\omega')-\varepsilon_0}{\omega'-\omega}\,d\omega'. \tag{4.2.29}$$

Such relations between real and imaginary parts as a consequence of causality are called dispersion relations. They have first been discovered in connection with optical dispersion as discussed here, by Kronig (1926) and Kramers (1927). It follows from (4.2.21) that $\mathrm{Re}\,\varepsilon(\omega)$ is even whereas $\mathrm{Im}\,\varepsilon(\omega)$ is odd. The dispersion integrals can then be transformed as follows

$$\mathrm{Re}\,\varepsilon(\omega) = \varepsilon_0 + \frac{1}{\pi}\,\mathrm{P}\int_0^\infty d\omega'\,\mathrm{Im}\,\varepsilon(\omega')\Big(\frac{1}{\omega'-\omega} - \frac{1}{-\omega'-\omega}\Big)$$

$$= \varepsilon_0 + \frac{2}{\pi}\int_0^\infty d\omega'\,\frac{\omega'\mathrm{Im}\,\varepsilon(\omega')}{\omega'^2-\omega^2} \tag{4.2.30}$$

$$\mathrm{Im}\,\varepsilon(\omega) = -\frac{1}{\pi}\,\mathrm{P}\int_0^\infty d\omega'\,(\mathrm{Re}\,\varepsilon(\omega')-\varepsilon_0)\Big(\frac{1}{\omega'-\omega} + \frac{1}{-\omega'-\omega}\Big)$$

$$= -\frac{2\omega}{\pi}\,\mathrm{P}\int_0^\infty d\omega'\,\frac{\mathrm{Re}\,\varepsilon(\omega')-\varepsilon_0}{\omega'^2-\omega^2}. \tag{4.2.31}$$

To make the foregoing discussion more concrete, we present a **simple classical theory of dispersion**. A more refined treatment must be quantum mechanical. We want to calculate $\varepsilon_r(\omega)$ for atoms that are represented by oscillators of harmonically bound charges. The equation of motion of such a charge q in the presence of an electric field is

$$\boldsymbol{x}''(t) + \gamma\boldsymbol{x}'(t) + \omega_0^2\boldsymbol{x}(t) = \frac{q}{m}\boldsymbol{E}(t). \tag{4.2.32}$$

γ is the damping constant of the oscillator. Such a damping must be included here, in order to have dissipation of mechanical energy into heat. The real damping mechanism is the coupling of the electronic motion to lattice vibrations. By Fourier transformation

$$\hat{\boldsymbol{x}}(\omega) = \frac{1}{\sqrt{2\pi}}\int dt\,\boldsymbol{x}(t)e^{i\omega t}$$

we get

$$-\omega^2\hat{\boldsymbol{x}}(\omega) - i\omega\gamma\hat{\boldsymbol{x}}(\omega) + \omega_0^2\hat{\boldsymbol{x}}(\omega) = \frac{q}{m}\hat{\boldsymbol{E}}(\omega),$$

or

$$\hat{\boldsymbol{x}}(\omega) = \frac{q}{m}\frac{\hat{\boldsymbol{E}}(\omega)}{\omega_0^2-\omega^2-i\gamma\omega}. \tag{4.2.33}$$

Assuming N atoms per unit volume, this leads to the following dipole moment or polarization

$$\boldsymbol{P}(\omega) = Nq\boldsymbol{x}(\omega) = \frac{Nq^2}{m}\frac{1}{\omega_0^2-\omega^2-i\gamma\omega}\hat{\boldsymbol{E}}(\omega). \tag{4.2.34}$$

The factor in front of $\boldsymbol{E}$ is essentially the electrical susceptibility $\varepsilon_0\chi_e$ (4.2.6). Hence,

$$\varepsilon_r(\omega) = 1 + \chi_e = 1 + \frac{Nq^2}{\varepsilon_0 m}\frac{1}{\omega_0^2-\omega^2-i\gamma\omega}. \tag{4.2.35}$$

In reality there exist more resonance frequencies represented by different sorts of oscillators. Then we get the more general dispersion formula

$$\varepsilon_r(\omega) = 1 + \frac{Nq^2}{\varepsilon_0 m} \sum_j \frac{f_j}{\omega_j^2 - \omega^2 - i\gamma_j\omega}, \tag{4.2.36}$$

where f_j is the so-called oscillator strength. One easily verifies that this function is regular in the upper half-plane, the poles lie in the lower half-plane in accordance with causality, because $\gamma_j > 0$. Furthermore, condition (4.2.23) is fulfilled, so that the dispersion relations (4.2.28, 29) are valid. The explicit verification is left as Problem 2.

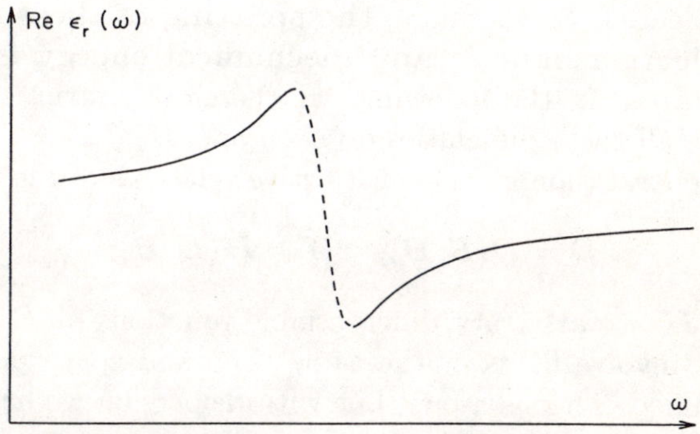

Fig. 13. Anomalous and normal dispersion

In Fig.13 the behavior of $\mathrm{Re}\,\varepsilon_r$ in the vicinity of a resonance $\omega = \omega_j$ is shown. The real part is decreasing near ω_j which is called anomalous dispersion. Away from the resonance, $\mathrm{Re}\,\varepsilon_r$ increases with increasing frequency. This is normal dispersion. Below each resonance frequency ω_j, there is a sharp maximum of $\varepsilon_r(\omega)$. We will discuss in Sect.4.4 some interesting physical consequences of this behavior. It follows from the dispersion relation (4.2.30) that this qualitative behavior is generally valid: **In the neighbourhood of a sharp maximum of** $\mathrm{Im}\,\varepsilon_r$ (absorption line, resonant absorption), on finds **anomalous dispersion**.

4.3 Macroscopic Conservation Laws, Boundary Conditions

We now come to the discussion of the conservation laws in phenomenological electrodynamics. Although energy and momentum conservation are very general principles of physics, it is completely undecided whether the macroscopic Maxwell's equations (4.2.1-4) imply energy and momentum conservation laws in a similar manner as the microscopic ones. This question depends on the constitutive relations. If those relations include dissipation, then Maxwell's equations alone do not yield the conservation laws. Additional field equations for the matter must be used. This makes the problem complicated. Furthermore, **the splitting of the energy**, for example, **into electromagnetic and mechanical energy is** to a certain extent **ambiguous**. In the following, we therefore restrict ourselves to a situation where all these difficulties do not appear.

We allow general (nonlinear) constitutive relations of the form

$$\boldsymbol{D} = \boldsymbol{D}(\boldsymbol{E}, \boldsymbol{B}), \quad \boldsymbol{H} = \boldsymbol{H}(\boldsymbol{E}, \boldsymbol{B}), \tag{4.3.1}$$

where $\boldsymbol{D}$ and $\boldsymbol{H}$ are arbitrary differentiable functions of $\boldsymbol{E}$ and $\boldsymbol{B}$. Since these relations involve fields at the same time and space point only, dispersion is excluded. That is a pity, but with dispersion we have dissipation and the problem is more complicated. In addition, since there is no explicit space dependence in (4.3.1), the medium must be homogeneous.

To derive energy conservation, we multiply (4.2.1) by $\boldsymbol{E}$ and (4.2.2) by $\boldsymbol{H}$ and add the two equations:

$$\boldsymbol{E} \cdot \boldsymbol{D}' + \boldsymbol{H} \cdot \boldsymbol{B}' = \boldsymbol{E} \cdot \operatorname{curl} \boldsymbol{H} - \boldsymbol{H} \cdot \operatorname{curl} \boldsymbol{E} - \boldsymbol{j}_c \cdot \boldsymbol{E}, \tag{4.3.2}$$

where the prime denotes the time derivative. The l.h.s. is the total time derivative of an energy density $u(\boldsymbol{D}, \boldsymbol{B})$, if and only if

$$\frac{\partial u}{\partial \boldsymbol{D}} = \boldsymbol{E}, \quad \frac{\partial u}{\partial \boldsymbol{B}} = \boldsymbol{H}. \tag{4.3.3}$$

This means that u depends on time only through the fields $\boldsymbol{D}$ and $\boldsymbol{B}$ and not explicitly. The system is homogeneous in time. Then (4.3.2) implies **energy conservation**

$$\frac{du}{dt} = \operatorname{div}(\boldsymbol{H} \wedge \boldsymbol{E}) - \boldsymbol{j}_c \cdot \boldsymbol{E}, \tag{4.3.4}$$

with the energy flow density

$$\boldsymbol{S} = \boldsymbol{E} \wedge \boldsymbol{H}. \tag{4.3.5}$$

The last term is Joule's heat

$$\boldsymbol{j}_c \cdot \boldsymbol{E} = E_i \sigma_{ik} E_k, \tag{4.3.6}$$

where σ_{ik} is the conductivity tensor (see (4.2.13)). According to (4.3.3), the energy density $u(\boldsymbol{D}, \boldsymbol{B})$ only exists, if $\boldsymbol{E}$ and $\boldsymbol{H}$ are potential fields with respect to $\boldsymbol{D}$ and $\boldsymbol{B}$, where u is the potential. Consequently, the curl must vanish:

$$\frac{\partial E_i}{\partial D_k} - \frac{\partial E_k}{\partial D_i} = 0, \quad \frac{\partial H_i}{\partial B_k} - \frac{\partial H_k}{\partial B_i} = 0, \quad \frac{\partial E_i}{\partial B_k} - \frac{\partial H_k}{\partial D_i} = 0, \tag{4.3.7}$$

for all $i, k = 1, 2, 3$. These are restrictions on the material equations (4.3.1). For example, for nonisotropic linear relations, like $E_i = \varepsilon_{ik}^{-1} D_k$, it follows that the dielectric tensor must be symmetric and the permeability tensor also. If the integrability conditions (4.3.7) are satisfied, $u(\boldsymbol{D}, \boldsymbol{B})$ can be obtained from (4.3.3) by a path-independent line integral

$$u(\boldsymbol{D}, \boldsymbol{B}) = \int^{(\boldsymbol{D},\boldsymbol{B})} (\boldsymbol{E} \cdot d\boldsymbol{D} + \boldsymbol{H} \cdot d\boldsymbol{B}). \tag{4.3.8}$$

In the static situation, this energy density can be used for a thermodynamic description of the material in electromagnetic fields (see *R.Balian, From Microphysics to Macrophysics, Springer-Verlag 1991, Sect.6.6.5*).

To arrive at **momentum conservation**, we study the time derivative of the Poynting-like vector

$$\begin{aligned} \frac{\partial}{\partial t} \boldsymbol{D} \wedge \boldsymbol{B} &= \boldsymbol{D}' \wedge \boldsymbol{B} + \boldsymbol{D} \wedge \boldsymbol{B}' \\ &= (\operatorname{curl} \boldsymbol{H}) \wedge \boldsymbol{B} - \boldsymbol{D} \wedge \operatorname{curl} \boldsymbol{E} - \boldsymbol{j}_c \wedge \boldsymbol{B}, \end{aligned} \tag{4.3.9}$$

where the dynamical equations (4.2.1, 2) have been used. The i-th component of the first term on the r.h.s. is equal to

$$[(\operatorname{curl} \boldsymbol{H}) \wedge \boldsymbol{B}]_i = \Big(\frac{\partial H_i}{\partial x_n} - \frac{\partial H_n}{\partial x_i}\Big) B_n - \Big(\frac{\partial H_k}{\partial x_i} - \frac{\partial H_i}{\partial x_k}\Big) B_k,$$

where i, k, n is a cyclic permutation of 1,2,3 (no summation over double indices). This is equal to

$$= \sum_{k=1}^{3} \Big(\frac{\partial H_i}{\partial x_k} - \frac{\partial H_k}{\partial x_i}\Big) B_k = \sum_{k=1}^{3} \Big[\frac{\partial}{\partial x_k}(H_i B_k) - H_i \frac{\partial B_k}{\partial x_k} - \frac{\partial}{\partial x_i}(H_k B_k) + H_k \frac{\partial B_k}{\partial x_i}\Big],$$

The second term vanishes by (4.2.4) and in the last one we use (4.3.3)

$$= \frac{\partial}{\partial x_k}(H_i B_k) - \frac{\partial}{\partial x_i}(\boldsymbol{H} \cdot \boldsymbol{B}) + \frac{\partial u}{\partial B_k} \frac{\partial B_k}{\partial x_i}, \tag{4.3.10}$$

assuming summation over double indices. The second term in (4.3.9) is transformed in a similar way

$$[\boldsymbol{D} \wedge \operatorname{curl} \boldsymbol{E}]_i = \frac{\partial}{\partial x_i}(\boldsymbol{D} \cdot \boldsymbol{E}) - \frac{\partial}{\partial x_k}(D_k E_i) - \frac{\partial u}{\partial D_k}\frac{\partial D_k}{\partial x_i} + \rho_c E_i. \quad (4.3.11)$$

Substituting everything into (4.3.9), we end up with

$$\frac{\partial}{\partial t}(\boldsymbol{D} \wedge \boldsymbol{B})_i = \frac{\partial T_{ik}}{\partial x_k} - \rho_c E_i - (\boldsymbol{j}_c \wedge \boldsymbol{B})_i, \quad (4.3.12)$$

where

$$T_{ik} = E_i D_k + H_i B_k - \delta_{ik}(\boldsymbol{E} \cdot \boldsymbol{D} + \boldsymbol{H} \cdot \boldsymbol{B} - u) \quad (4.3.13)$$

is the momentum flow density. We have assumed here that u depends on $\boldsymbol{x}$ only implicitly through the fields $\boldsymbol{D}$ and $\boldsymbol{B}$. The system is homogeneous in space. In vacuum T_{ik} reduces of course to Maxwell's stress tensor (2.6.17). There are reasons to call this conservation law quasimomentum conservation instead of momentum conservation (see *V.L.Gurevich and A.Thellung, Physica A 188, 654 (1992)* and *H.Schoeller and A.Thellung, Annals of Phys. 220, 18 (1992)*). Note also that T_{ik} is not symmetric. For this reason, the derivation of angular momentum conservation is not straightforward (compare (2.6.25-29)).

We now turn to a very important subject, namely **the boundary conditions on discontinuity surfaces between different media**. Let us write Maxwell's equations (4.2.1-4) in integral form, using Stokes' theorem in (4.2.1-2)and Gauss' theorem in (4.2.3-4):

$$\frac{\partial}{\partial t}\int_{K_2} \boldsymbol{D} \cdot d\boldsymbol{\sigma} = \int_{\partial K_2} \boldsymbol{H} \cdot d\boldsymbol{s} - \int_{K_2} \boldsymbol{j}_c \cdot d\boldsymbol{\sigma} \quad (4.3.14)$$

$$\frac{\partial}{\partial t}\int_{K_2} \boldsymbol{B} \cdot d\boldsymbol{\sigma} = -\int_{\partial K_2} \boldsymbol{E} \cdot d\boldsymbol{s} \quad (4.3.15)$$

$$\int_{\partial K_3} \boldsymbol{D} \cdot d\boldsymbol{\sigma} = \int_{K_3} \rho_c \, d^3x \quad (4.3.16)$$

$$\int_{\partial K_3} \boldsymbol{B} \cdot d\boldsymbol{\sigma} = 0. \quad (4.3.17)$$

As in (1.4.35, 39) we choose a small 2-dimensional surface K_2 intersecting the boundary between the two media 1,2. Its 1-dimensional boundary ∂K_2 is called a Stokes contour (Fig.14). We let K_2 shrink to two lines of length d in the media 1 and 2. Since all fields remain bounded in this limit, it follows from (4.3.15) that

$$0 = -d(E_1^t - E_2^t), \quad (4.3.18)$$

hence, **the tangential component E^t of the electric field is continuous.**

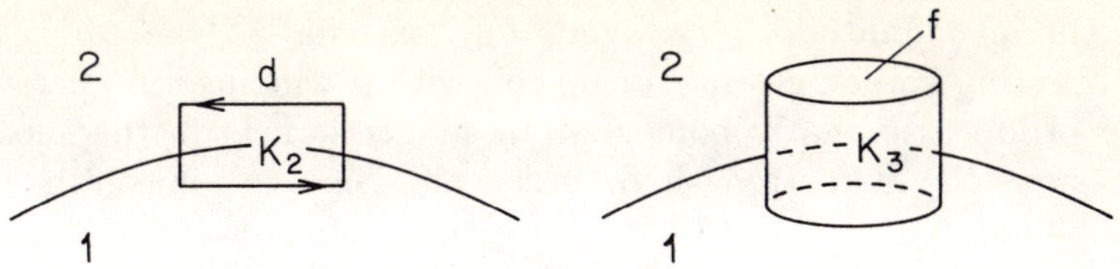

Fig. 14. Stokes' contour and Gauss' cylinder at the boundary between two media 1, 2.

Similarly, in (4.3.17) we consider a Gauss' cylinder K_3 whose height goes to zero, while the upper and lower plane surfaces f remain finite. Then (4.3.17) implies

$$f(B_1^n - B_2^n) = 0, \tag{4.3.19}$$

hence, **the normal component** B^n **of the B-field is continuous**. In the inhomogeneous equation (4.3.16), the r.h.s. goes only to 0, if there is no surface charge density s. Otherwise we have

$$\int\limits_{K_3} \rho_c \, d^3x \to s \cdot f.$$

The corresponding **boundary condition**

$$D_1^n - D_2^n = s = \varepsilon_1 E_1^n - \varepsilon_2 E_2^n \tag{4.3.20}$$

has been obtained already in (1.4.37). To discuss the last Maxwell's equation (4.3.14), we note that the same reasoning with the continuity equation

$$\partial_t \rho_c + \operatorname{div} \boldsymbol{j}_c$$

implies that the normal component j_c^n is continuous. But the tangential components j_c^t may be discontinuous, if there are surface currents

$$\int\limits_{K_2} \boldsymbol{j}_c \cdot d\boldsymbol{\sigma} \to k_\perp \cdot d. \tag{4.3.21}$$

The surface current density $k_\perp$ is parallel to the surface element $d\boldsymbol{\sigma}$ of K_2, thus it is perpendicular to $d\boldsymbol{s}$. The **fourth boundary condition**, following from (4.3.14), reads

$$H_1^t - H_2^t = k_\perp, \tag{4.3.22}$$

where $k_\perp$ is the surface current component perpendicular to H^t.

As a first application of the boundary conditions we want to discuss the **measurement of the magnetization** $M(H_i)$ as a function of the field H_i in the interior of a substance. The best possible arrangement would be to form a ring of length l_i of the material, which is not closed, but has a small slit of magnitude l_e (e stands for "external"). One winds a wire around this ring kernel, getting a ring coil with n windings. A current J in the wire produces a nearly homogeneous magnetic field in the substance. One measures the field strength B_e in the slit. Since B^n is continuous, we know

$$B_e = \mu_0 H_e = B_i, \tag{4.3.23}$$

assuming that we have vacuum in the slit. On the other hand, it follows from (4.3.14) that

$$\int_{\partial K_2} \boldsymbol{H} \cdot d\boldsymbol{s} = H_i l_i + H_e l_e = Jn.$$

This gives us the inner field

$$H_i = \frac{n}{l_i} J - \frac{l_e}{l_i} H_e. \tag{4.3.24}$$

The magnetization M follows from (4.3.23)

$$B_e = B_i = \mu_0 H_i + M = \mu_0 H_e.$$

Substituting H_e into (4.3.24) we shall obtain

$$H_i = \frac{n}{l_i} J - \frac{l_e}{l_i}\Big(H_i + \frac{M}{\mu_0}\Big).$$

This gives the following final result

$$H_i = \frac{nJ}{l} - \frac{l_e}{l}\frac{M}{\mu_0}, \tag{4.3.25}$$

where $l = l_e + l_i$ is the total length of the ring plus slit. The first term herein is the field of the coil alone. It is decreased by "demagnetization" represented by the second term. The demagnetization factor l_e/l depends on the geometry and can only by calculated in a simple way for a few arrangements. Many further applications of the boundary conditions will be discussed in the next chapter on optics.

Finally we want to return to **electro-hydrodynamics** which we have introduced in Sect.3.6 (3.6.49). This theory describes the electromagnetic field in interaction with a fluid of charged particles by means of a classical relativistic field theory. Since the atomistic structure of matter is already smoothed out in this fluid picture, it should be possible to derive the phenomenological Maxwell's equations without any averaging and, furthermore, the corresponding material equations should also come out. This we will now show in an important special case, which leads to the **electrodynamics of superconductors**. We study this field theory here for its principal interest. The complete phenomenological description of superconductors requires more refined treatment, and a relativistic theory is not necessary.

We start from the microscopic Maxwell's equation

$$\partial_\mu F^{\mu\nu} = \mu_0 j^\nu, \tag{4.3.26}$$

where the current density is given by (3.6.46) in terms of a four-velocity

$$j^\nu = c\rho_0 u^\nu, \quad \text{with} \quad u_\nu u^\nu = 1. \tag{4.3.27}$$

ρ_0 is the charge density in the rest frame of the fluid, and it is assumed here to be constant. For u^ν we use the relativistic Euler equation (3.6.49)

$$u_\mu \partial^\mu u^\nu + \frac{e}{mc} u_\mu F^{\mu\nu} = 0. \tag{4.3.28}$$

It can be written in antisymmetric form

$$u_\mu (\partial^\mu u^\nu - \partial^\nu u^\mu + \frac{e}{mc} F^{\mu\nu}) = 0, \tag{4.3.29}$$

because the subtracted term vanishes

$$u_\mu \partial^\nu u^\mu = \tfrac{1}{2} \partial^\nu u^2 = 0$$

due to (4.3.27). The bracket in (4.3.29) is the so-called vortex tensor

$$\Pi^{\mu\nu} = \partial^\mu u^\nu - \partial^\nu u^\mu + \frac{e}{mc} F^{\mu\nu}. \tag{4.3.30}$$

From now on we restrict ourselves to vortex-free motion where $\Pi_{\mu\nu} = 0$, which is certainly a solution of (4.3.29). The resulting equation

$$\partial^\mu u^\nu - \partial^\nu u^\mu + \frac{e}{mc} F^{\mu\nu} = 0 \tag{4.3.31}$$

is linear in u and can therefore be solved.

In considering (4.3.31) together with (4.3.26), we shall always use j^ν instead of u^ν. Multiplying (4.3.31) by mc/e, we get

$$\Lambda(\partial^\mu j^\nu - \partial^\nu j^\mu) + F^{\mu\nu} = 0, \tag{4.3.32}$$

where

$$\Lambda = \frac{m}{e\rho_0} \tag{4.3.33}$$

is called London's constant. Equations (4.3.32) are the **relativistic London's equations** for superconductors of type I. In real experiments with superconductors one needs an external macroscopic current density j^ν_{ext} to generate electromagnetic fields that drive the system. j_{ext} is only different from 0 outside the superconductor. Instead of (4.3.26) we now have

$$\partial_\mu F^{\mu\nu} = \mu_0 (j^\nu + j^\nu_{\text{ext}}). \tag{4.3.34}$$

We differentiate (4.3.32) with respect to x^μ and insert (4.3.26):

$$\Lambda(\partial_\mu \partial^\mu j^\nu - \partial^\nu \partial_\mu j^\mu) + \mu_0 (j^\nu + j^\nu_{\text{ext}}) = 0. \tag{4.3.35}$$

Since the second term vanishes by microscopic current conservation, we shall obtain

$$(\Box + \kappa^2) j^\nu(x) = -\kappa^2 j^\nu_{\text{ext}}(x), \tag{4.3.36}$$

with

$$\kappa^2 = \frac{\mu_0}{\Lambda}. \tag{4.3.37}$$

We discuss this equation with help of the 4-dimensional Fourier transform

$$\hat{f}(k) = (2\pi)^{-2} \int d^4x \, e^{ikx} f(x), \quad \text{where} \quad kx = k^\mu x_\mu. \tag{4.3.38}$$

The wave operator in (4.3.36) is transformed into $-k^2$, thus

$$(-k^2 + \kappa^2) \hat{j}^\nu(k) = -\kappa^2 \hat{j}^\nu_{\text{ext}}(k), \tag{4.3.39}$$

and

$$\hat{j}^\nu(k) = -\frac{\kappa^2}{-k^2 + \kappa^2} \hat{j}^\nu_{\text{ext}}(k). \tag{4.3.40}$$

Now we are able to eliminate the microscopic current density j^ν from Maxwell's equation (4.3.34) after Fourier transformation:

$$-ik_\mu \hat{F}^{\mu\nu} = \mu_0 \frac{k^2}{k^2 - \kappa^2} \hat{j}^\nu_{\text{ext}}(k),$$

or

$$-ik_\mu \left(1 - \frac{\kappa^2}{k^2}\right) \hat{F}^{\mu\nu} = \mu_0 \hat{j}^\nu_{\text{ext}}(k). \tag{4.3.41}$$

In this form only the macroscopic external current density j^ν_{ext} appears as source of the electromagnetic field. This equation must then be considered as the phenomenological inhomogeneous Maxwell's equation. $(1 - \kappa^2/k^2)\hat{F}^{\mu\nu}$ on the l.h.s. involves the phenomenological fields $\boldsymbol{D}$, $\boldsymbol{B}$. Consequently, we may conclude

$$\hat{\boldsymbol{B}}(k) = \mu(k) \hat{\boldsymbol{H}}(k), \quad \text{with} \tag{4.3.42}$$

$$\mu(k) = \frac{\mu_0}{1 - \kappa^2/k^2}. \tag{4.3.43}$$

Since $\mu(k) \to 0$ for $k \to 0$, the material becomes a **perfect diamagnet** in the static and spatially homogeneous situation. The B-field must vanish inside the superconductor, only in a small layer at the surface it is different from zero, as we will shortly see. This is the so-called **Meissner effect**. We are also interested in the conductivity inside the material. From (4.3.40) we find

$$\hat{j}^\nu_{\text{ext}}(k) = -\frac{k^2 - \kappa^2}{\mu_0 k^2} ik_\mu \hat{F}^{\mu\nu}(k). \tag{4.3.44}$$

Inside the superconductor there flows the microscopic current, only, $j^\nu_{\text{ext}}(x)$ vanishes there. Substituting (4.3.44) into (4.3.40) we shall obtain

$$\hat{j}^\nu(k) = \frac{\kappa^2}{\mu_0 k^2} i k_\mu \hat{F}^{\mu\nu}(k). \tag{4.3.45}$$

From this equation we can identify the conductivity. In the homogeneous situation $k = (\omega/c, \mathbf{0})$, we get

$$\hat{\boldsymbol{j}}(\omega) = \frac{ic\kappa^2}{\mu_0 \omega} \hat{\boldsymbol{E}}(\omega). \tag{4.3.46}$$

This is a purely imaginary conductivity

$$\sigma(\omega) = \frac{ic\kappa^2}{\mu_0 \omega}. \tag{4.3.47}$$

There are no losses (dissipation) in the material. For $\omega \to 0$ we have a **perfect conductor**, $\sigma \to \infty$. These are the typical properties of superconductors.

We want to discuss the static situation in more detail. In (4.3.40)

$$\hat{j}^\nu(\boldsymbol{k}) = -\frac{\kappa^2}{\boldsymbol{k}^2 + \kappa^2} \hat{j}^\nu_{\text{ext}}(\boldsymbol{k}) \tag{4.3.48}$$

we perform the inverse Fourier transformation, using the convolution theorem

$$j^\nu(\boldsymbol{x}) = -\kappa^2 \int G(\boldsymbol{x} - \boldsymbol{y}) j^\nu_{\text{ext}}(\boldsymbol{y})\, d^3y. \tag{4.3.49}$$

Here $G(\boldsymbol{x})$ is given by the following 3-dimensional Fourier integral

$$G(\boldsymbol{x}) = (2\pi)^{-3/2} \int d^3k \, \frac{e^{i\boldsymbol{k}\cdot\boldsymbol{x}}}{\boldsymbol{k}^2 + \kappa^2} = \frac{e^{-\kappa|\boldsymbol{x}|}}{4\pi|\boldsymbol{x}|}. \tag{4.3.50}$$

This is a Yukawa potential; its range

$$\frac{1}{\kappa} = \sqrt{\frac{\Lambda}{\mu_0}} = \lambda \tag{4.3.51}$$

is the **London penetration depth**. It follows from (4.3.49) that the microscopic current $j^\nu(\boldsymbol{x})$ and the B-field are only different from zero in a surface layer of extension λ.

Finally we discuss the time-dependent situation. Even without external sources ($j_{\text{ext}} = 0$) non-trivial solutions are now possible. According to (4.3.36) they follow from the Klein-Gordon equation

$$(\Box + \kappa^2) j^\nu = 0. \tag{4.3.52}$$

In the spatially homogeneous case we have simply to consider

$$\frac{1}{c^2}\frac{\partial^2}{\partial t^2}j^\nu = -\kappa^2 j^\nu.$$

The solutions are

$$j^\nu(t) \sim \sin\kappa ct\,,\ \cos\kappa ct. \tag{4.3.53}$$

Here the so-called **plasma frequency**

$$\omega_p = c\kappa = \sqrt{\frac{c^2\mu_0}{\Lambda}} = \sqrt{\frac{e\rho_0}{\varepsilon_0 m}}$$

appears. Expressing the charge density ρ_0 by the particle density n, $\rho_0 = en$, we get

$$\omega_p = \sqrt{\frac{ne^2}{\varepsilon_0 m}}. \tag{4.3.54}$$

The solutions (4.3.53) describe plasma oscillations in any charged fluid. To illustrate this, let us consider a linear arrangement of point charges e with distance d between them. If one charge is displaced by a distance x, the Coulomb forces of the two neighbours no longer cancel, there remains a net restoring force given by

$$\frac{e^2}{4\pi\varepsilon_0}\left(\frac{1}{(d+x)^2} - \frac{1}{(d-x)^2}\right) = \frac{e^2}{4\pi\varepsilon_0 d^2}\left(\frac{1}{1+2x/d} - \frac{1}{1-2x/d}\right).$$

for small $x \ll d$. We set this force equal to mass times acceleration and expand further:

$$-\frac{e^2}{\varepsilon_0 \pi d^3}x = mx''. \tag{4.3.55}$$

This is an harmonic oscillator with characteristic frequency

$$\omega_p = \sqrt{\frac{e^2}{\varepsilon_0 \pi d^3 m}}. \tag{4.3.56}$$

Substituting the particle density

$$n = \frac{1}{\pi d^3},$$

this is indeed identical with (4.3.54). Unfortunately, plasma oscillations cannot be observed in superconductors. If one tries to excite them, one breaks the electron pairs which are essential for superconductivity. But they can be produced in ordinary metals (see (4.4.71)).

4.4 Motion of Particles Through Matter

The matter under consideration will be not magnetic

$$\boldsymbol{H} = \frac{1}{\mu_0}\boldsymbol{B}, \tag{4.4.1}$$

but we allow for dispersion in the D-field according to (4.2.18)

$$\boldsymbol{D}(t, \boldsymbol{x}) = \varepsilon_0 \boldsymbol{E} + \int_0^\infty R(\tau)\boldsymbol{E}(t-\tau, \boldsymbol{x})\, d\tau \stackrel{\text{def}}{=} \varepsilon \boldsymbol{E}, \tag{4.4.2}$$

in order to have energy dissipation (see (4.2.32)). The integral operator ε commutes with all derivatives, for example

$$\partial_t \boldsymbol{D} = \partial_t(\varepsilon \boldsymbol{E}) = \varepsilon \partial_t \boldsymbol{E}. \tag{4.4.3}$$

In the phenomenological Maxwell's equations

$$\frac{\partial}{\partial t}\varepsilon \boldsymbol{E} = \frac{1}{\mu_0}\operatorname{curl} \boldsymbol{B} - \boldsymbol{j}_c \tag{4.4.4}$$

$$\frac{\partial \boldsymbol{B}}{\partial t} = -\operatorname{curl} \boldsymbol{E} \tag{4.4.5}$$

$$\operatorname{div} \varepsilon \boldsymbol{E} = \rho_c \tag{4.4.6}$$

$$\operatorname{div} \boldsymbol{B} = 0 \tag{4.4.7}$$

we consider a point particle with charge q moving with constant velocity $\boldsymbol{v}$ as the source. It gives rise to a charge density

$$\rho_c = q\delta(\boldsymbol{x} - \boldsymbol{v}t) \tag{4.4.8}$$

and current density

$$\boldsymbol{j}_c = q\boldsymbol{v}\delta(\boldsymbol{x} - \boldsymbol{v}t). \tag{4.4.9}$$

As usual it follows from the homogeneous equations (4.4.5, 7) that we can introduce the **potentials**

$$\boldsymbol{B} = \operatorname{curl} \boldsymbol{A} \tag{4.4.10}$$

$$\boldsymbol{E} = -\operatorname{grad} V - \frac{\partial \boldsymbol{A}}{\partial t}. \tag{4.4.11}$$

Substituting into (4.4.6), we get

$$-\varepsilon \bigtriangleup V - \varepsilon \frac{\partial}{\partial t}\operatorname{div} \boldsymbol{A} = \rho_c. \tag{4.4.12}$$

With the definition of ε (4.4.2), this is now an integro-differential equation. We want again to use the **Lorentz gauge**

$$\frac{1}{c^2}\frac{\partial V}{\partial t} + \operatorname{div} \boldsymbol{A} = 0, \tag{4.4.13}$$

where

$$\frac{1}{c^2} \stackrel{\text{def}}{=} \mu_0 \varepsilon \tag{4.4.14}$$

is, after Fourier transformation, the frequency dependent phase velocity. Then we get from (4.4.12)

$$\varepsilon\left(\frac{1}{c^2}\frac{\partial^2 V}{\partial t^2} - \triangle V\right) = q\delta(\boldsymbol{x} - \boldsymbol{v}t). \tag{4.4.15}$$

Similarly, substituting the potentials into (4.4.4), we shall obtain

$$-\varepsilon \operatorname{grad} \partial_t V - \varepsilon \frac{\partial^2 \boldsymbol{A}}{\partial t^2} = \frac{1}{\mu_0}(\operatorname{grad} \operatorname{div} \boldsymbol{A} - \triangle \boldsymbol{A}) - \boldsymbol{j}_c, \tag{4.4.16}$$

and, using the Lorentz condition (4.4.13),

$$\frac{1}{c^2}\frac{\partial^2 \boldsymbol{A}}{\partial t^2} - \triangle \boldsymbol{A} = \mu_0 q \boldsymbol{v} \delta(\boldsymbol{x} - \boldsymbol{v}t). \tag{4.4.17}$$

It is clear that these integro-differential equations must be solved by Fourier transformation. Let us first carry out the Fourier transformation in space, using

$$F_x[\delta(\boldsymbol{x} - \boldsymbol{v}t)] = \frac{1}{(2\pi)^{3/2}} e^{-i\boldsymbol{k}\boldsymbol{v}t}, \tag{4.4.18}$$

then we get

$$\varepsilon\left(\frac{1}{c^2}\frac{\partial^2 \hat{V}}{\partial t^2} + \boldsymbol{k}^2 \hat{V}\right) = \frac{q}{(2\pi)^{3/2}} e^{-i\boldsymbol{k}\boldsymbol{v}t} \tag{4.4.19}$$

$$\frac{1}{c^2}\frac{\partial^2 \hat{\boldsymbol{A}}}{\partial t^2} + \boldsymbol{k}^2 \hat{\boldsymbol{A}} = \mu_0 \frac{q\boldsymbol{v}}{(2\pi)^{3/2}} e^{-i\boldsymbol{k}\boldsymbol{v}t}. \tag{4.4.20}$$

These are inhomogeneous linear equations, so that we may add an arbitrary solution of the homogeneous equations to any solution. This would correspond to the superposition of a free electromagnetic wave in the medium, which we disregard. Let us now perform the temporal Fourier transform

$$\frac{1}{\sqrt{2\pi}} \int_{-\infty}^{+\infty} e^{-i\boldsymbol{k}\boldsymbol{v}t} e^{i\omega t} dt = \sqrt{2\pi}\,\delta(\omega - \boldsymbol{k}\boldsymbol{v}), \tag{4.4.21}$$

understood of course in the distributive sense. Using the abbreviation

$$\boldsymbol{k} \cdot \boldsymbol{v} = \omega_0(\boldsymbol{k}), \tag{4.4.22}$$

the two equations (4.4.19, 20) assume the following form:

$$\varepsilon(\omega)\Big(-\frac{\omega^2}{c^2(\omega)}+\boldsymbol{k}^2\Big)\hat{V}(\omega,\boldsymbol{k})=\frac{q}{2\pi}\delta(\omega-\omega_0) \tag{4.4.23}$$

$$\Big(-\frac{\omega^2}{c^2}+\boldsymbol{k}^2\Big)\hat{\boldsymbol{A}}(\omega,\boldsymbol{k})=\mu_0\frac{q\boldsymbol{v}}{2\pi}\delta(\omega-\omega_0). \tag{4.4.24}$$

This gives the Fourier transformed potentials

$$\hat{V}(\omega,\boldsymbol{k})=\frac{q}{2\pi}\frac{\delta(\omega-\omega_0)}{\varepsilon(\omega)(\boldsymbol{k}^2-\omega^2/c^2)} \tag{4.4.25}$$

$$\hat{\boldsymbol{A}}(\omega,\boldsymbol{k})=\mu_0\frac{q\boldsymbol{v}}{2\pi}\frac{\delta(\omega-\omega_o)}{\boldsymbol{k}^2-\omega^2/c^2}=\frac{\boldsymbol{v}}{c^2}\hat{V}(\omega,\boldsymbol{k}). \tag{4.4.26}$$

From the potentials we get the fields (4.4.11, 12). After Fourier transformation they are given by

$$\hat{\boldsymbol{E}}(\omega,\boldsymbol{k})=-i\boldsymbol{k}\hat{V}+i\omega\hat{\boldsymbol{A}} \tag{4.4.27}$$

$$\hat{\boldsymbol{B}}(\omega,\boldsymbol{k})=i\boldsymbol{k}\wedge\hat{\boldsymbol{A}}. \tag{4.4.28}$$

We first **calculate the electric field**

$$\hat{\boldsymbol{E}}(\omega,\boldsymbol{k})=\frac{q}{2\pi}i\frac{\omega_0/c^2-\boldsymbol{k}}{\varepsilon(\omega)(\boldsymbol{k}^2-\omega_0c^2)}\delta(\omega-\omega_0). \tag{4.4.29}$$

Let us assume the velocity $\boldsymbol{v}$ to be parallel to the x-axis, thus $\omega_0=k_1v$ (4.4.22). We want to compute the field at a point $\boldsymbol{x}_0=(0,a,0)$ on the y-axis. The field has cylindrical symmetry around the x-axis. The inverse Fourier transform in $\boldsymbol{x}$

$$\boldsymbol{E}(\omega,\boldsymbol{x}_0)=\frac{1}{(2\pi)^{3/2}}\int d^3k\,\hat{\boldsymbol{E}}(\omega,\boldsymbol{k})e^{iak_2} \tag{4.4.30}$$

gives

$$\begin{aligned}E_1(\omega,\boldsymbol{x}_0)&=\frac{iq}{(2\pi)^{5/2}}\int d^3k\,\frac{k_1v^2/c^2-k_1}{\varepsilon(\boldsymbol{k}^2-\omega_0^2/c^2)}\,\delta(\omega-k_1v)e^{iak_2}\\&=\frac{iq}{(2\pi)^{5/2}}\frac{\omega}{v^2\varepsilon}\Big(\frac{v^2}{c^2}-1\Big)\int dk_2\,e^{iak_2}\int\frac{dk_3}{k_2^2+k_3^2+\kappa^2},\end{aligned} \tag{4.4.31}$$

where

$$\kappa^2=\frac{\omega^2}{v^2}-\frac{\omega_0^2}{c^2}=\frac{\omega^2}{v^2}\Big(1-\frac{v^2}{c^2}\Big)=\frac{\omega^2}{v^2}\Big(1-\varepsilon_r(\omega)\frac{v^2}{c_0^2}\Big), \tag{4.4.32}$$

and c_0 is the vacuum light velocity. Since the last integral is equal to $\pi/\sqrt{k_2^2+\kappa^2}$, we arrive at

$$E_1(\omega,\boldsymbol{x}_0)=\frac{iq}{2(2\pi)^{3/2}}\frac{\omega}{v^2\varepsilon}\Big(\frac{v^2}{c^2}-1\Big)\int dk_2\,\frac{e^{iak_2}}{\sqrt{k_2^2+\kappa^2}}. \tag{4.4.33}$$

Here the last integral gives the modified Bessel function $2K_0(\kappa a)$, hence

$$E_1(\omega, \boldsymbol{x}_0) = \frac{iq}{(2\pi)^{3/2}} \frac{\omega}{v^2 \varepsilon(\omega)} \Big(\frac{v^2}{c^2(\omega)} - 1\Big) K_0(\kappa a). \tag{4.4.34}$$

The modified Bessel function is exponentially decreasing

$$K_0(\kappa a) \to \sqrt{\frac{\pi}{2\kappa a}} e^{-\kappa a}, \quad a \to \infty, \tag{4.4.35}$$

for real κ. According to (4.4.2),

$$\kappa = \frac{\omega}{v} \sqrt{1 - \varepsilon_r(\omega) \frac{v^2}{c_0^2}} \tag{4.4.36}$$

is real for

$$v < \frac{c_0}{\sqrt{\varepsilon_r(\omega)}} = c, \tag{4.4.37}$$

up to the small imaginary part in ε_r which is negligible. Relativity theory demands $v < c_0$, however,

$$v > \frac{c_0}{\sqrt{\varepsilon_r(\omega)}} \tag{4.4.38}$$

is possible, if $\varepsilon_r(\omega) > 1$. In this case the particle moves more rapidly then the phase velocity of light in the medium, and this really happens. As discussed below, the fields then do not decrease exponentially, but represent radiation that is emitted (Cherenkov radiation).

The y-component of the E-field follows from (4.4.29, 30)

$$E_2(\omega, \boldsymbol{x}_0) = \frac{iq}{(2\pi)^{5/2}} \int d^3k \, \frac{-k_2}{\varepsilon(\boldsymbol{k}^2 - \omega_0^2/c^2)} \frac{i}{v} \delta(k_1 - \frac{\omega}{v}) e^{iak_2}$$

$$= -\frac{iq}{(2\pi)^{5/2}} \frac{1}{\varepsilon v} \int dk_2 \, k_2 e^{iak_2} \int \frac{dk_3}{k_2^2 + k_3^2 + \frac{\omega^2}{v^2} - \frac{\omega_0^2}{c^2}}. \tag{4.4.39}$$

Since the last integral is again equal to $\pi/\sqrt{k_2^2 + \kappa^2}$, we arrive at

$$E_2(\omega, \boldsymbol{x}_0) = -\frac{q}{(2\pi)^{3/2} 2\varepsilon v} \int dk_2 \, \frac{ik_2}{\sqrt{k_2^2 + \kappa^2}} e^{iak_2}. \tag{4.4.40}$$

The Fourier integral occurring here is the derivative of the previous one (4.4.33), it is thus equal to

$$\frac{\partial}{\partial a} 2K_0(\kappa a) = -2\kappa K_1(\kappa a),$$

hence

$$E_2(\omega, \boldsymbol{x}_0) = \frac{q}{(2\pi)^{3/2}} \frac{\kappa}{\varepsilon v} K_1(\kappa a). \tag{4.4.41}$$

The **magnetic field** follows from

$$\hat{\boldsymbol{B}} = i\boldsymbol{k} \wedge \hat{\boldsymbol{A}} = i\boldsymbol{k} \wedge \frac{\boldsymbol{v}}{c^2}\hat{V}. \tag{4.4.42}$$

Thus the 3-component

$$\hat{B}_3(\omega, \boldsymbol{k}) = -ik_2\frac{v}{c^2}\hat{V} = \frac{v}{c^2}\hat{E}_2(\omega, \boldsymbol{k}) \tag{4.4.43}$$

is essentially equal to the 2-component of the electric field:

$$B_3(\omega, \boldsymbol{x}_0) = \frac{v}{c^2}E_2(\omega, \boldsymbol{x}_0) = \frac{q}{(2\pi)^{3/2}}\frac{\kappa}{c^2\varepsilon}K_1(\kappa a). \tag{4.4.44}$$

Since B_1 vanishes, the y-component of the Poynting vector is given by

$$S_2 = -\frac{1}{\mu_0}E_1 B_3. \tag{4.4.45}$$

The quantity of interest is the energy loss per unit time N_a of the particle that is transferred to the medium at distances $> a$. This is equal to the energy flow through the surface of a cylinder of radius a:

$$N_a = \frac{dW_a(t)}{dt} = 2\pi a \int_{-\infty}^{+\infty} S_2\, dx = -\frac{2\pi a}{\mu_0}\int_{-\infty}^{+\infty} E_1 B_3\, dx. \tag{4.4.46}$$

Even more important is the **energy loss per unit length of the particle trajectory**. This is obtained by putting $x = vt$, i.e. by multiplying (4.4.46) by $1/v$

$$\frac{dW_a(x)}{dx} = -\frac{2\pi a}{\mu_0 v}\int_{-\infty}^{+\infty} (E_1 B_3)(t, \boldsymbol{x}')\, dx', \tag{4.4.47}$$

where $\boldsymbol{x}' = (x', a, 0)$. Instead of integrating over x' for fixed t, we set $x' = vt' + x_0$ and integrate over t' for fixed x_0

$$\frac{dW_a(x)}{dx} = -\frac{2\pi a}{\mu_0}\int_{-\infty}^{+\infty} (E_1 B_3)(t', \boldsymbol{x}_0)\, dt'. \tag{4.4.48}$$

Using Parseval's equality of Fourier transformation (3.3.20), this is equal to

$$\frac{dW_a(x)}{dx} = -\frac{2\pi a}{\mu_0}\int_{-\infty}^{+\infty} (E_1 B_3^*)(\omega, \boldsymbol{x}_0)\, d\omega. \tag{4.4.49}$$

To simplify the notation, we use the same symbols for the Fourier transformed quantities but with different arguments. Since $\boldsymbol{E}(t, \boldsymbol{x})$, $\boldsymbol{B}(t, \boldsymbol{x})$ are real, the Fourier transforms fulfill $E_j(-\omega) = E_j(\omega)^*$ etc. Hence,

$$\frac{dW_a(x)}{dx} = -\frac{4\pi a}{\mu_0}\mathrm{Re}\int_0^\infty (E_1 B_3^*)(\omega, \boldsymbol{x}_0)\,d\omega$$

$$= -\frac{4\pi a}{\mu_0}\frac{q^2}{(2\pi)^3}\frac{1}{v^2}\mathrm{Re}\int_0^\infty i\frac{\omega}{\varepsilon c^2}\Big(\frac{v^2}{c^2}-1\Big)\frac{\kappa^*}{\varepsilon}K_0(\kappa a)K_1(\kappa^* a)\,d\omega. \quad (4.4.50)$$

Since $c^2\varepsilon = c_0^2\varepsilon_0$, we have

$$\frac{v^2}{c^2} = \frac{v^2}{c_0^2}\varepsilon_r \stackrel{\text{def}}{=} \beta^2\varepsilon_r, \quad (4.4.51)$$

thus

$$\frac{dW_a(x)}{dx} = \frac{1}{4\pi\varepsilon_0}\frac{2}{\pi}\frac{aq^2}{v^2}\mathrm{Re}\int_0^\infty i\omega\kappa^*\Big(\frac{1}{\varepsilon_r}-\beta^2\Big)K_0(\kappa a)K_1(\kappa^* a)\,d\omega. \quad (4.4.52)$$

This **formula** was first derived **by Fermi** in 1940.

First we want to discuss particles with small velocities $\beta \ll 1$. Then

$$\kappa = \frac{\omega}{v}\sqrt{1-\varepsilon_r\beta^2} \quad (4.4.53)$$

is real, up to a small negligible imaginary part, and the main contribution to (4.4.52) comes from $\mathrm{Im}\,1/\varepsilon_r$. We use the dispersion formula (4.2.36)

$$\varepsilon_r = 1 + \frac{Ne^2}{\varepsilon_0 m}\sum_j \frac{f_j}{\omega_j^2-\omega^2-i\gamma_j\omega}. \quad (4.4.54)$$

Sometimes we will neglect the sum, if it is small compared to 1 for the interesting frequencies, for example

$$\mathrm{Im}\,\frac{1}{\varepsilon_r} = -\frac{\mathrm{Im}\,\varepsilon_r}{|\varepsilon_r|^2} \approx -\mathrm{Im}\,\varepsilon_r = -\frac{Ne^2}{\varepsilon_0 m}\sum_j f_j\frac{\gamma_j\omega}{(\omega_j^2-\omega^2)^2+\gamma_j^2\omega^2}. \quad (4.4.55)$$

Using

$$\kappa \approx \frac{\omega}{v}\sqrt{1-\beta^2} \stackrel{\text{def}}{=} \frac{\omega}{v\gamma}, \quad \text{where} \quad \gamma = \frac{1}{\sqrt{1-v^2/c_0^2}}, \quad (4.4.56)$$

we shall obtain

$$\frac{dW_a(x)}{dx} = \frac{1}{4\pi\varepsilon_0}\frac{2}{\pi}\frac{q^2}{v^2}\frac{Ne^2}{\varepsilon_0 m}\sum_j f_j\int_0^\infty d\omega\,\frac{\omega a}{v\gamma}K_0\Big(\frac{\omega a}{v\gamma}\Big)$$

$$\times K_1\Big(\frac{\omega a}{v\gamma}\Big)\frac{\gamma_j\omega^2}{(\omega_j^2-\omega^2)^2+\gamma_j^2\omega^2}. \quad (4.4.57)$$

For small γ_j, the last fraction is peaked at ω_j, hence

$$\frac{dW_a(x)}{dx} \approx \frac{1}{(4\pi\varepsilon_0)^2}\frac{8q^2}{v^2}\frac{Ne^2}{m}\sum_j f_j\xi_j K_0(\xi_j)K_1(\xi_j)$$

$$\times \int_0^\infty d\omega \, \frac{\gamma_j\omega^2}{(\omega_j^2-\omega^2)^2+\gamma_j^2\omega^2}, \tag{4.4.58}$$

where we have introduced the dimensionless quantities

$$\xi_j = \frac{\omega_j a}{v\gamma}. \tag{4.4.59}$$

Since the last integral is equal to $\pi/2$, we arrive at

$$\frac{dW_a(x)}{dx} = \frac{1}{(4\pi\varepsilon_0)^2}\frac{q^2}{v^2}\frac{4\pi Ne^2}{m}\sum_j f_j\xi_j K_0(\xi_j)K_1(\xi_j). \tag{4.4.60}$$

The resonance frequencies ω_j are mainly in the optical region and in the infrared. To control the total energy loss, we choose a as small as possible, say of the order of the atomic size. Then the quantities (4.4.59) are small compared to 1. We use the following asymptotic expressions for the modified Bessel functions

$$K_0(\xi) \to \log\frac{2}{\xi} - 0.5772\ldots = \log\frac{1.123}{\xi}$$

$$K_1(\xi) \to \frac{1}{\xi}, \quad \text{for} \quad \xi \to 0, \tag{4.4.61}$$

where $\gamma = 0.5772\ldots$ is Euler's constant. The sum in (4.4.60) is now equal to

$$\sum_j f_j \log\frac{1.123}{\xi_j} = \sum f_j \log\frac{1.123\,v\gamma}{\omega_j a} \stackrel{\text{def}}{=} Z\log\frac{1.123\,v\gamma}{\langle\omega\rangle a}. \tag{4.4.62}$$

Here we have introduced the mean value

$$\sum_j f_j \log\omega_j \stackrel{\text{def}}{=} Z\log\langle\omega\rangle, \tag{4.4.63}$$

where Z is the number of electrons per atom of the medium. In fact, the oscillator strengths obey the following sum rule

$$\sum_j f_j = Z, \tag{4.4.64}$$

which follows from quantum mechanics. Inserting all this into (4.4.60), we get the final formula for the **energy loss of slow particles**

$$\frac{dW_a(x)}{dx} = \frac{1}{(4\pi\varepsilon_0)^2} 4\pi NZ \frac{e^2 q^2}{mv^2} \log \frac{1.123 v\gamma}{\langle\omega\rangle a}. \tag{4.4.65}$$

This shows a rather complicated dependence on the particle velocity v and the electronic structure of the medium through $\langle\omega\rangle$.

Next we turn to rapidly moving, **relativistic particles**. In the the limit $v \to c < c_0$ we set

$$\kappa^2 = \frac{\omega^2}{v^2}(1 - \varepsilon_r \beta^2) \approx \frac{\omega^2}{c^2}(1 - \varepsilon_r(\omega)). \tag{4.4.66}$$

For optical frequencies ω and a of the order of the atomic size, we have $\kappa a \ll 1$. Using the above asymptotic expressions (4.4.61) for the modified Bessel functions in the Fermi formula (4.4.52), we shall obtain

$$\frac{dW_a(x)}{dx} = \frac{1}{4\pi\varepsilon_0} \frac{2}{\pi} \frac{q^2}{c^2} \mathrm{Re} \int_0^\infty d\omega\, i\omega \left(\frac{1}{\varepsilon_r(\omega)} - 1 \right) \log \frac{1.123}{\kappa(\omega) a}. \tag{4.4.67}$$

Since $\varepsilon_r(\omega)$ is regular in the upper half plane $\mathrm{Im}\,\omega > 0$ due to causality (4.2.20), we may change the path of integration and integrate along the imaginary axis $\omega = i\omega'$ and along a large quarter circle. On the imaginary axis we have

$$\varepsilon_r(i\omega') = 1 + \frac{Ne^2}{\varepsilon_0 m} \sum_j \frac{f_j}{\omega_j^2 + \omega'^2 + \gamma_j \omega'}, \tag{4.4.68}$$

according to (4.4.54). This is real and $1 - \varepsilon_r(i\omega') < 0$, thus $\kappa^2 > 0$ (4.4.66) and κ is also real. Consequently, the integral from 0 to $i\infty$ in (4.4.67) becomes purely imaginary and the real part vanishes.

There remains the integral over the large quarter circle

$$\omega = \omega' e^{i\varphi}, \quad \varphi = \frac{\pi}{2} \ldots 0, \quad \omega' \to \infty, \tag{4.4.69}$$

only. Here, in the limit $\omega' \to \infty$, we find

$$\varepsilon_r(\omega) \to 1 - \frac{Ne^2}{\varepsilon_0 m} \frac{1}{\omega'^2 \exp(2i\varphi)} \cdot Z, \tag{4.4.70}$$

where (4.4.64) has been used, and

$$\kappa^2 \to \frac{NZe^2}{\varepsilon_0 mc^2} = \frac{\omega_p^2}{c^2}. \tag{4.4.71}$$

The plasma frequency (4.3.54) appears here. Obviously, the incident particle excites plasma oscillations in the medium and this is the way to observe them. For the reciprocal dielectric constant we get on the large quarter circle

$$\frac{1}{\varepsilon_r} \to 1 + \frac{NZe^2}{\varepsilon_0 m\omega'^2 \exp(2i\varphi)} = 1 + \frac{\omega_p^2}{\omega'^2} e^{-2i\varphi}. \tag{4.4.72}$$

Using

$$d\omega = i\omega' e^{i\varphi} d\varphi$$

in the integral (4.4.67), we observe that the φ-dependence and ω' drop out completely, so that the integral and the limit $\omega' \to \infty$ are trivial

$$\frac{dW_a(x)}{dx} = \frac{1}{4\pi\varepsilon_0}\frac{q^2}{c^2}\omega_p^2 \log \frac{1.123c}{\omega_p a}. \tag{4.4.73}$$

The **energy loss per unit length depends only on the plasma frequency of the medium**, i.e. on its electron density $\omega_p \sim \sqrt{NZ}$. No details of the electronic structure, as $\langle\omega\rangle$ (4.4.63) for example, enter. The electronic structure is unimportant if the moving particle has very high energy. In both cases (4.4.65, 73) the result depends on the distance a from the particle trajectory. That means that the energy is deposited in the neighborhood of the trajectory. There is no emission of radiation. For $a \to 0$ we observe a logarithmic divergence. It signals that in the forward direction this classical calculation breaks down. Here one must use quantum mechanical scattering theory.

We now turn to **Cherenkov radiation**. We have discussed in Sect.4.2 (Fig.13, (4.2.36)) that $\varepsilon_r(\omega)$ has sharp maxima just below the resonance frequencies ω_j. The small imaginary part of ε_r will be neglected in the following. It gives rise to an additional small absorption that is of no interest here. It is possible that $\varepsilon_r(\omega)$ becomes so big that

$$v > c = \frac{c_0}{\sqrt{\varepsilon_r(\omega)}}, \tag{4.4.74}$$

or

$$\varepsilon_r \beta^2 > 1, \quad \beta = \frac{v}{c_0}. \tag{4.4.75}$$

Then

$$\kappa = \frac{\omega}{v}\sqrt{1 - \varepsilon_r \beta^2} = \pm i|\kappa| \tag{4.4.76}$$

becomes imaginary. The sign herein must be chosen to be $\kappa = -i|\kappa|$, the plus sign is unphysical, as can be seen in the final result (4.4.78). The distance a from the trajectory we choose now so big that $|\kappa a| \gg 1$. Then we can use the limiting expressions for the modified Bessel functions

$$K_{0,1}(\kappa a) \to \sqrt{\frac{\pi}{2\kappa a}} e^{-i\kappa a}, \quad |\kappa a| \to \infty. \tag{4.4.77}$$

These functions now oscillate in contrast to the exponential decrease (4.4.35) found previously. This signals the Cherenkov radiation. The two oscillating factors cancel in the Fermi formula (4.4.52)

$$\frac{dW_a(x)}{dx} = \frac{1}{4\pi\varepsilon_0}\frac{2}{\pi}\frac{aq^2}{v^2}\mathrm{Re} \int\limits_{\varepsilon_r(\omega>\beta^{-2}} d\omega\,(-\omega)|\kappa|\Big(\frac{1}{\varepsilon_r(\omega)} - \beta^2\Big)\frac{\pi}{2|\kappa|a}$$

$$= \frac{1}{4\pi\varepsilon_0}\frac{q^2}{c_0^2} \int\limits_{\varepsilon_r>\beta^{-2}} d\omega\,\omega\Big(1 - \frac{1}{\beta^2\varepsilon_r(\omega)}\Big). \tag{4.4.78}$$

Also the distance a has dropped out, so that the energy is really transported to infinity. That is Cherenkov radiation. According to (4.4.78) the emission takes place in the small bands below the resonance frequencies where $\varepsilon_r(\omega) > 1/\beta^2$. With the plus sign in (4.4.76), the resulting energy loss would come out negative, i.e. an energy gain, which is impossible.

It is interesting to determine the direction of the radiation. Since the radiation is transverse in the wave zone and the particle moves in the 1-direction, the radiation is emitted under an angle ϑ, given by

$$\tan\vartheta = -\frac{E_1}{E_2} \tag{4.4.79}$$

Using our previous results (4.4.34, 41) with $\kappa = -i|\kappa|$ and $K_0 \approx K_1$ in the asymptotic region, we immediately get

$$\tan\vartheta = \frac{\omega}{|\kappa|v}\Big(\frac{v^2}{c^2} - 1\Big) = \sqrt{\varepsilon_r\beta^2 - 1}. \tag{4.4.80}$$

This leads to

$$\cos\vartheta = \frac{1}{\sqrt{1+\tan^2\vartheta}} = \frac{1}{\beta\sqrt{\varepsilon_r}} = \frac{c}{v} = \frac{c_0}{v\sqrt{\varepsilon_r}} < 1. \tag{4.4.81}$$

There exists a simple explanation of this result: The radiation generated by the particle propagates with velocity c in all directions. The particle moves with velocity $v > c$ along the hypotenuse of an right-angled triangle (Fig.15). In the direction ϑ the radiation can constructively interfere. It forms a wave front along **Mach's cone**. The same phenomenon occurs in supersonic flow. For $v < c$ no interference is possible.

Cherenkov radiation is a common phenomenon in connection with high energy particle beams. For illustration we make some rough estimates. The refractive index of air is

$$n = \sqrt{\varepsilon_r} \approx 1.0003,$$

and it increases with frequency (normal dispersion). To get Cherenkov radiation, we must have

$$v > \frac{c_0}{1.0003} \approx c_0(1 - 0.0003). \tag{4.4.82}$$

Let us imagine a beam of high energy protons with energy

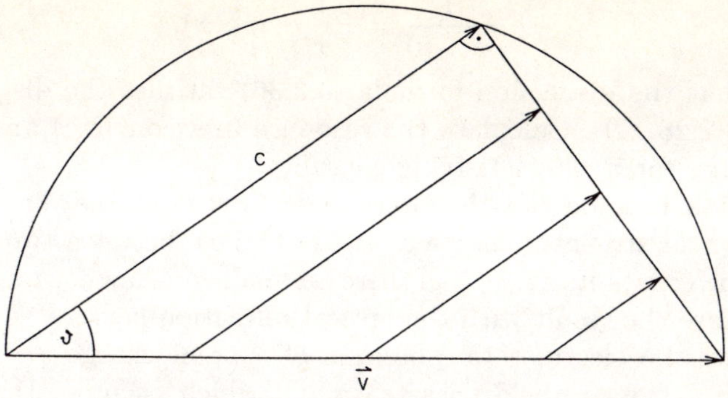

Fig. 15. Mach's cone for Cherenkov radiation

$$E = \frac{Mc_0^2}{\sqrt{1 - v^2/c_0^2}}, \quad Mc_0^2 = 1\,\mathrm{GeV}.$$

To satisfy (4.4.82) we must have

$$\frac{v}{c_0} = \sqrt{1 - \frac{1}{E}} \approx 1 - \frac{1}{2E^2(\mathrm{GeV})^2} > 1 - 3 \cdot 10^{-4}$$

that means

$$E > \frac{100}{\sqrt{6}}\,\mathrm{GeV} = 41\,\mathrm{GeV}.$$

Such a beam is indeed visible in air. But only a short part can be seen because the Cherenkov radiation is emitted under the Mach angle (4.4.81). Since ϑ increases with frequency ω, the color of the radiation changes along the beam in a typical manner from yellow to blue.

4.5 Problems

1. Dispersion relations with subtractions: Let $f(\omega)$ be an analytic function, regular in the upper half-plane $\mathrm{Im}\,\omega > 0$, and satisfying

$$|f(\omega)| \le C|\omega|^n \quad \mathrm{for} |\omega| \to \infty. \tag{4.5.1}$$

   Derive a dispersion relation for $f(\omega)$ by starting from a dispersion relation for the function

$$\frac{f(\omega)}{(\omega - \omega_0 + i\varepsilon)^{n+1}}, \quad \omega_0 \quad \text{real}, \quad \varepsilon > 0. \tag{4.5.2}$$

   Take the distributive limit $\varepsilon \to 0$ using the n-th distributive derivative of the relation

$$\frac{1}{x+i0} = \mathrm{P}\frac{1}{x} - i\pi\delta(x).$$

2. Show that the dispersion formula (4.2.36) satisfies the dispersion relations (4.2.28, 29). Calculate the response function $R(\tau)$ and show that it vanishes for $\tau < 0$, satisfying causality.
3. Refraction law for electric force lines: How does the direction of the electric field strength change at the boundary between two media with dielectric constants ε_1, ε_2 ? Is there an analogous law for magnetic fields ? Compare the result with the optical refraction law.
4. Calculate the electrostatic potential of a point charge q in vacuum at distance d from a dielectrics ($\varepsilon_r \neq 1$), which occupies the half space $x_1 < 0$. Discuss the limit $\varepsilon_r \to \infty$. Hint: Write the potential for $x_1 > 0$ and $x_1 < 0$ by means of mirror charges, determine their magnitude from the boundary conditions at $x_1 = 0$.
5. Prove the distributive relation

$$\triangle \log |\boldsymbol{x}| = 2\pi\delta(\boldsymbol{x}), \quad \boldsymbol{x} \in \mathbb{R}^2 \tag{4.5.3}$$

by using Green's theorem.
6. A homogeneously charged, infinitely long, rectilinear wire in vacuum has a distance d from a dielectric half space with dielectric constant ε. Calculate the electric field. How big is the force per unit length on the wire ?
7. A sphere with dielectric constant ε is placed in vacuum in an electric field that is homogeneous ($= E_\infty$) at large distances. Expand the potential inside and outside of the sphere in terms of Legendre polynomials. Determine the coefficients from the boundary conditions at the surface. Discuss the electric field inside and outside of the sphere and calculate the polarization.
8. Dielectric ellipsoid: An ellipsoid with dielectric constant ε is placed in vacuum in an asymptotically homogeneous electric field ($= E_\infty$) in the 1-direction.
   a) Write the homogeneous field in ellipsoidal coordinates u_1, u_2, u_3 introduced in Problem 4 of Chap.1.
   b) In the Laplace equation in ellipsoidal coordinates use the following ansatz

$$V(u_1, u_2, u_3) = -E_i x_1 f(u_1) \tag{4.5.4}$$

   for the outer potential ($u_1 > 0$). Determine $f(u_1)$ and the constant E_i in terms of E_∞.
   c) Show that this outer solution matches to the homogeneous inner field $V_i(\boldsymbol{x}) = -E_i x_1$ and determine the last constant of integration in the solution.
9. Dispersive delay of pulsar signals: A pulsar is a rotating neutron star which emits radio frequency pulses as a lighthouse. Because of dispersion in interstellar space, the pulses from different frequencies arrive at

delayed times. Determine the time delay for two frequencies ω_1 and ω_2 assuming the simple dispersion formula

$$\varepsilon_r(\omega) = 1 - \frac{\omega_p^2}{\omega^2}, \tag{4.5.5}$$

where ω_p is the plasma frequency which is assumed to be constant. Why is (4.5.5) reasonable? This method is used to estimate the distance of pulsars.

5. Optics

What is light? Since Maxwell and Hertz the answer seems to be simple: Light is electromagnetic radiation with wave length between 400 and 800 nanometers. This is just one "octave", so that the eye is much less sensitive in frequency than the ear which is able to distinguish almost eight octaves. However, the above answer is only true for "ideal" light. By that we mean a plane wave solution (4.2.10)

$$\boldsymbol{E}(t,\boldsymbol{x}) = \boldsymbol{f}(n\boldsymbol{e}\cdot\boldsymbol{x} - c_0 t) \tag{5.1}$$

of Maxwell's equations. Here $\boldsymbol{E}$ is the electric field strength, $\boldsymbol{e}$ is the propagation direction, $|\boldsymbol{e}| = 1$, and n is the refractive index (4.2.9)

$$n = \sqrt{\varepsilon_r \mu_r} \approx \sqrt{\varepsilon_r}. \tag{5.2}$$

In real light the function $\boldsymbol{f}$ changes discontinuously, because this light is emitted by many atoms which radiate independently. The length l over which $\boldsymbol{f}$ is continuous is called coherence length. The coherence length of thermal light sources amounts to only a few wave lengths λ. Better are lasers with $l \approx 1000\,\lambda$, because here many atoms are radiating coherently. But this is still not ideal light. Nevertheless, the classical optical phenomena discussed in the following sections are practically the same for real and ideal light, because they refer to macroscopic observations which average out the microscopic irregularities.

5.1 Reflection and Refraction

We consider a plane interface $z = 0$ between two media 1 and 2 with refractive indices n_1 and n_2. Absorption is neglected. The incident wave in medium 1 is assumed to be linearly polarized

$$\boldsymbol{E} = (0, E_y, 0) \tag{5.1.1}$$

where

$$E_y = f(n_1 \boldsymbol{e}\cdot\boldsymbol{x} - c_0 t). \tag{5.1.2}$$

The wave is transversal, consequently, the propagation direction $\boldsymbol{e}$ must be in the xz-plane (Fig.16)

$$e = (\sin\alpha, 0, -\cos\alpha), \tag{5.1.3}$$

so that

$$E_y = f(n_1(x\sin\alpha - z\cos\alpha) - c_0 t). \tag{5.1.4}$$

The B-field follows from the induction law

$$\partial_t \boldsymbol{B} = -\mathrm{curl}\,\boldsymbol{E} = \Big(\frac{\partial E_y}{\partial z}, 0, -\frac{\partial E_y}{\partial x}\Big) = \mu_0 \partial_t \boldsymbol{H}. \tag{5.1.5}$$

The B-field is also in the xz-plane, perpendicular to $\mathbf{e}$. The direction of $\boldsymbol{B}$ is called the direction of polarization for reasons which will become clear below. The xz-plane is the plane of incidence, so that we are considering the case of polarization in the plane of incidence.

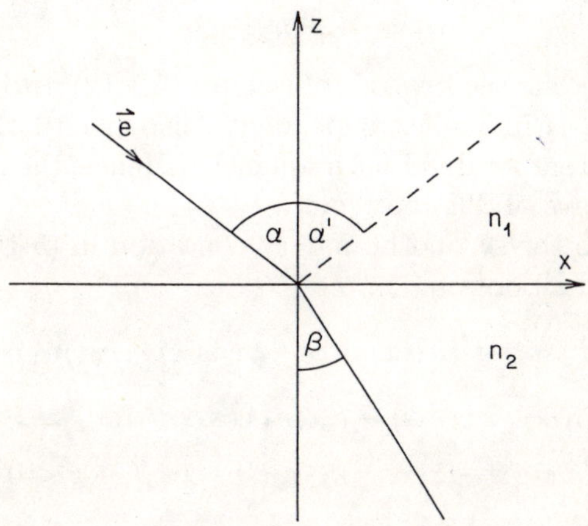

Fig. 16. Reflection and Refraction

At the interface $z = 0$ the boundary conditions (4.3.18) and (4.3.22) must be satisfied, i.e. the tangential components

$$E_t = E_y, \quad H_t = H_x \quad \text{must be continuous.} \tag{5.1.6}$$

We have assumed insulators here, so that no surface current is possible, metals will be considered later. In addition, there are two further boundary conditions (4.3.19, 20), saying that the normal components B_n and D_n are also continuous. It is not clear how so many conditions can actually be satisfied at a simple interface. It could be that this is impossible because the simple assumption of discontinuous jumps in ε and μ is wrong. We therefore make a fairly general ansatz assuming a reflected wave g in medium 1 and a refracted wave f_2 in medium 2:

$$E_y = f(n_1(x\sin\alpha - z\cos\alpha) - c_0 t) + g(n_1(x\sin\alpha' + z\cos\alpha') - c_0 t) \tag{5.1.7}$$

$$\text{if} \quad z > 0,$$

$$= f_2(n_2(x\sin\beta - z\cos\beta) - c_0 t), \quad \text{if} \quad z < 0. \tag{5.1.8}$$

Continuity at $z = 0$ demands

$$f(n_1 x \sin\alpha - c_0 t) + g(n_1 x \sin\alpha' - c_0 t) = f_2(n_2 x \sin\beta - c_0 t) \tag{5.1.9}$$

for all x, t and all f. Since everything depends linearly of f, we conclude

$$g = rf, \quad f_2 = d \cdot f, \tag{5.1.10}$$

where r is called the reflection coefficient and

$$1 + r = d, \tag{5.1.11}$$

furthermore

$$\alpha' = \alpha \tag{5.1.12}$$

$$n_1 \sin\alpha = n_2 \sin\beta. \tag{5.1.13}$$

We have derived the simple **law of reflection** (5.1.12) **and the refraction law** (5.1.13). If n_2 is frequency dependent (dispersion), then the angle of refraction β is different for different frequencies. Hence, the light is spectrally decomposed into rays of different color.

We now turn to the second boundary condition in (5.1.6). From (5.1.5) and (5.1.7, 8) we shall obtain

$$\mu_0 \partial_t H_x = -f'(n_1(x\sin\alpha - z\cos\alpha) - c_0 t) n_1 \cos\alpha$$

$$+ g'(n_1(x\sin\alpha + z\cos\alpha) - c_0 t) n_1 \cos\alpha, \quad \text{for} \quad z > 0, \tag{5.1.14}$$

$$= -f_2'(n_2(x\sin\beta - z\cos\beta) - c_0 t) n_2 \cos\beta, \quad z < 0. \tag{5.1.15}$$

By integrating in t we find

$$\mu_0 H_x = \frac{n_1}{c_0}\cos\alpha\, f - \frac{n_1}{c_0}\cos\alpha\, g, \quad \text{if} \quad z > 0, \tag{5.1.16}$$

$$= \frac{n_2}{c_0}\cos\beta\, f_2, \quad \text{if} \quad z < 0. \tag{5.1.17}$$

Continuity at $z = 0$ together with (5.1.10, 11) implies

$$n_1 \cos\alpha\,(1 - r) = n_2 \cos\beta\, d = (1 + r) n_2 \cos\beta,$$

or

$$\frac{1+r}{1-r} = \frac{n_1\cos\alpha}{n_2\cos\beta} = \frac{\sin\beta\cos\alpha}{\sin\alpha\cos\beta}, \tag{5.1.18}$$

and

$$r = \frac{\sin\beta\cos\alpha - \cos\beta\sin\alpha}{\sin\beta\cos\alpha + \cos\beta\sin\alpha} = \frac{\sin(\beta - \alpha)}{\sin(\beta + \alpha)}. \tag{5.1.19}$$

This is the **first formula of Fresnel, expressing the reflection coefficient by the angles of incidence and refraction**. The above ansatz (5.1.7, 8) is now completely fixed, but we have still to verify two further boundary conditions. $D_n = D_z = 0$ is trivially continuous, and it is easy to show that $B_n = B_z$ is also continuous (Problem 1). In case of flat incidence $\alpha \to \pi/2$, we have $\sin(\alpha \pm \beta) \to \mp \cos\beta$, and, hence, $|r| \to 1$ (assuming $n_1 < n_2$, so that β is real). That means almost complete reflection. This is the reason why one sees nice mirror images at the border of a water.

Let us now consider the polarization perpendicular to the plane of incidence. Compared with the foregoing case, we must interchange $\boldsymbol{E}$ and $\boldsymbol{B}$. Instead of (5.1.7, 8) we now assume

$$B_y = f(n_1(x\sin\alpha - z\cos\alpha) - c_0 t) + g(n_1(x\sin\alpha' + z\cos\alpha') - c_0 t), \quad z > 0, \tag{5.1.20}$$

$$= f_2(n_2(x\sin\beta - z\cos\beta) - c_0 t), \quad \text{if} \quad z < 0. \tag{5.1.21}$$

Again the continuity of $H_t = H_y$ implies

$$g = rf, \quad f_2 = df, \quad 1 + r = d, \tag{5.1.22}$$

and the reflection and refraction laws (5.1.12, 13) follow. The E-field is obtained from Ampère's law

$$\partial_t \boldsymbol{D} = \varepsilon \partial_t \boldsymbol{E} = n^2 \varepsilon_0 \partial_t \boldsymbol{E} = \operatorname{curl} \boldsymbol{H} = \mu_0 \operatorname{curl} \boldsymbol{B}. \tag{5.1.23}$$

Compared with (5.1.16, 17), the signs come out differently

$$E_x = -\frac{\cos\alpha}{n_1 c_0} f + \frac{\cos\alpha}{n_1 c_0} g, \quad z > 0, \tag{5.1.24}$$

$$= -\frac{\cos\beta}{n_2 c_0} f_2, \quad \text{if} \quad z < 0. \tag{5.1.25}$$

Continuity at $z = 0$ implies

$$-\frac{\cos\alpha}{n_1}(1 - r) = -\frac{\cos\beta}{n_2} d = -\frac{\cos\beta}{n_1}(1 + r),$$

or

$$\frac{1+r}{1-r} = \frac{n_1\cos\alpha}{n_1\cos\beta} = \frac{\sin\alpha\cos\alpha}{\sin\beta\cos\beta} \tag{5.1.26}$$

and

$$\begin{aligned} r_\perp &= \frac{\sin\alpha\cos\alpha - \sin\beta\cos\beta}{\sin\alpha\cos\alpha + \sin\beta\cos\beta} = \frac{\sin 2\alpha - \sin 2\beta}{\sin 2\alpha + \sin 2\beta} \\ &= \frac{2\sin(\alpha - \beta)\cos(\alpha + \beta)}{2\sin(\alpha + \beta)\cos(\alpha - \beta)} = \frac{\tan(\alpha - \beta)}{\tan(\alpha + \beta)}. \end{aligned} \tag{5.1.27}$$

This is the **second formula of Fresnel for the reflection coefficient for polarization perpendicular to the plane of incidence**.

This second formula has an interesting application. For $\alpha + \beta = \pi/2$ we have $r_\perp = 0$. For this special angle, called **Brewster angle**, there is no reflected wave of perpendicular polarization. Consequently, if an unpolarized wave is coming in at the Brewster angle, the reflected wave is completely linearly polarized with $\boldsymbol{B}$ in the plane of incidence. This is the reason why the direction of polarization is defined as the direction of $\boldsymbol{B}$. There is a simple microscopic explanation of this phenomenon (Fig.17). The incident wave with $\boldsymbol{E}$ in the plane of incidence excites the microscopic dipoles in the medium, which reemit the radiation. As we have found in Chap.3 (3.4.57), these dipoles cannot radiate in their direction of oscillation. Consequently, at the Brewster angle, no reflected wave is possible, if the E-field oscillates in the plane of incidence.

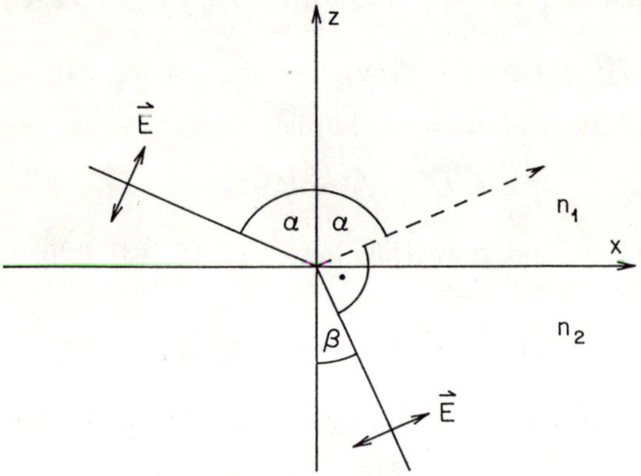

Fig. 17. Linear polarization in reflection at the Brewster angle

Let us now assume that $n_1 > n_2$, that means the radiation enters from the medium with higher index of refraction. From the refraction law (5.1.13)

$$\frac{n_1}{n_2} \sin\alpha = \sin\beta \tag{5.1.28}$$

we notice that for α sufficiently big (near $\pi/2$) the l.h.s. becomes bigger than 1. Therefore, we do not have a real solution of Maxwell's equations. We emphasize again that only real solutions of Maxwell's equations have a physical interpretation. What shall we do? The way out here is very simple: We take the real part of the complex solution. This is obviously a real solution and in addition, it satisfies the boundary conditions. However, to continue to complex arguments, we now assume monochromatic light. This was not necessary before, if dispersion is neglected.

For polarization in the plane of incidence, $\boldsymbol{E} = (0, E_y, 0)$, we make the ansatz

$$E_y = \operatorname{Re} A \exp i\omega\Big[\frac{n_1}{c_0}(x\sin\alpha - z\cos\alpha) - t\Big]$$

$$+ r \exp i\omega\Big[\frac{n_1}{c_0}(x\sin\alpha + z\cos\alpha) - t\Big], \quad \text{if} \quad z > 0 \tag{5.1.29}$$

$$= \operatorname{Re} Ad \exp i\omega\Big[\frac{n_2}{c_0}(x\sin\beta - z\cos\beta) - t\Big], \quad \text{if} \quad z < 0, \tag{5.1.30}$$

instead of (5.1.7, 8). As above the boundary conditions imply (5.1.28) and (5.1.19)

$$r_{\|} = \frac{\sin\beta\cos\alpha - \cos\beta\sin\alpha}{\sin\beta\cos\alpha + \cos\beta\sin\alpha}. \tag{5.1.31}$$

For $\sin\beta > 1$, β becomes complex

$$\beta = \frac{\pi}{2} - i\vartheta, \tag{5.1.32}$$

thus

$$\sin\beta = \frac{1}{2i}\left(e^{i\beta} - e^{-i\beta}\right) = \frac{1}{2}\left(e^{\vartheta} + e^{-\vartheta}\right) = \cosh\vartheta > 1 \tag{5.1.33}$$

$$\cos\beta = \frac{1}{2}\left(e^{i\beta} + e^{-i\beta}\right) = \frac{i}{2}\left(e^{\vartheta} - e^{-\vartheta}\right) = i\sinh\vartheta. \tag{5.1.34}$$

Substituting this into (5.1.31), we find

$$r_{\|} = \frac{\cosh\vartheta\cos\alpha - i\sinh\vartheta\sin\alpha}{\cosh\vartheta\cos\alpha + i\sinh\vartheta\sin\alpha} = e^{i\delta}. \tag{5.1.35}$$

This is a complex number of absolute value $|r_{\|}| = 1$. Consequently, **the entire intensity is reflected**, only with a phase shift δ. This is called **total internal reflection**. One easily observes this perfect reflection when diving under water or if one is looking to the surface of the water from below in an aquarium. Although there is no refracted wave, a second wave must be somewhere, in order to fulfill all boundary conditions. Indeed, we have $d \neq 0$ in (5.1.30) for $z < 0$ and the wave in medium 2 is given by

$$E_y = \mathrm{Re}\, Ad\exp\left[i\omega\left(\frac{n_2}{c_0}x\cosh\vartheta - t\right)\right]\exp\left(\omega\frac{n_2}{c_0}z\sinh\vartheta\right). \tag{5.1.36}$$

Without loss of generality we may assume that A is real, thus

$$E_y = Ad\cos\left[\omega\left(\frac{n_2}{c_0}x\cosh\vartheta - t\right)\right]\exp\left(\omega\frac{n_2}{c_0}z\sinh\vartheta\right). \tag{5.1.37}$$

Due to the cos-factor, this is a wave propagating in the x-direction, but it is damped in the z-direction. This is a transversally damped wave. It penetrates only a few wavelengths $\sim c_0/\omega = \lambda/2\pi$ into medium 2. For polarization perpendicular to the plane of incidence, one obtains in the same manner

$$r_{\perp} = \frac{\sin\alpha\cos\alpha - \sin\beta\cos\beta}{\sin\alpha\cos\alpha + \sin\beta\cos\beta} = \frac{\sin\alpha\cos\alpha - i\sinh\vartheta\cosh\vartheta}{\sin\alpha\cos\alpha + i\sinh\vartheta\cosh\vartheta}$$

$$= \exp i\delta_{\perp}, \tag{5.1.38}$$

with similar consequences as above. Total reflection has important applications nowadays for the development of light conductors in the field of opto-electronics. It prevents losses at the boundaries of a glass fibre where the propagating laser light is totally reflected.

What we have discussed until now applies to insulators, not to metals. The reason is that we have used the boundary condition (4.3.22) that H_t is continuous. This is only true if there are no currents flowing. We **now consider metals** and allow a macroscopic current density $\boldsymbol{j}_m$ according to Ohm's law (4.2.11)

$$\boldsymbol{j}_m(t,\boldsymbol{x}) = \sigma \boldsymbol{E}(t,\boldsymbol{x}). \tag{5.1.39}$$

Then the phenomenological Maxwell's equation

$$\partial_t \boldsymbol{D} = \operatorname{curl} \boldsymbol{H} - \boldsymbol{j}_m \tag{5.1.40}$$

assumes the following form

$$\varepsilon \partial_t \boldsymbol{E} + \sigma \boldsymbol{E} = \operatorname{curl} \boldsymbol{H}. \tag{5.1.41}$$

We again assume monochromatic radiation and first consider a complex solution

$$\boldsymbol{E} \sim e^{-i\omega t}, \tag{5.1.42}$$

taking the real part afterwards. Since

$$\partial_t \boldsymbol{E} = -i\omega \boldsymbol{E}, \quad \boldsymbol{E} = \frac{i}{\omega}\partial_t \boldsymbol{E}, \tag{5.1.43}$$

equation (5.1.41) can be written in the same form as for insulators

$$\left(\varepsilon + i\frac{\sigma}{\omega}\right)\partial_t \boldsymbol{E} = \operatorname{curl} \boldsymbol{H}. \tag{5.1.44}$$

Only the dielectric constant has become complex

$$\tilde{\varepsilon} = \varepsilon + i\frac{\sigma}{\omega}. \tag{5.1.45}$$

Then, all previous results can be taken over, namely the reflection and refraction laws, as well as Fresnel's formulae. The only modification is that we have to take real parts as above for total reflection.

From the complex dielectric constant $\tilde{\varepsilon}$ we get a complex refractive index

$$\tilde{n} = \sqrt{\tilde{\varepsilon}} \overset{\text{def}}{=} n(1+i\kappa), \tag{5.1.46}$$

where

$$n = \sqrt{\frac{1}{2}\left(\sqrt{\varepsilon^2 + \sigma^2/\omega^2} + \varepsilon\right)}, \quad \kappa = \frac{\sigma}{2\omega n^2} = \frac{\sigma}{\omega\left(\sqrt{\varepsilon^2 + \sigma^2/\omega^2} + \varepsilon\right)}. \tag{5.1.47}$$

This follows by decomposing the square root (5.1.46) into real and imaginary parts, where σ is assumed to be real. For a perfect conductor $\sigma \to \infty$ we get

$n \to \infty$ and $\kappa \to 1$. Let us now examine the reflection coefficient (5.1.19) for polarization parallel to the plane of incidence

$$r_{\parallel} = \frac{\sin(\beta - \alpha)}{\sin(\alpha + \beta)}. \tag{5.1.48}$$

We restrict ourselves to the worst case of small α and β, i.e. almost perpendicular incidence, where insulators show the lowest reflectivity. Then we get

$$r_{\parallel} \approx \frac{\beta - \alpha}{\beta + \alpha} = \frac{1 - \alpha/\beta}{1 + \alpha/\beta} = -r_{\perp}, \tag{5.1.49}$$

and for perpendicular polarization (5.1.27) we have the negative result. Here we use the approximate refraction law

$$\frac{\alpha}{\beta} \approx n(1 + i\kappa). \tag{5.1.50}$$

Then we arrive at

$$r_{\parallel} = \frac{1 - n - in\kappa}{1 + n + in\kappa} = -r_{\perp}. \tag{5.1.51}$$

Consequently, the reflectivity

$$|r_{\parallel}|^2 = |r_{\perp}|^2 = \frac{(n-1)^2 + n^2\kappa^2}{(n+1)^2 + n^2\kappa^2} = 1 - O\Big(\frac{1}{n}\Big) \tag{5.1.52}$$

is very big for good conductors ($n \gg 1$) for all optical frequencies. For this reason we make our mirrors with silver.

There arises the question, what happens with the refracted wave. Assuming the polarization in the plane of incidence, we obtain this wave for $z < 0$ from (5.1.30)

$$E_y = \operatorname{Re} Ad \exp i\omega\Big[\frac{\tilde{n}}{c_0}(x \sin\beta - z\cos\beta) - t\Big]. \tag{5.1.53}$$

Here we have to use the (complex) refraction law

$$\tilde{n} \sin\beta = n_1 \sin\alpha. \tag{5.1.54}$$

Since the r.h.s. is real and has absolute value < 1, we conclude

$$\tilde{n}\cos\beta = \tilde{n}\sqrt{1 - \sin^2\beta} = \sqrt{\tilde{n}^2 - n_1^2 \sin^2\alpha}. \tag{5.1.55}$$

For large n the second term under the square root is negligible, thus

$$\tilde{n}\cos\beta \approx n(1 + i\kappa). \tag{5.1.56}$$

Substituting (5.1.54, 56) into (5.1.53), we arrive at

$$E_y = \operatorname{Re} Ad \exp\Big[i\omega\Big(\frac{n_1}{c_0}x\sin\alpha - \frac{n}{c_0}z - t\Big)\Big]\exp\Big(\omega\frac{n}{c_0}\kappa z\Big). \tag{5.1.57}$$

The x-dependent term in the first exponent can be dropped compared to the second term with the large factor n. We see that there is indeed a wave in the z-direction that penetrates into the metal. However, this wave is exponentially damped by the last exponential factor. It decreases to zero after a few wavelengths. In contrast to total internal reflection by insulators, where the refracted wave is transversally damped (5.1.37), we have longitudinal damping in the propagation direction here. The missing energy is converted into Joule's heat (Problem 3).

5.2 Light Scattering

The most common optical observations, namely that the sky is blue and that the sun goes down red, are most complicated to understand. In fact, the famous poet Goethe was completely misled when he took these observations as the basis of his "Farbenlehre". The reason why the simple facts are so difficult to explain is that for light scattering the wave nature of radiation is essential. But even with this knowledge, the problem remains mysterious: How can light be scattered in the atmosphere, if the gas molecules are only 10^{-7} cm small and the wavelength of light is a thousand times bigger? Obviously, scattering does not take place at individual molecules. There exist inhomogeneities on a larger length scale due to density fluctuations in the air. Those are responsible for light scattering. So it was not before the beginning of this century, when finally Einstein cleared up the subject completely.

We start again from the phenomenological Maxwell's equations for insulators

$$\partial_t \boldsymbol{D} = \operatorname{curl} \boldsymbol{H} \tag{5.2.1}$$

$$\partial_t \boldsymbol{B} = -\operatorname{curl} \boldsymbol{E}, \tag{5.2.2}$$

which are not magnetic: $\boldsymbol{B} = \mu_0 \boldsymbol{H}$. We consider monochromatic light

$$\boldsymbol{D}, \boldsymbol{H} \sim e^{-i\omega t}. \tag{5.2.3}$$

It is again convenient to calculate with complex quantities and to take real or imaginary parts at the end. Then the time dependence in (5.2.1, 2) drops out

$$-i\omega \boldsymbol{D} = \operatorname{curl} \boldsymbol{H} \tag{5.2.4}$$

$$\mu_0 i\omega \boldsymbol{H} = \operatorname{curl} \boldsymbol{E}. \tag{5.2.5}$$

Here we eliminate the H-field

$$\operatorname{curl}\operatorname{curl} \boldsymbol{E} = \mu_0 i\omega \operatorname{curl} \boldsymbol{H} = \mu_0 \omega^2 \boldsymbol{D}. \tag{5.2.6}$$

In addition, we need some constitutive equation

$$D_j = \varepsilon_{jk} E_k, \tag{5.2.7}$$

allowing a non-isotropic dielectric constant. In reality the **dielectric constant** of a gas or liquid is not constant, but **shows temporal and spatial variations**. The latter are due to small inhomogeneities, we therefore write

$$\varepsilon_{jk} = \varepsilon(\omega)\delta_{jk} + \eta_{jk}(\omega, \boldsymbol{x}), \tag{5.2.8}$$

where $\eta_{jk} \ll \varepsilon$.

We look for a scattering solution of (5.2.6) in the following form

$$\boldsymbol{D} = \boldsymbol{D}^{\text{in}} + \boldsymbol{D}' \tag{5.2.9}$$

$$\boldsymbol{E} = \boldsymbol{E}^{\text{in}} + \boldsymbol{E}'. \tag{5.2.10}$$

Here $\boldsymbol{D}^{\text{in}}$ and $\boldsymbol{E}^{\text{in}}$ represent the undisturbed incoming wave which satisfies the equation

$$\boldsymbol{D}^{\text{in}} = \varepsilon(\omega)\boldsymbol{E}^{\text{in}}. \tag{5.2.11}$$

The primed quantities represent the scattered wave satisfying

$$\operatorname{curl}\operatorname{curl}\boldsymbol{E}' = \mu_0\omega^2\boldsymbol{D}', \tag{5.2.12}$$

as a consequence of (5.2.6). Substituting (5.2.9, 10) into (5.2.7) and taking (5.2.11) into account, we find

$$D'_j = \varepsilon E'_j + \eta_{jk}E^{\text{in}}_k + \eta_{jk}E'_k. \tag{5.2.13}$$

The last term is a product of two small quantities. Calculating in first order of perturbation theory, this term will be neglected, thus

$$E'_j = \frac{1}{\varepsilon}D'_j - \frac{1}{\varepsilon}\eta_{jk}E^{\text{in}}_k. \tag{5.2.14}$$

This enables us to eliminate $\boldsymbol{E}'$ in (5.2.12)

$$\operatorname{curl}\operatorname{curl}\boldsymbol{E}' = \frac{1}{\varepsilon}\Big[\operatorname{curl}\operatorname{curl}\boldsymbol{D}' - \operatorname{curl}\operatorname{curl}(\eta\boldsymbol{E}^{\text{in}})\Big]. \tag{5.2.15}$$

Here we use

$$\operatorname{curl}\operatorname{curl}\boldsymbol{D}' = \operatorname{grad}\operatorname{div}\boldsymbol{D}' - \triangle\boldsymbol{D}' \tag{5.2.16}$$

where

$$\operatorname{div}\boldsymbol{D} = 0 = \operatorname{div}\boldsymbol{D}_{\text{in}}, \tag{5.2.17}$$

because we are considering insulators, so that the conduction charge density vanishes. Substituting into (5.2.12), we shall obtain

$$-\triangle\boldsymbol{D}' = \mu_0\omega^2\varepsilon\boldsymbol{D}' + \operatorname{curl}\operatorname{curl}(\eta\boldsymbol{E}^{\text{in}}), \tag{5.2.18}$$

or

$$\triangle\boldsymbol{D}' + \boldsymbol{k}^2\boldsymbol{D}' = -\operatorname{curl}\operatorname{curl}(\eta\boldsymbol{E}^{\text{in}}), \tag{5.2.19}$$

where

$$|\boldsymbol{k}| = \sqrt{\mu_0 \varepsilon}\omega = \frac{\omega}{c}. \tag{5.2.20}$$

The direction of the wave vector $\boldsymbol{k}$ will be specified later. This equation (5.2.19) is the so-called Helmholtz equation. The incoming wave $\boldsymbol{E}^{\text{in}}$ on the r.h.s. is considered to be given. We therefore deal with the **inhomogeneous Helmholtz equation.**

Similarly to the treatment of the wave equation in Chap.3.3, we have to calculate the **Green's function** $G(\boldsymbol{x})$. It is defined as the distributive solution of the equation with a δ-source

$$\Delta G + \boldsymbol{k}^2 G(\boldsymbol{x}) = \delta(\boldsymbol{x}). \tag{5.2.21}$$

We perform spatial Fourier transformation

$$(-\boldsymbol{p}^2 + \boldsymbol{k}^2)\hat{G}(\boldsymbol{p}) = (2\pi)^{-3/2}. \tag{5.2.22}$$

A formal solution of this equation is

$$\hat{G}_{\text{formal}}(\boldsymbol{p}) = (2\pi)^{-3/2} \frac{1}{\boldsymbol{k}^2 - \boldsymbol{p}^2}. \tag{5.2.23}$$

But this is not a distribution, because it is not integrable in the neighbourhood of $\boldsymbol{p} = \boldsymbol{k}$. A well-defined distributive solution of (5.2.22) is given by

$$\hat{G}(\boldsymbol{p}) = (2\pi)^{-3/2} \frac{1}{\boldsymbol{k}^2 - \boldsymbol{p}^2 + i0}. \tag{5.2.24}$$

This distribution is defined as follows (see (4.2.26)): For any test function $\varphi \in S(\mathbb{R}^3)$ we have

$$\langle \hat{G}, \varphi \rangle = (2\pi)^{-3/2} \lim_{\varepsilon \to 0} \int d^3p \, \frac{\varphi(\boldsymbol{p})}{\boldsymbol{k}^2 - \boldsymbol{p}^2 + i\varepsilon}. \tag{5.2.25}$$

By inverse Fourier transformation we get the desired Green's function

$$G(\boldsymbol{x}) = \text{w-} \lim_{\varepsilon \to 0} \frac{1}{(2\pi)^3} \int d^3p \, \frac{e^{i\boldsymbol{p}\boldsymbol{x}}}{\boldsymbol{k}^2 - \boldsymbol{p}^2 + i\varepsilon}. \tag{5.2.26}$$

We calculate the integral (5.2.26) in spherical coordinates taking the polar axis parallel to $\boldsymbol{x}$:

$$\begin{aligned} G(\boldsymbol{x}) &= \text{w-} \lim_{\varepsilon \to 0} \frac{1}{(2\pi)^2} \int_0^\infty dp \, \frac{p^2}{k^2 - p^2 + i\varepsilon} \int_{-1}^{+1} d\cos\vartheta \, \exp(ipx\cos\vartheta) \\ &= \text{w-} \lim_{\varepsilon \to 0} \frac{1}{(2\pi)^2} \frac{1}{2ix} \int_{-\infty}^{+\infty} dp \, \frac{p}{k^2 - p^2 + i\varepsilon} (e^{ipx} - e^{-ipx}). \end{aligned} \tag{5.2.27}$$

These integrals can be easily computed by contour integration: In the integral with $\exp ipx$ one closes the contour in the upper half-plane. The result comes from the residue of the pole at $p = k + i\varepsilon'$, where $\varepsilon' \approx \varepsilon/2k$, and similarly for the other integral. Taking the limit $\varepsilon \to 0$, we finally get

$$G(\boldsymbol{x}) = \frac{1}{(2\pi)^2}\frac{2\pi i}{2ix}\frac{k}{(-2k)}e^{ikx}\cdot 2$$

$$= -\frac{1}{4\pi}\frac{e^{ik|\boldsymbol{x}|}}{|\boldsymbol{x}|}. \tag{5.2.28}$$

For $k = 0$ this agrees, of course, with the Green's function of the Laplace equation (0.2.11) and for $k = i\kappa$ it becomes the Yukawa potential (4.3.50).

The Green's function $G(\boldsymbol{x})$ together with the time-dependent factor (5.2.3)

$$e^{-i\omega t}\frac{e^{ik|\boldsymbol{x}|}}{|\boldsymbol{x}|}, \quad k = |\boldsymbol{k}|, \tag{5.2.29}$$

represent an outgoing spherical wave. Physically speaking, the solution of the inhomogeneous Helmholtz equation (5.2.19) expresses the scattering wave $\boldsymbol{D}'$ in terms of spherical waves. $\boldsymbol{D}'(\boldsymbol{x})$ is given by the convolution

$$\boldsymbol{D}'(\boldsymbol{x}) = \frac{1}{4\pi}\int \operatorname{curl}\operatorname{curl}_y(\eta\boldsymbol{E}^{\text{in}})\frac{e^{ik|\boldsymbol{x}-\boldsymbol{y}|}}{|\boldsymbol{x}-\boldsymbol{y}|}\,d^3y. \tag{5.2.30}$$

The integral goes only over the small region of the spatial inhomogeneity, but the scattering wave is observed at large distances $|\boldsymbol{x}| \gg |\boldsymbol{y}|$. We therefore use the usual expansion

$$|\boldsymbol{x}-\boldsymbol{y}| = r - \frac{\boldsymbol{x}\cdot\boldsymbol{y}}{r}\Big(1 + O\Big(\frac{y}{r}\Big)\Big), \quad r = |\boldsymbol{x}|, \tag{5.2.31}$$

and introduce the wave vector

$$\boldsymbol{k} = k\,\frac{\boldsymbol{x}}{r} \tag{5.2.32}$$

in the direction where the scattered light is observed. The **scattered wave** is then given by

$$\boldsymbol{D}'(\boldsymbol{x}) = \frac{e^{ikr}}{4\pi r}\int \operatorname{curl}\operatorname{curl}(\eta\boldsymbol{E}^{\text{in}})e^{-i\boldsymbol{k}\cdot\boldsymbol{y}}\,d^3y. \tag{5.2.33}$$

It **is proportional to the Fourier transform of the inhomogeneity** that produces the scattering. This is typical for single scattering, where the incident wave is scattered only once.

Let us compute the integrand

$$\Big[\operatorname{curl}\operatorname{curl}(\eta\boldsymbol{E}^{\text{in}})\Big]\,e^{-i\boldsymbol{k}\cdot\boldsymbol{y}} = \operatorname{curl}\Big[e^{-i\boldsymbol{k}\cdot\boldsymbol{y}}\operatorname{curl}(\eta\boldsymbol{E}^{\text{in}})\Big]$$

$$+\,i\boldsymbol{k}e^{-i\boldsymbol{k}\cdot\boldsymbol{y}}\wedge\operatorname{curl}(\eta\boldsymbol{E}^{\text{in}}).$$

The first term gives rise to a surface integral in (5.2.33), that vanishes. The second term can be written as follows

$$= \ldots + i\boldsymbol{k} \wedge \left[\operatorname{curl}\left(e^{-i\boldsymbol{k}\cdot\boldsymbol{y}}(\eta\boldsymbol{E}^{\text{in}})\right) + i\boldsymbol{k} \wedge (\eta\boldsymbol{E}^{\text{in}})e^{-i\boldsymbol{k}\cdot\boldsymbol{y}}\right]. \tag{5.2.34}$$

Again the first term gives a surface integral, so that we arrive at

$$\boldsymbol{D}'(\boldsymbol{x}) = -\frac{e^{ikr}}{4\pi r}\boldsymbol{k} \wedge (\boldsymbol{k} \wedge \boldsymbol{F}), \tag{5.2.35}$$

where

$$\boldsymbol{F}(\boldsymbol{k}) = \int (\eta\boldsymbol{E}^{\text{in}})(\boldsymbol{y})\, e^{-i\boldsymbol{k}\cdot\boldsymbol{y}}\, d^3y \tag{5.2.36}$$

is the above mentioned Fourier transform of the scattering inhomogeneity.

For the electric field we find

$$\boldsymbol{E}'(\boldsymbol{x}) = \frac{1}{\varepsilon}\boldsymbol{D}'(\boldsymbol{x}) = -\frac{e^{ikr}}{4\pi\varepsilon r}\,\boldsymbol{k} \wedge (\boldsymbol{k} \wedge \boldsymbol{F}(\boldsymbol{k})). \tag{5.2.37}$$

According to (5.2.20) we have

$$k = \frac{\omega}{c_0}\sqrt{\varepsilon_r}, \tag{5.2.38}$$

so that the electric field of the scattering light increases as ω^2. This is the so-called Rayleigh scattering. The magnetic field is given by

$$\boldsymbol{H}' = -\frac{i}{\mu_0\omega}\operatorname{curl}\boldsymbol{E}' = \frac{1}{\mu_0\omega}\boldsymbol{k} \wedge \boldsymbol{E}'. \tag{5.2.39}$$

For the physical discussion we have still to take the reals parts of the scattering solutions. We are especially interested in the energy flow (4.3.5)

$$\boldsymbol{S}'(t, \boldsymbol{x}) = \operatorname{Re}\boldsymbol{E}'(t, \boldsymbol{x}) \wedge \operatorname{Re}\boldsymbol{H}'(t, \boldsymbol{x}). \tag{5.2.40}$$

The t-dependence herein is given by the exponential factor (5.2.3) $\exp(-i\omega t)$. Expressing the real parts by means of the complex conjugated fields, we shall obtain

$$\begin{aligned} \boldsymbol{S}'(t, \boldsymbol{x}) &= \frac{1}{4}\,(\boldsymbol{E}' + \boldsymbol{E}'^*) \wedge (\boldsymbol{H}' + \boldsymbol{H}'^*) \\ &= \frac{1}{4}\,(\boldsymbol{E}' \wedge \boldsymbol{H}' + \boldsymbol{E}'^* \wedge \boldsymbol{H}'^* + \boldsymbol{E}' \wedge \boldsymbol{H}'^* + \boldsymbol{E}'^* \wedge \boldsymbol{H}'). \end{aligned} \tag{5.2.41}$$

The time dependence drops out in the last two terms, while periodic oscillation $\sim \exp(\pm 2i\omega t)$ remains in the first two terms. These oscillating terms give no contribution in the temporal average. The mean energy flow density is therefore given by

$$\begin{aligned} \boldsymbol{S}'(\boldsymbol{x}) &= \frac{1}{2}\operatorname{Re}(\boldsymbol{E}' \wedge \boldsymbol{H}'^*) = \frac{1}{2\mu_0\omega}\operatorname{Re}\boldsymbol{E}' \wedge (\boldsymbol{k} \wedge \boldsymbol{E}'^*) \\ &= \frac{1}{2\mu_0\omega}\,|\boldsymbol{E}'|^2\boldsymbol{k}. \end{aligned} \tag{5.2.42}$$

Hence, the scattered **intensity is proportional to the absolute square of the complex electric field strength in the temporal mean.**

The discussion of light scattering is now easy. We assume that a linearly polarized plane wave

$$\boldsymbol{E}^{\text{in}} = \boldsymbol{E}_0 e^{i\boldsymbol{k}_{\text{in}}\cdot\boldsymbol{x}} \tag{5.2.43}$$

is coming in. It gives rise to the following scattering inhomogeneity (5.2.36)

$$F_j(\boldsymbol{k}) = E_{0l}\int d^3y\,\eta_{jl}(\omega,\boldsymbol{y})e^{-i(\boldsymbol{k}-\boldsymbol{k}_{\text{in}})\boldsymbol{y}}. \tag{5.2.44}$$

Using (5.2.37), the intensity of the scattered light is proportional to

$$|\boldsymbol{E}'(\boldsymbol{x})|^2 = \frac{k^4}{16\pi^2\varepsilon^2}\frac{|\boldsymbol{F}|^2}{r^2}\sin^2\vartheta, \tag{5.2.45}$$

in the temporal mean. The angle ϑ is measured between $\boldsymbol{F}$ and the direction of observation $\boldsymbol{k} = k\boldsymbol{x}/r$. The result is still to be averaged over the microscopic (density) fluctuations in the gas. We denote this averaging by an overline

$$\overline{|\boldsymbol{F}(\boldsymbol{k})|^2_{lm}} = E^*_{0l}E_{0m}\int d^3y_1\int d^3y_2\,\overline{\eta_{jl}(\boldsymbol{y}_1)^*\eta_{jm}(\boldsymbol{y}_2)}\,e^{i\boldsymbol{q}(\boldsymbol{y}_1-\boldsymbol{y}_2)}. \tag{5.2.46}$$

Here $\boldsymbol{q} = \boldsymbol{k} - \boldsymbol{k}_{\text{in}}$ is the so-called momentum transfer in the scattering process. Since the length scale of density fluctuations is small compared with the wave length of optical light $\lambda \sim 1/|\boldsymbol{k}|$, the exponential in (5.2.46) can be approximated by 1. (This is not true at the critical point of the gas, see below.) Furthermore, the density fluctuations in a gas at rest are isotropic, so that

$$\eta_{jl}(\boldsymbol{y}) = \delta_{jl}\eta(\boldsymbol{y}). \tag{5.2.47}$$

Then the integral in (5.2.46) simplifies to

$$\int d^3y_1\int d^3y_2\,\overline{\eta_{jl}(\boldsymbol{y}_1)^*\eta_{jm}(\boldsymbol{y}_2)} = \delta_{lm}\overline{\left(\int d^3y\,\eta(\boldsymbol{y})\right)^2} \stackrel{\text{def}}{=} \delta_{lm}\langle\eta^2\rangle. \tag{5.2.48}$$

This gives the following final result

$$|\boldsymbol{E}'|^2 = \frac{k^4}{16\pi^2\varepsilon^2}\frac{|\boldsymbol{E}_0|^2}{r^2}\langle\eta^2\rangle\sin^2\vartheta. \tag{5.2.49}$$

The angle ϑ is measured with respect to the direction of $\boldsymbol{E}_0$.

The **scattered light intensity strongly increases with frequency** $\sim \omega^4$. Since the daylight from the sky consists only of scattered light, the blue frequencies get amplified. On the other hand, if the direct sunlight passes through a thick or dim atmosphere, the blue frequencies are scattered away so that the sun looks red. This **explains the blue sky and the red sunset**. Another interesting point is the polarization of the scattered light. If one looks at the sky in the direction perpendicular to the direct sun rays, the observing direction $\boldsymbol{k}$ agrees with one direction of polarization $\boldsymbol{E}_0$ in the incoming light, i.e. $\vartheta = 0$. Hence, there is no scattered light with the corresponding polarization (parallel to the incoming rays). The scattered light perpendicular to the incoming rays is completely polarized (in the third orthogonal direction). The sky has a characteristic polarization pattern. Our eyes are not sensitive to it (except with polarizing sun glasses), but the bees can see it and use it for navigation.

The total scattered intensity I is obtained from (5.2.49) by multiplying with r^2 and integrating over the solid angle. A measure for the total scattering is the extinction coefficient

$$h \overset{\text{def}}{=} \frac{I}{|\boldsymbol{S}_0|^2 V} = \frac{k^4}{16\pi^2\varepsilon^2}\frac{\langle\eta^2\rangle}{V} 2\pi \int_{-1}^{+1} (1 - \cos^2\vartheta)\, d\cos\vartheta. \tag{5.2.50}$$

Here V is the volume of the gas exposed to radiation. Since the last ϑ-integral is equal to 4/3 and

$$k = \frac{\omega}{c_0}\sqrt{\varepsilon_r}, \tag{5.2.51}$$

we get

$$h = \frac{\omega^4}{6\pi\varepsilon_0^2 c_0^4}\frac{\langle\eta^2\rangle}{V}. \tag{5.2.52}$$

The fluctuations of the dielectric constant are caused by density and temperature fluctuations

$$\eta = \left(\frac{\partial\varepsilon}{\partial\rho}\right)_T \delta\rho + \left(\frac{\partial\varepsilon}{\partial T}\right)_\rho \delta T. \tag{5.2.53}$$

Since the density and temperature fluctuations are independent

$$\langle\delta\rho\,\delta T\rangle = 0,$$

we find for the averaged square

$$\langle\eta^2\rangle = \left(\frac{\partial\varepsilon}{\partial\rho}\right)_T^2 \langle(\delta\rho)^2\rangle + \left(\frac{\partial\varepsilon}{\partial T}\right)_\rho^2 \langle(\delta T)^2\rangle. \tag{5.2.54}$$

The square fluctuations are known from statistical mechanics:

$$\langle(\delta\rho)^2\rangle = V k_B T \rho \left(\frac{\partial\rho}{\partial p}\right)_T \tag{5.2.55}$$

$$\langle(\delta T)^2\rangle = V\frac{(k_B T)^2}{\rho C_V}, \tag{5.2.56}$$

where k_B is Boltzmann's constant, p the pressure and C_V the specific heat at constant volume of the gas. Substituting all this into (5.2.52) we arrive at the following final formula

$$h = \frac{\omega^4}{6\pi\varepsilon_0^2 c_0^4}\left[\left(\frac{\partial\varepsilon}{\partial\rho}\right)_T^2 k_B T\rho\left(\frac{\partial\rho}{\partial p}\right)_T + \left(\frac{\partial\varepsilon}{\partial T}\right)_\rho^2 \frac{(k_B T)^2}{\rho C_V}\right]. \tag{5.2.57}$$

The light scattering becomes very big at the critical point of the gas, where

$$\left(\frac{\partial p}{\partial V}\right)_T = 0. \tag{5.2.58}$$

This is a consequence of the fact that the compressibility

$$\left(\frac{\partial\rho}{\partial p}\right)_T = \left(\frac{\partial}{\partial p}\frac{N}{V}\right)_T = -\frac{N}{V^2}\left(\frac{\partial V}{\partial p}\right)_T$$

diverges (N is the particle number). Then the first term in the square bracket in (5.2.57) becomes infinite. Although at the critical point, the length scale of density fluctuations is no longer small compared to the wave length, the qualitative behavior is still correct: The gas shows extremely big light scattering that can be seen with the naked eye. This beautiful phenomenon is called **critical opalescence**. At the critical point the gas becomes dim like milk or like a moon-stone or an opal.

5.3 Geometrical Optics

Geometrical optics is known from elementary courses in experimental physics. If it is reasonable at all, and, of course it is, it must be an approximation to electrodynamical wave optics. We are going to show that the relation between the two is the same as classical mechanics is an approximation to wave or quantum mechanics.

Let us start again from Maxwell's equations for a non-magnetic insulator

$$\partial_t \boldsymbol{D} = \operatorname{curl} \boldsymbol{H} \tag{5.3.1}$$

$$\partial_t \boldsymbol{B} = -\operatorname{curl} \boldsymbol{E} \tag{5.3.2}$$

$$\operatorname{div} \boldsymbol{D} = 0, \quad \operatorname{div} \boldsymbol{B} = 0. \tag{5.3.3}$$

The constitutive equations are

$$\boldsymbol{B} = \mu_0 \boldsymbol{H}, \quad \boldsymbol{D} = \varepsilon(\boldsymbol{x})\boldsymbol{E}. \tag{5.3.4}$$

In the equation (5.3.3)

$$0 = \operatorname{div} \boldsymbol{D} = \varepsilon \operatorname{div} \boldsymbol{E} + \boldsymbol{E} \cdot \operatorname{grad} \varepsilon \tag{5.3.5}$$

we neglect the term with $\operatorname{grad}\varepsilon$, assuming that $\varepsilon(\boldsymbol{x})$ is slowly varying over a length of the order of the wavelength λ of the radiation. This is consistent because we will consider the short wavelength limit $\lambda \to 0$ below. Then (5.3.5) implies

$$\operatorname{div} \boldsymbol{E} = 0. \tag{5.3.6}$$

Taking the curl of (5.3.2)

$$\operatorname{curl}\operatorname{curl} \boldsymbol{E} + \operatorname{curl} \partial_t \boldsymbol{B} = \boldsymbol{0}, \tag{5.3.7}$$

and using (5.3.1) we find

$$\boldsymbol{0} = \operatorname{curl}\operatorname{curl} \boldsymbol{E} + \mu_0 \partial_t^2 \boldsymbol{D} = \operatorname{grad}\operatorname{div} \boldsymbol{E} - \triangle \boldsymbol{E} + \mu_0 \varepsilon \partial_t^2 \boldsymbol{E}, \tag{5.3.8}$$

or

$$- \triangle \boldsymbol{E} + \varepsilon_r(\boldsymbol{x}) \partial_t^2 \boldsymbol{E} = \boldsymbol{0}. \tag{5.3.9}$$

We assume a periodic time dependence

$$\boldsymbol{E} = \boldsymbol{E}_0(\boldsymbol{x}) e^{-i\omega t} \tag{5.3.10}$$

and calculate with complex quantities as in the last section. Introducing the wave number

$$k = \frac{\omega}{c_0} = \frac{2\pi}{\lambda}, \tag{5.3.11}$$

we get the following time independent equation

$$\frac{1}{k^2} \triangle \boldsymbol{E}_0(\boldsymbol{x}) + \varepsilon_r(\boldsymbol{x}) \boldsymbol{E}_0(\boldsymbol{x}) = \boldsymbol{0}. \tag{5.3.12}$$

This equation looks like the time-independent Schrödinger equation, where $\varepsilon_r(\boldsymbol{x})$ plays the rôle of the (negative) potential and $\boldsymbol{E}_0(\boldsymbol{x})$ is the wave function. We consider one component, say $E(\boldsymbol{x})$, of $\boldsymbol{E}_0$ in the following. **In geometrical optics one is interested in the asymptotic theory for small wavelength $\lambda \to 0$, i.e. $k \to \infty$.** Then the factor in front of the Laplace operator goes to 0. The corresponding factor in the Schrödinger equation is $\hbar^2/2m$, where $\hbar$ is Planck's constant (divided by 2π). The corresponding limit is the semi-classical limit of quantum mechanics. It leads to classical mechanics. In both cases the wave phenomena disappear in the short wavelength limit.

To solve (5.3.12) one makes the so-called **eikonal or semi-classical ansatz**

$$E(\boldsymbol{x}) = \varphi(\boldsymbol{x}) \exp[ikS(\boldsymbol{x})]. \tag{5.3.13}$$

$S(\boldsymbol{x})$ is called the eikonal. It gives the phase, while $\varphi(\boldsymbol{x})$ gives the amplitude of the radiation; both are real quantities. We compute

$$\triangle E = \triangle\varphi\, e^{ikS} + 2ik \operatorname{grad}\varphi \cdot \operatorname{grad} S\, e^{ikS} + \varphi \triangle e^{ikS} \tag{5.3.14}$$

and

$$\begin{aligned} \triangle e^{ikS} &= \operatorname{div}\operatorname{grad} e^{ikS} = ik \operatorname{div}\left(\operatorname{grad} S e^{ikS}\right) \\ &= ik e^{ikS} \triangle S - k^2 (\operatorname{grad} S)^2 e^{ikS}. \end{aligned} \tag{5.3.15}$$

Inserting everything into (5.3.12), we shall obtain

$$e^{ikS}\Big[\frac{1}{k^2}\triangle\varphi + \frac{2i}{k}\operatorname{grad}\varphi\cdot\operatorname{grad}S + \frac{i}{k}\varphi\triangle S$$

$$-(\operatorname{grad}S)^2\varphi + \varepsilon_r\varphi\Big] = 0. \tag{5.3.16}$$

For $k \to \infty$ we first get the equation

$$(\operatorname{grad}S)^2 - \varepsilon_r(\boldsymbol{x}) = 0 \tag{5.3.17}$$

for the eikonal $S(\boldsymbol{x})$. If this is solved, the amplitude φ follows from the equation $O(1/k)$

$$2\operatorname{grad}\varphi\cdot\operatorname{grad}S + \varphi\triangle S = 0. \tag{5.3.18}$$

The **eikonal equation** (5.3.17) is a first order partial differential equation. It is known from mechanics that such an equation is equivalent to a system of ordinary differential equations. The corresponding partial differential equation in mechanics is the (time-independent) Hamilton-Jacobi equation

$$H\Big(\frac{\partial S}{\partial \boldsymbol{x}}, \boldsymbol{x}\Big) = E, \tag{5.3.19}$$

where E is the energy (=0 in (5.3.17)). The gradient

$$\boldsymbol{p} = \frac{\partial S}{\partial \boldsymbol{x}} \tag{5.3.19}$$

is the momentum of a particle and

$$H = \boldsymbol{p}^2 - \varepsilon_r(\boldsymbol{x}), \tag{5.3.20}$$

according to (5.3.17), is the Hamiltonian function. The particle trajectories parallel to $\boldsymbol{p}$ (5.3.19) are orthogonal to the surfaces $S(\boldsymbol{x}) = \text{const}$. In optics the latter are the surfaces of constant phase and their orthogonal trajectories are the light rays. Hence, the light rays correspond to the particle trajectories in mechanics. Geometrical optics is indeed a mechanical theory.

A first order partial differential equation like (5.3.17) or (5.3.18) can be handled by the **method of characteristics**. Let us consider a general equation

$$F(S, \boldsymbol{p}, \boldsymbol{x}) = 0, \quad \boldsymbol{p} = \frac{\partial S}{\partial \boldsymbol{x}} \tag{5.3.21}$$

for $S(\boldsymbol{x})$. A solution can be specified by initial values as follows: Let S be given on a 2-dimensional initial surface

$$\boldsymbol{x} = \boldsymbol{x}(s_1, s_2), \quad S = S_0(s_1, s_2), \tag{5.3.22}$$

where s_1, s_2 are real parameters, then we want to calculate the solution $S(\boldsymbol{x})$ in $\mathbb{R}^3$ as far as possible. Such a solution is called an integral surface $S = S(\boldsymbol{x})$. Let $P_1 = (\boldsymbol{x}_1, S_1) \in \mathbb{R}^4$ be a point on the integral surface, then the tangent plane in P_1 to the surface has the directional coefficients

$$\frac{\partial S}{\partial \boldsymbol{x}}(\boldsymbol{x}_1) = \boldsymbol{p}. \tag{5.3.23}$$

In an arbitrary point $(\boldsymbol{x}_2, S_2)$ (not on a fixed integral surface), the possible tangent planes fulfill the condition

$$F(S_2, \boldsymbol{p}, \boldsymbol{x}_2) = 0, \tag{5.3.24}$$

so that there is a 2-dimensional family $\boldsymbol{p}(t_1, t_2)$ of possible directions. The corresponding tangent planes form the so-called Monge's cone. A particular surface must be tangent to the cone at every point.

We want to determine the straight lines $\boldsymbol{x}(\tau), S(\tau)$ on the cone through the vertex. These are found by intersecting neighbouring tangent planes. Since

$$\frac{dS}{d\tau} = \frac{\partial S}{\partial x_k}\frac{dx_k}{d\tau} = \boldsymbol{p}(t_1, t_2) \cdot \frac{d\boldsymbol{x}}{d\tau}, \tag{5.3.25}$$

we get by differentiation with respect to $t_j, j = 1, 2$

$$0 = \frac{\partial \boldsymbol{p}}{\partial t_j} \cdot \frac{d\boldsymbol{x}}{d\tau}. \tag{5.3.26}$$

On the other hand, differentiating (5.3.24), we find

$$\frac{\partial F}{\partial \boldsymbol{p}} \cdot \frac{\partial \boldsymbol{p}}{\partial t_j} = 0, \quad j = 1, 2, \tag{5.3.27}$$

thus the directions $d\boldsymbol{x}/d\tau$ of the lines are parallel to the gradient $\partial F/\partial \boldsymbol{p}$. By suitable choice of the parameter τ we may achieve

$$\frac{d\boldsymbol{x}}{d\tau} = \frac{\partial F}{\partial \boldsymbol{p}}. \tag{5.3.28}$$

Then (5.3.25) implies

$$\frac{dS}{d\tau} = \boldsymbol{p} \cdot \frac{d\boldsymbol{x}}{d\tau} = \boldsymbol{p} \cdot \frac{\partial F}{\partial \boldsymbol{p}}. \tag{5.3.29}$$

These two differential equations determine the straight lines on the Monge cone, if the directions $p(\tau)$ are known.

We want now to determine $\boldsymbol{p}(\tau)$ in such a way that the corresponding curves $\boldsymbol{x}(\tau), S(\tau)$ lie on the same integral surface. To find

$$\frac{d\boldsymbol{p}}{d\tau} = \frac{\partial \boldsymbol{p}}{\partial x_j}\frac{dx_j}{d\tau} = \frac{\partial \boldsymbol{p}}{\partial x_j}\frac{\partial F}{\partial p_j}, \tag{5.3.30}$$

we first notice that

$$\frac{\partial p_k}{\partial x_j} = \frac{\partial^2 S}{\partial x_j \partial x_k} = \frac{\partial p_j}{\partial x_k}. \tag{5.3.31}$$

Next we differentiate the original equation (5.3.21) under the assumption that all quantities ultimately depend on $\boldsymbol{x}(\tau)$:

$$\frac{\partial F}{\partial x_k} + \frac{\partial F}{\partial S}\frac{\partial S}{\partial x_k} + \frac{\partial F}{\partial p_j}\frac{\partial p_j}{\partial x_k} = 0. \tag{5.3.32}$$

Taking (5.3.31) into account in the last term, we find

$$\frac{\partial F}{\partial p_j}\frac{\partial p_k}{\partial x_j} = -\frac{\partial F}{\partial x_k} - p_k\frac{\partial F}{\partial S}. \tag{5.3.33}$$

Using this in (5.3.30), we get the desired equation for the directions $\boldsymbol{p}(\tau)$

$$\frac{d\boldsymbol{p}}{d\tau} = -\frac{\partial F}{\partial \boldsymbol{x}} - \boldsymbol{p}\frac{\partial F}{\partial S}. \tag{5.3.34}$$

Equations (5.3.28, 29) and (5.3.34) form the characteristic system of ordinary differential equations. A solution gives the characteristic curves (characteristics) $\boldsymbol{x}(\tau), S(\tau)$ plus a direction $\boldsymbol{p}(\tau)$. These quantities together are sometimes called characteristic strip.

We now suppose that the characteristic system has been solved and we want to construct the integral surface through the initial values (5.3.22). We complete these initial values to an initial strip by computing $\boldsymbol{p}(s_1, s_2)$ from the equations

$$\frac{\partial S_0}{\partial s_k} = \frac{\partial S}{\partial \boldsymbol{x}} \cdot \frac{\partial \boldsymbol{x}}{\partial s_k} = \boldsymbol{p}(s_1, s_2) \cdot \frac{\partial \boldsymbol{x}}{\partial s_k}, \quad k = 1, 2 \tag{5.3.35}$$

$$F\Big(\boldsymbol{x}(s_1, s_2), S_0(s_1, s_2), \boldsymbol{p}(s_1, s_2)\Big) = 0. \tag{5.3.36}$$

The equations (5.3.35) are called strip relations. Let now

$$\boldsymbol{x}(\tau; s_1, s_2), \quad S(\tau; s_1, s_2), \quad \boldsymbol{p}(\tau; s_1, s_2)$$

be the solution of the characteristic system (5.3.28, 29, 34) through the initial strip ($\tau = 0$). Eliminating the three parameters τ, s_1, s_2 by the three-vector $\boldsymbol{x}$, we obtain the desired integral surface $S(\boldsymbol{x}), \boldsymbol{p}(\boldsymbol{x})$. This elimination is possible, if the Jacobian determinant

$$D = \frac{\partial(x_1, x_2, x_3)}{\partial(\tau, s_1 s_2)} = \begin{vmatrix} \frac{\partial x_1}{\partial \tau} & \frac{\partial x_2}{\partial \tau} & \frac{\partial x_3}{\partial \tau} \\ \frac{\partial x_1}{\partial s_1} & \frac{\partial x_2}{\partial s_1} & \frac{\partial x_3}{\partial s_1} \\ \frac{\partial x_1}{\partial s_2} & \frac{\partial x_2}{\partial s_2} & \frac{\partial x_3}{\partial s_2} \end{vmatrix}$$

$$= \begin{vmatrix} \frac{\partial F}{\partial p_1} & \frac{\partial F}{\partial p_2} & \frac{\partial F}{\partial p_3} \\ \frac{\partial x_1}{\partial s_1} & \frac{\partial x_2}{\partial s_1} & \frac{\partial x_3}{\partial s_1} \\ \frac{\partial x_1}{\partial s_2} & \frac{\partial x_2}{\partial s_2} & \frac{\partial x_3}{\partial s_2} \end{vmatrix} \neq 0 \tag{5.3.37}$$

is different from zero. If this is the case initially, i.e. at $\tau = 0$, then we have $D \neq 0$ in a certain τ-interval by continuity. If, on the other hand, $D = 0$, then the corresponding integral surface is called characteristic.

Now we return to the **eikonal equation** (5.3.17)

$$F(\boldsymbol{x}, S, \boldsymbol{p}) = \boldsymbol{p}^2 - \varepsilon_r(\boldsymbol{x}) = 0. \tag{5.3.38}$$

This equation is simple because the function S does not appear in F, only its derivative

$$\boldsymbol{p} = \frac{\partial S}{\partial \boldsymbol{x}}. \tag{5.3.39}$$

Its absolute value

$$|\boldsymbol{p}| = \sqrt{\varepsilon_r(\boldsymbol{x})} = n(\boldsymbol{x}) \tag{5.3.40}$$

is just the index of refraction. The **characteristic system** reads

$$\frac{d\boldsymbol{x}}{d\tau} = \frac{\partial F}{\partial \boldsymbol{p}} = 2\boldsymbol{p} \tag{5.3.41}$$

$$\frac{dS}{d\tau} = \boldsymbol{p} \cdot \frac{\partial F}{\partial \boldsymbol{p}} = 2\boldsymbol{p}^2 = 2\varepsilon_r(\boldsymbol{x}) \tag{5.3.42}$$

$$\frac{d\boldsymbol{p}}{d\tau} = -\frac{\partial F}{\partial \boldsymbol{x}} = \frac{\partial \varepsilon_r(\boldsymbol{x})}{\partial \boldsymbol{x}}. \tag{5.3.43}$$

By (5.3.41), $d\boldsymbol{x}/d\tau$ is parallel to $\boldsymbol{p}$, hence, the **characteristics in 3-dimensional space are the light rays.**

It is convenient to introduce a unit vector $\boldsymbol{e}$ in the direction of the rays

$$\boldsymbol{e} = \frac{\boldsymbol{p}}{|\boldsymbol{p}|} = \frac{1}{n}\boldsymbol{p}. \tag{5.3.44}$$

Then the last characteristic equation (5.3.43) can be written as follows

$$\frac{d\boldsymbol{p}}{d\tau} = \frac{d}{d\tau}(n\boldsymbol{e}) = \operatorname{grad} \varepsilon_r = 2n \operatorname{grad} n. \tag{5.3.45}$$

Using instead of the arbitrary parameter τ the arc length along the light rays

$$ds = \sqrt{\left(\frac{d\boldsymbol{x}}{d\tau}\right)^2} d\tau = 2|\boldsymbol{p}|\, d\tau = 2n\, d\tau, \tag{5.3.46}$$

we obtain the **differential equation of the light path**

$$\frac{d}{ds}(n\boldsymbol{e}) = \operatorname{grad} n(\boldsymbol{x}). \tag{5.3.47}$$

These are in fact only two independent equations because $|\boldsymbol{e}| = 1$. If this equation is integrated, the eikonal $S(\boldsymbol{x})$ can be found from

$$n\boldsymbol{e} = \boldsymbol{p} = \operatorname{grad} S \tag{5.3.48}$$

by a line integral

$$S(2) - S(1) = \int_1^2 n\boldsymbol{e} \cdot d\boldsymbol{x}, \tag{5.3.49}$$

which is path independent. The integral (5.3.49) is equal to

$$S(2) - S(1) = \int_1^2 n\, ds \tag{5.3.50}$$

along the ray, i.e. always integrated in the direction $\boldsymbol{e}(\boldsymbol{x})$. Since $\boldsymbol{e}\cdot d\boldsymbol{x} \le |d\boldsymbol{x}|$, the integral along any other path is bigger. This is **Fermat's principle**:

$$\int_1^2 n\,ds = \text{minimal}, \tag{5.3.51}$$

i.e. the light rays are the geodesics in $\mathbb{R}^3$ with the metric given by $n(\boldsymbol{x})$. Using

$$n(\boldsymbol{x}) = \frac{c_0}{c(\boldsymbol{x})}, \quad \text{and} \quad \frac{ds}{dt} = c(\boldsymbol{x}),$$

Fermat's principle

$$c_0 \int_1^2 \frac{ds}{c(\boldsymbol{x})} = c_0 \int_1^2 dt = \text{minimal} \tag{5.3.52}$$

says that the light always propagates in such a way that the propagation time is minimal.

We have deduced Fermat's principle from the differential equations (5.3.47) of the light path. The two are actually equivalent, because (5.3.47) are the Euler equations for the variational principle

$$\int_1^2 n\,ds = \int_1^2 n(\boldsymbol{x})\sqrt{\left(\frac{d\boldsymbol{x}}{d\tau}\right)^2}\,d\tau = \text{stationary}. \tag{5.3.53}$$

To verify this, we denote the integrand in (5.3.53) by $L(\boldsymbol{x}, \boldsymbol{x}')$,

$$\boldsymbol{x}' = \frac{d\boldsymbol{x}}{d\tau},$$

L plays the rôle of the Lagrangian function in mechanics. Then (5.3.53) is equivalent to the Euler-Lagrange equations

$$\frac{d}{d\tau}\frac{\partial L}{\partial x_j'} - \frac{\partial L}{\partial x_j} = 0. \tag{5.3.54}$$

In fact

$$\frac{d}{d\tau}\frac{n}{\sqrt{\boldsymbol{x}'^2}}\frac{dx_j}{d\tau} - \sqrt{\boldsymbol{x}'^2}\frac{\partial n}{\partial x_j} = 0, \tag{5.3.55}$$

and substituting

$$\frac{\boldsymbol{x}'}{|\boldsymbol{x}'|} = \boldsymbol{e}, \quad |\boldsymbol{x}'| = 2|\boldsymbol{p}| = 2n,$$

we get

$$\frac{d}{d\tau}(n\boldsymbol{e}) - 2n\,\mathrm{grad}\,n = 0. \tag{5.3.56}$$

Taking (5.3.46) into account, this is identical with (5.3.47).

In order that the method of characteristics is applicable, the Jacobian determinant D (5.3.37) must be different from zero. In the exceptional case

$$D = \begin{vmatrix} p_1 & p_2 & p_3 \\ \dfrac{\partial x_1}{\partial s_1} & \dfrac{\partial x_2}{\partial s_1} & \dfrac{\partial x_3}{\partial s_1} \\ \dfrac{\partial x_1}{\partial s_2} & \dfrac{\partial x_2}{\partial s_2} & \dfrac{\partial x_3}{\partial s_2} \end{vmatrix} = 0 \tag{5.3.57}$$

the light rays ($\parallel \boldsymbol{p}$) and the tangent vectors $d\boldsymbol{x}/ds_1$, $d\boldsymbol{x}/ds_2$ are linearly dependent. This characteristic integral surface is called a **caustic**. More than one characteristic (light ray) goes through a point on a caustic. As we will see below, the intensity then becomes big and geometrical optics breaks down. In a good optical system the caustic essentially shrinks to the focal points.

To make this discussion still more concrete, we consider the characteristic system (5.3.41, 42, 43) in some detail. For constant $\varepsilon_r(\boldsymbol{x})$, $\boldsymbol{p}(\tau)$ is also constant and

$$S(\tau) = 2\varepsilon_r\tau + S_0 \tag{5.3.58}$$

$$\boldsymbol{x}(\tau) = 2\boldsymbol{p}\tau + \boldsymbol{x}_0, \tag{5.3.59}$$

so that the light rays are straight lines. In the initial-value problem the eikonal $S = S_0(s_1, s_2)$ is given on a two-dimensional surface $\boldsymbol{x} = \boldsymbol{x}_0(s_1, s_2)$. For example, an incoming wave front $S_0 =$ const. is specified. Then we must determine $\boldsymbol{p}(s_1, s_2)$ from the strip conditions (5.3.35)

$$\frac{\partial S_0(s_1, s_2)}{\partial s_k} = \boldsymbol{p} \cdot \frac{\partial \boldsymbol{x}_0(s_1, s_2)}{\partial s_k}, \quad k = 1, 2 \tag{5.3.60}$$

and from (5.3.42)

$$\boldsymbol{p}^2(s_1, s_2) = \varepsilon_r. \tag{5.3.61}$$

From (5.3.59)

$$\boldsymbol{x}(\tau; s_1, s_2) = 2\boldsymbol{p}(s_1, s_2)\tau + \boldsymbol{x}_0(s_1, s_2) \tag{5.3.62}$$

we obtain

$$\tau = \tau(\boldsymbol{x}), \quad s_k = s_k(\boldsymbol{x}), \quad k = 1, 2, \tag{5.3.63}$$

and then

$$S(\boldsymbol{x}) = 2\varepsilon_r\tau(\boldsymbol{x}) + S_0(s_1(\boldsymbol{x}), s_2(\boldsymbol{x})). \tag{5.3.64}$$

If we eliminate τ and $\boldsymbol{p}$ from (5.3.61) and (5.3.62) without using the strip condition (5.3.60), we get the possible light paths

$$\tau = \frac{|\boldsymbol{x} - \boldsymbol{x}_0|}{2|\boldsymbol{p}|} = \frac{|\boldsymbol{x} - \boldsymbol{x}_0|}{2\sqrt{\varepsilon_r}}, \tag{5.3.65}$$

$$S - S_0 = \sqrt{\varepsilon_r}|\boldsymbol{x} - \boldsymbol{x}_0|, \tag{5.3.66}$$

or

$$(S - S_0)^2 = \varepsilon_r(\boldsymbol{x} - \boldsymbol{x}_0)^2. \tag{5.3.67}$$

This is the light cone in $\mathbb{R}^4$, sometimes called conoid here. The points of constant phase S =const lie on spheres. The actual wave front, obtained by taking the strip conditions (5.3.60) into account, are envelopes of these spherical waves. This is **Huygens' principle**, which, thus, **is a principle of geometrical optics, not of wave optics**.

After discussion of the eikonal equation, we briefly turn to the amplitude equation (5.3.18)

$$2\text{grad}\,\varphi \cdot \text{grad}\,S + \varphi \bigtriangleup S = 0. \tag{5.3.68}$$

If S is known, this is again a first order partial differential equation, so that the methods of characteristics is once more applicable. Introducing

$$\text{grad}\,\varphi = \boldsymbol{q}, \quad \text{and} \tag{5.3.69}$$

$$F(\boldsymbol{x}, \varphi, \boldsymbol{q}) = 2\boldsymbol{q} \cdot \text{grad}\,S + \varphi \bigtriangleup S = 0, \tag{5.3.70}$$

the characteristic system reads

$$\frac{d\boldsymbol{x}}{d\tau} = \frac{\partial F}{\partial \boldsymbol{q}} = 2\text{grad}\,S(\boldsymbol{x}) = 2\boldsymbol{p} \tag{5.3.71}$$

$$\frac{d\boldsymbol{q}}{d\tau} = -\frac{\partial F}{\partial \boldsymbol{x}} - \boldsymbol{q}\frac{\partial F}{\partial \varphi} = -2\text{grad}\,(\boldsymbol{q} \cdot \text{grad}\,S) - \varphi\text{grad} \bigtriangleup S - \boldsymbol{q} \bigtriangleup S \tag{5.3.72}$$

$$\frac{d\varphi}{d\tau} = \boldsymbol{q} \cdot \frac{\partial F}{\partial \boldsymbol{q}} = 2\boldsymbol{q} \cdot \text{grad}\,S = -\varphi \bigtriangleup S = -\varphi\text{div}\,\boldsymbol{p}. \tag{5.3.73}$$

The first equation (5.3.71) gives the light rays $\boldsymbol{x} = \boldsymbol{x}(\tau)$ as above (5.3.41). Dividing the last equation by φ, we find for the amplitude

$$\frac{d}{d\tau}\log\varphi = -\text{div}\,\boldsymbol{p}, \quad \text{or} \tag{5.3.74}$$

$$\log\frac{\varphi}{\varphi_0} = -\int_0^\tau (\text{div}\,\boldsymbol{p})(\boldsymbol{x}(\tau'))\,d\tau', \tag{5.3.75}$$

where the integral runs along the light ray. **On the caustic** there is more then one ray through a point $\boldsymbol{x}$. Consequently, $\boldsymbol{p}(\boldsymbol{x})$ may change without change of $\boldsymbol{x}$, i.e. div $\boldsymbol{p}$ is singular. Then the amplitude φ diverges and **geometrical optics breaks down**.

We now want to apply these results to the **quantitative discussion of optical systems**. We start from the characteristic equation (5.3.42)

$$\frac{dS}{d\tau} = 2n^2(\boldsymbol{x}), \tag{5.3.76}$$

and integrate it along the light ray from $\boldsymbol{x}(0) = \boldsymbol{x}$ to $\boldsymbol{x}(\tau) = \boldsymbol{x}'$:

$$S(\boldsymbol{x}, \boldsymbol{x}') = 2 \int_0^\tau n^2(\boldsymbol{x}(\tau'))\, d\tau'. \tag{5.3.77}$$

This equation shows that

$$(\text{grad}_{\boldsymbol{x}'} S)^2 = n^2(\boldsymbol{x}') \quad \text{and} \tag{5.3.78}$$

$$(\text{grad}_{\boldsymbol{x}} S)^2 = n^2(\boldsymbol{x}). \tag{5.3.79}$$

We introduce the directions of the rays in $\boldsymbol{x}$, $\boldsymbol{x}'$

$$\mathbf{e} = -\frac{1}{n(\boldsymbol{x})} \frac{\partial S}{\partial \boldsymbol{x}}, \quad \mathbf{e}' = \frac{1}{n(\boldsymbol{x}')} \frac{\partial S}{\partial \boldsymbol{x}'}, \tag{5.3.80}$$

respectively. We are interested in these ray directions $\mathbf{e}, \mathbf{e}'$ as functions of $\boldsymbol{x}, \boldsymbol{x}'$. This characterizes the properties of the optical system.

In the total differential

$$\begin{aligned} dS &= \frac{\partial S}{\partial \boldsymbol{x}'} \cdot d\boldsymbol{x}' + \frac{\partial S}{\partial \boldsymbol{x}} \cdot d\boldsymbol{x} \\ &= n(\boldsymbol{x}')\mathbf{e}' \cdot d\boldsymbol{x}' - n(\boldsymbol{x})\mathbf{e} \cdot d\boldsymbol{x} \end{aligned} \tag{5.3.81}$$

we perform a Legendre transformation

$$\begin{aligned} dS = {} & d(n(\boldsymbol{x}')\mathbf{e}' \cdot \boldsymbol{x}') - \boldsymbol{x}' \cdot d(n(\boldsymbol{x}')\mathbf{e}') \\ & -d(n(\boldsymbol{x})\mathbf{e} \cdot \boldsymbol{x}) + \boldsymbol{x} \cdot d(n(\boldsymbol{x})\mathbf{e}). \end{aligned} \tag{5.3.82}$$

Let us introduce the so-called angular eikonal ψ

$$\psi = n(\boldsymbol{x}')\mathbf{e}' \cdot \boldsymbol{x}' - n(\boldsymbol{x})\mathbf{e} \cdot \boldsymbol{x} - S. \tag{5.3.83}$$

Its differential is equal to

$$d\psi = \boldsymbol{x}' \cdot d(n(\boldsymbol{x}')\mathbf{e}') - \boldsymbol{x} \cdot d(n(\boldsymbol{x})\mathbf{e}). \tag{5.3.84}$$

We now assume that the points $\boldsymbol{x}, \boldsymbol{x}'$ are in vacuum, so that $n(\boldsymbol{x}) = n(\boldsymbol{x}') = 1$. Then

$$\begin{aligned} d\psi &= \boldsymbol{x}' \cdot d\mathbf{e}' - \boldsymbol{x} \cdot d\mathbf{e} \\ &= x' de'_x + y' de'_y + z' de'_z - x de_x - y de_y - z de_z, \end{aligned} \tag{5.3.85}$$

and we may take the directions (angles) as the independent variables.

Since

$$e'_x = \sqrt{1 - e'^2_y - e'^2_z}, \quad e_x = \sqrt{1 - e^2_y - e^2_z},$$

we can eliminate these two variables:

$$\begin{aligned} de'_x &= -\frac{1}{e'_x}(e'_y de'_y + e'_z de'_z) \\ de_x &= -\frac{1}{e_x}(e_y de_y + e_z de_z). \end{aligned} \tag{5.3.86}$$

Then we find for the differential of the angular eikonal (5.3.85)

$$d\psi = \left(y' - \frac{e'_y}{e'_x}x'\right)de'_y + \left(z' - \frac{e'_z}{e'_x}x'\right)de'_z$$

$$-\left(y - \frac{e_y}{e_x}x\right)de_y - \left(z - \frac{e_z}{e_x}x\right)de_z. \tag{5.3.87}$$

This gives us the partial derivatives of the angular eikonal

$$y' - \frac{e'_y}{e'_x}x' = \frac{\partial\psi}{\partial e'_y}, \quad z' - \frac{e'_z}{e'_x}x' = \frac{\partial\psi}{\partial e'_z} \tag{5.3.88}$$

$$y - \frac{e_y}{e_x}x = -\frac{\partial\psi}{\partial e_y}, \quad z - \frac{e_z}{e_x}x = -\frac{\partial\psi}{\partial e_z}. \tag{5.3.89}$$

We now restrict to **systems with small aperture**, where $|e_y|, |e_z| \ll e_x \approx 1$ and $|e'_y|, |e'_z| \ll e'_x \approx \pm 1$. The plus sign corresponds to lens systems, while mirrors have $e'_x \approx -1$. In addition we assume that the system is axial symmetric and that the x-axis is the optical axis. Then we expand the angular eikonal

$$\psi(\mathbf{e}', \mathbf{e}) = \psi(e'_y, e'_z, e_y, e_z)$$

into a power series. ψ is axial symmetric under rotation around the x-axis, consequently, it can only depend on the invariants $\mathbf{e}^2, \mathbf{e}'^2$ and $\mathbf{e}\cdot\mathbf{e}'$:

$$\psi = \psi_0 + \frac{g}{2}(e_y^2 + e_z^2) + f(e_y e'_y + e_z e'_z) + \frac{h}{2}(e'^2_y + e'^2_z) + O(e^4). \tag{5.3.90}$$

We insert this into (5.3.88, 89), assuming a lens system ($e'_x = 1$),

$$y' = e'_y x' + f e_y + h e'_y = f e_y + (x' + h)e'_y$$

$$z' = e'_z x' + f e_z + h e'_z = f e_z + (x' + h)e'_z, \tag{5.3.91}$$

$$y = e_y x - g e_y - f e'_y = -f e'_y + (x - g)e_y$$

$$z = e_z x - g e_z - f e'_z = -f e'_z + (x - g)e_z. \tag{5.3.92}$$

For given $\boldsymbol{x}, \boldsymbol{x}'$, these are four linear equations for e_y, e_z, e'_y, e'_z. In general, there is a unique solution, so that there exists one ray going from $\boldsymbol{x}$ to $\boldsymbol{x}'$, only. This does not give an optical imaging. The condition for this is that rays intersecting at point $\boldsymbol{x}$ (so-called homocentral rays) do again intersect at point $\boldsymbol{x}'$ in the object space, after going through the system. This is only possible if the pair of equations (5.3.92) is a multiple of the pair of equations (5.3.91):

$$\frac{f}{x - g} = -\frac{x' + h}{f} = \frac{y'}{y} = \frac{z'}{z}. \tag{5.3.93}$$

This implies

$$(x - g)(x' + h) = -f^2. \tag{5.3.94}$$

It follows from (5.3.93) that the point $x = g$ is mapped to infinity $x' = \infty$, and that the point $x' = -h$ corresponds to $x = \infty$. These are the two main focal points F, F' of the system. It is convenient to introduce two new coordinate systems in the object and image space, respectively, with origins F and F'. Denoting these new coordinates by capital letters, we have

$$X = x - g, \quad X' = x' + h \tag{5.3.95}$$
$$Y = y, \quad Y' = y'$$
$$Z = z, \quad Z' = z'.$$

Then (5.3.94, 93) leads to **Newton's equations of optical imaging**

$$X \cdot X' = -f^2 \tag{5.3.96}$$

$$\frac{Y'}{Y} = \frac{Z'}{Z} = \frac{f}{X} = -\frac{X'}{f}. \tag{5.3.97}$$

Here f is the focal distance and the ratio $Y'/Y = Z'/Z$ is the lateral magnification. The directions of the rays can be found from (5.3.91)

$$Y' = fe_y + X'e'_y$$
$$Z' = fe_z + X'e'_z. \tag{5.3.98}$$

A glance to (5.3.96, 97) shows that for $X = f$, $X' = -f$ there is no lateral magnification, $Y'/Y = 1$. The image is upright and of equal magnitude. These are the two principal planes of the system. The points $(X = f,\ 0,\ 0)$ and $(X' = -f,\ 0,\ 0)$ are the so-called principal points. For these points (5.3.98) implies

$$0 = fe_y - fe'_y, \quad \text{i.e.} \quad e'_y = e_y$$
$$0 = fe_z - fe'_z, \quad \text{i.e.} \quad e'_z = e_z. \tag{5.3.99}$$

Hence the rays through the principal points must be parallel. These facts allow a simple geometric construction of the optical image, if the main planes and the focal points are given.

We want to illustrate this construction in the simple but important example of a convex lens ($f > 0$), used as a magnifier (Fig.18). The lens is represented by its two principal planes H and H'. The focal points are F and F'. The object to be imaged is placed between F and H. A ray parallel to the x-axis, which corresponds to an image at infinity, is refracted into a ray through the focal point F'. A ray through the main point $(f, 0, 0)$ becomes a parallel ray through the other main point $(-f, 0, 0)$. The two resulting rays do not intersect in the image space on the right-hand side of the lens. There is no real image. But there is a virtual image at the backward intersection. It seems to the observer as if the rays are coming from the magnified upright image. This explains how a magnifier is working. The "rays" used in this construction are not the real light rays, they only serve for the purpose of finding a geometric solution of the image equations.

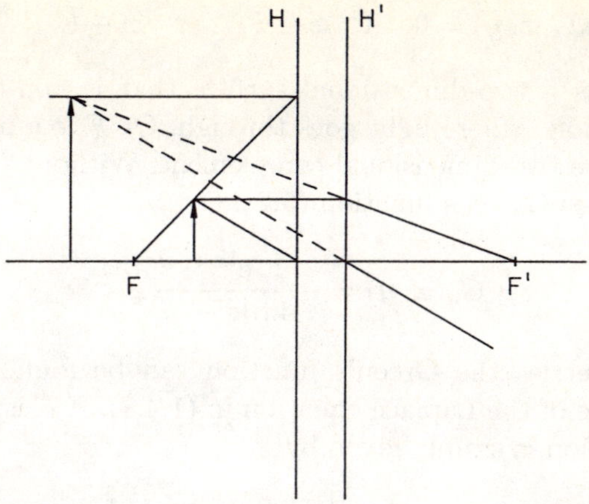

Fig. 18. Image construction for the magnifier

5.4 Diffraction

Diffraction phenomena cannot be explained in the limit $\lambda \to 0$ of geometrical optics, but we still may assume that the wavelength λ is small compared to a typical geometrical length d in the arrangement. We are studying **small deviations from geometrical optics**, which, nevertheless, can be quite spectacular. As in the last section, we assume monochromatic light

$$E \sim e^{-i\omega t} u(\boldsymbol{x}), \tag{5.4.1}$$

and we calculate with complex quantities. The wave equation for E then implies the Helmholtz equation

$$\triangle u + k^2 u = 0, \quad k = \frac{\omega}{c} \tag{5.4.2}$$

for $u(\boldsymbol{x})$. As in geometrical optics, we consider one (mean) component of $\boldsymbol{E}$ (or $\boldsymbol{B}$), only. This is scalar wave optics, where all polarization phenomena are neglected. More seriously, the boundary conditions on surfaces can only be satisfied approximately, and the error of this approximation is not under full control, in general.

We have to solve a boundary value problem for the Helmholtz equation (5.4.2), which we will carry out by the method of Green's function, similarly to Sect.1.4 in the case of the Laplace equation. Let $G(\boldsymbol{x}, \boldsymbol{y})$ be the distributive solution of

$$\triangle_x G(\boldsymbol{x}, \boldsymbol{y}) + k^2 G(\boldsymbol{x}, \boldsymbol{y}) = \delta(\boldsymbol{x} - \boldsymbol{y}), \tag{5.4.3}$$

if $\boldsymbol{x}$ and $\boldsymbol{y}$ are in the region Ω, and

$$G(\boldsymbol{x},\boldsymbol{y}) = 0 \quad \text{if} \quad \boldsymbol{x} \in F, \quad \text{or} \quad \boldsymbol{x} \in \bar{F}. \tag{5.4.4}$$

Here F represents a two-dimensional surface that is non-transparent for light and $\bar{F}$ is a hole, where light goes through. Or F is a non-transparent obstacle and $\bar{F}$ the two-dimensional complement. Without the obstacle, we would have the free Green's function (5.2.28)

$$G_0(\boldsymbol{x},\boldsymbol{y}) = -\frac{1}{4\pi}\frac{e^{ik|\boldsymbol{x}-\boldsymbol{y}|}}{|\boldsymbol{x}-\boldsymbol{y}|}. \tag{5.4.5}$$

For simple geometries the Green's function can be found by the image method, as in case of the Laplace equation in (1.4.8). If F is a plane screen, the Green's function is simply given by

$$G(\boldsymbol{x},\boldsymbol{y}) = -\frac{1}{4\pi}\frac{e^{ik|\boldsymbol{x}-\boldsymbol{y}|}}{|\boldsymbol{x}-\boldsymbol{y}|} + \frac{1}{4\pi}\frac{e^{ik|\boldsymbol{x}-\boldsymbol{y}^*|}}{|\boldsymbol{x}-\boldsymbol{y}^*|}. \tag{5.4.6}$$

Here $\boldsymbol{y}^*$ is the mirror point of $\boldsymbol{y}$ with respect to F. The Green's function is symmetric

$$G(\boldsymbol{y},\boldsymbol{x}) = G(\boldsymbol{x},\boldsymbol{y}). \tag{5.4.7}$$

We suppose the radiation coming in from the half-space $\bar{\Omega}$, and we want to calculate the intensity behind the obstacle in Ω (Fig.19). As in electrostatics, we apply Green's theorem to the region Ω

$$\int_{\Omega} (u(\boldsymbol{y}) \,\triangle_y\, G(\boldsymbol{x},\boldsymbol{y}) - G(\boldsymbol{x},\boldsymbol{y}) \,\triangle\, u(\boldsymbol{y}))\, d^3y$$

$$= \int_{\partial\Omega} (u(\boldsymbol{y})\operatorname{grad}_y G(\boldsymbol{x},\boldsymbol{y}) - G(\boldsymbol{x},\boldsymbol{y})\operatorname{grad} u(\boldsymbol{y}))\, d\boldsymbol{\sigma}_y. \tag{5.4.8}$$

The last term vanishes because $\boldsymbol{y} \in \partial\Omega$ (5.4.4). Since G and u fulfill the Helmholtz equation, we arrive at

$$u(\boldsymbol{x}) - k^2 \int u(\boldsymbol{y}) G(\boldsymbol{x},\boldsymbol{y})\, d^3y + \int Gk^2 u\, d^3y$$

$$= \int_{\partial\Omega} u(\boldsymbol{y})\operatorname{grad}_y G(\boldsymbol{x},\boldsymbol{y})\, d\boldsymbol{\sigma}_y. \tag{5.4.9}$$

The two integrals on the l.h.s. cancel. Hence, as in electrostatics, we have succeeded in expressing $u(\boldsymbol{x})$ by its values on the boundaries.

Let us choose the screen in the yz-plane $x_1 = 0$ and assume the origin within the hole. For $\boldsymbol{y} \in \Omega$, not on the boundary $\partial\Omega$, we then have

$$r = |\boldsymbol{x}-\boldsymbol{y}| = \sqrt{(x_1-y_1)^2 + (x_2-y_2)^2 + (x_3-y_3)^2} \tag{5.4.10}$$

$$r^* = |\boldsymbol{x}-\boldsymbol{y}^*| = \sqrt{(x_1+y_1)^2 + (x_2-y_2)^2 + (x_3-y_3)^2}. \tag{5.4.11}$$

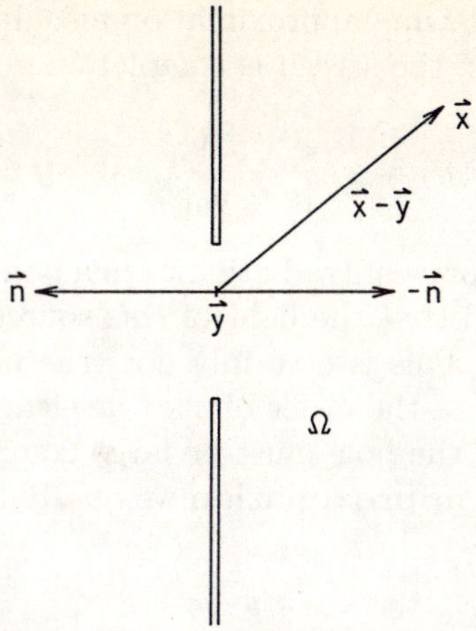

Fig. 19. Diffraction at a plane screen with a hole

Using

$$\frac{\partial r}{\partial y_1} = \frac{y_1 - x_1}{r}, \quad \frac{\partial r^*}{\partial y_1} = \frac{y_1 + x_1}{r^*}, \tag{5.4.12}$$

we compute the normal derivative

$$\frac{\partial G}{\partial n_y} = -\frac{\partial G}{\partial y_1}$$

$$= \frac{1}{4\pi}\Big(\frac{\partial}{\partial r}\frac{e^{ikr}}{r}\Big)\Big(\frac{y_1 - x_1}{r} - \frac{y_1 + x_1}{r^*}\Big). \tag{5.4.13}$$

For $\boldsymbol{y}$ on the boundary $\partial\Omega$, i.e. $y_1 \to 0$, $r^* \to r$, we get

$$\frac{\partial G}{\partial n_y}\Big|_{\partial\Omega} = -\frac{2}{4\pi}\frac{e^{ikr}}{r}\Big(ik - \frac{1}{r}\Big)\frac{x_1}{r}. \tag{5.4.14}$$

In the asymptotic region $kr \gg 1$ on the other hand, we find

$$\frac{\partial G}{\partial n_y}\Big|_{\partial\Omega} = -\frac{ik}{2\pi}\frac{e^{ik|\boldsymbol{x}-\boldsymbol{y}|}}{|\boldsymbol{x}-\boldsymbol{y}|}\cos(-\boldsymbol{n}, \boldsymbol{x}-\boldsymbol{y}). \tag{5.4.15}$$

Substituting this into (5.4.9), we obtain

$$u(\boldsymbol{x}) = -\frac{i}{\lambda}\int\limits_{\partial\Omega}\frac{e^{ik|\boldsymbol{x}-\boldsymbol{y}|}}{|\boldsymbol{x}-\boldsymbol{y}|}\cos(-\boldsymbol{n}, \boldsymbol{x}-\boldsymbol{y})u(\boldsymbol{y})\,d\sigma_y. \tag{5.4.16}$$

This is **Kirchhoff's formula for diffraction**. The cos-dependence herein is also called Lambert's area law.

Kirchhoff's formula does still not solve the problem, because the unknown amplitude $u(\boldsymbol{y})$ appears under the integral on the r.h.s. For $u(\boldsymbol{y})$

one now makes the following approximation also due to Kirchhoff: One sets $u(\boldsymbol{y}) = 0$ for $\boldsymbol{y} \in \bar{F}$, i.e. the screen is completely impenetrable, and

$$u(\boldsymbol{y}) = u_0 \frac{e^{ik|\boldsymbol{y}-\boldsymbol{y}_0|}}{|\boldsymbol{y}-\boldsymbol{y}_0|}, \quad \text{if} \quad \boldsymbol{y} \in F. \tag{5.4.17}$$

This last expression represents radiation from a point light source at $\boldsymbol{y}_0 \in \bar{\Omega}$. It is therefore assumed that the light of this source arrives in the aperture without perturbation. This is certainly not true near the boundary of the hole on a length scale of the order of the wavelength λ. This is the reason why the diameter d of the hole must be large compared to λ.

With **Kirchhoff's approximation** we obtain for the amplitude behind the screen

$$u(\boldsymbol{x}) = -\frac{i}{\lambda} u_0 \int\limits_F \frac{e^{ik|\boldsymbol{x}-\boldsymbol{y}|}}{|\boldsymbol{x}-\boldsymbol{y}|} \frac{e^{ik|\boldsymbol{y}-\boldsymbol{y}_0|}}{|\boldsymbol{y}-\boldsymbol{y}_0|} \cos(-\boldsymbol{n}, \boldsymbol{x}-\boldsymbol{y})\, d\sigma_y. \tag{5.4.18}$$

As a first simple application of this formula we calculate the amplitude at distance ρ on the axis behind a **circular obstacle of radius** a (Fig.20). If the light source is far away, $|y_0| \to \infty$, the y-dependence in the second spherical wave can be neglected. Writing the integral in planar polar coordinates, we get

$$u(\rho) \sim \frac{e^{ik|\boldsymbol{y}_0|}}{|\boldsymbol{y}_0|} \int\limits_a^\infty \frac{e^{ikr}}{r} \frac{\rho}{r} 2\pi z\, dz,$$

where r is given in (5.4.10). Using $r = \sqrt{\rho^2 + z^2}$ as new integration variable, with lower limit $R = \sqrt{\rho^2 + a^2}$, the resulting integral can be transformed by partial integration

$$u(\rho) \sim \rho \int\limits_r^\infty \frac{e^{ikr}}{r}\, dr = \frac{\rho}{ik} \left[\frac{e^{ikr}}{r} + \int \frac{e^{ikr}}{r^2}\, dr \right]_R^\infty.$$

If $kR \gg 1$, the remaining integral is small compared to the first term

$$u(\rho) \sim -\frac{\rho}{ikR} e^{ikR} \Big(1 + O\Big(\frac{1}{kR}\Big)\Big),$$

as can be seen by another partial integration. Taking the absolute square, we obtain for the ratio of the diffracted to the incoming intensity

$$\frac{I}{I_0} = \frac{\rho^2}{R^2} = \frac{\rho^2}{\rho^2 + a^2}. \tag{5.4.19}$$

Surprisingly enough, **the intensity on the axis** behind the obstacle is nowhere 0 (for $\rho > 0$), and it **increases to the undisturbed intensity** I_0 **for large** $\rho \gg a$. This is in sharp contrast to geometrical optics.

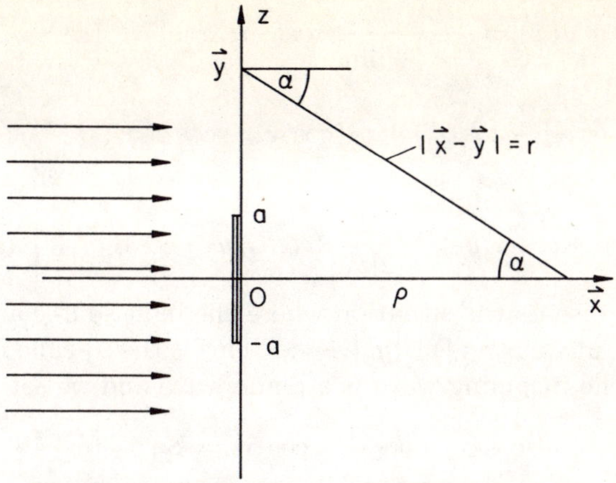

Fig. 20. Diffraction at a circular obstacle

If both $|\boldsymbol{x}|$ and $|\boldsymbol{y}_0|$ are large compared to $|\boldsymbol{y}| \approx d$, formula (5.4.18) can be simplified as follows

$$u(\boldsymbol{x}) = -\frac{i}{\lambda}\frac{u_0}{|\boldsymbol{x}||\boldsymbol{y}_0|}\cos(-\boldsymbol{n},\boldsymbol{x})\int\limits_F e^{ik(|\boldsymbol{x}-\boldsymbol{y}|+|\boldsymbol{y}-\boldsymbol{y}_0|)}d\sigma_y.$$

The exponent can then be expanded for large $|\boldsymbol{x}|$

$$|\boldsymbol{x}-\boldsymbol{y}| = \sqrt{x_1^2+(x_2-y_2)^2+(x_3-y_3)^2}$$

$$= \sqrt{\boldsymbol{x}^2-2(x_2y_2+x_3y_3)+y_2^2+y_3^2}$$

$$= |\boldsymbol{x}|\Big[1-2\Big(\frac{x_2}{|\boldsymbol{x}|^2}y_2+\frac{x_3}{|\boldsymbol{x}|^2}y_3\Big)+\frac{y_2^2+y_3^2}{|\boldsymbol{x}|^2}\Big]^{1/2}$$

$$= |\boldsymbol{x}|-\frac{x_2}{|\boldsymbol{x}|}y_2-\frac{x_3}{|\boldsymbol{x}|}y_3+\frac{y_2^2+y_3^2}{2|\boldsymbol{x}|}-\frac{4(x_2y_2+x_3y_3)^2}{8|\boldsymbol{x}|^3}$$

$$= |\boldsymbol{x}|-\cos\alpha_2' y_2-\cos\alpha_3' y_3+\frac{1}{2|\boldsymbol{x}|}\Big[y_2^2+y_3^2-(\cos\alpha_2' y_2+\cos\alpha_3' y_3)^2\Big]. \quad (5.4.20)$$

Here α_2, α_3 are the angles between $\boldsymbol{x}$ and the 2- and 3-axis, respectively. In the same way we find

$$|\boldsymbol{y}-\boldsymbol{y}_0| = |\boldsymbol{y}_0|+\cos\alpha_2 y_2+\cos\alpha_3 y_3+\frac{1}{2|\boldsymbol{y}_0|}\Big[y_2^2+y_3^2-(\cos\alpha_2 y_2+\cos\alpha_3 y_3)^2\Big], \quad (5.4.21)$$

where $\cos\alpha_2, \cos\alpha_3$ are the directional cosines of the incoming light. Substituting all that into (5.4.19), we end up with

$$u(\boldsymbol{x}) = -\frac{i}{\lambda} \frac{u_0}{|\boldsymbol{x}||\boldsymbol{y}_0|} \cos(-\boldsymbol{n}, \boldsymbol{x}) e^{ik(|\boldsymbol{x}|+|\boldsymbol{y}_0|)}$$

$$\cdot \int_F \exp ik \Big[y_2(\cos\alpha_2 - \cos\alpha_2') + y_3(\cos\alpha_3 - \cos\alpha_3') + \Big(\frac{1}{|\boldsymbol{x}|} + \frac{1}{|\boldsymbol{y}_0|}\Big) \frac{y_2^2 + y_3^2}{2}$$

$$- \frac{1}{2|\boldsymbol{x}|}(\cos\alpha_2' y_2 + \cos\alpha_3' y_3)^2 - \frac{1}{2|\boldsymbol{y}_0|}(\cos\alpha_2 y_2 + \cos\alpha_3 y_3)^2 \Big] \, dy_2 dy_3. \quad (5.4.22)$$

Let us first discuss the situation where the light source and the point of observation are at infinity: $|\boldsymbol{x}|, |\boldsymbol{y}_0| \to \infty$. This is the so-called **Frauenhofer diffraction**. The incoming wave is a plane wave and we set

$$k \cos\alpha_2 = k_2, \quad k \cos\alpha_2' = k_2', \quad \text{etc.}$$

According to (5.2.42) the diffracted intensity I is proportional to $|u|^2$, hence

$$I \sim \Big| \int_F \exp i[(k_2 - k_2')y_2 + (k_3 - k_3')y_3] \, dy_2 dy_3 \Big|^2. \quad (5.4.23)$$

This is a two-dimensional Fourier integral. If we add the corresponding contribution from the aperture $\bar{F}$, we get a two-dimensional δ-distribution

$$\int_F + \int_{\bar{F}} = \delta^2(k_2 - k_2', k_3 - k_3'). \quad (5.4.24)$$

Since this vanishes for $k_2' \neq k_2$ or $k_3' \neq k_3$, the diffracted intensities for the two complementary arrangements must be the same

$$I_F = I_{\bar{F}}. \quad (5.4.25)$$

That means, the diffraction pattern behind a screen with a spherical hole agrees with that of a small spherical obstacle without a screen. This is Babinet's principle.

We want to evaluate (5.4.23) for a **rectangular aperture** $2a \times 2b$:

$$I \sim \Big| \int_{-a}^{a} dy_2 \, e^{i(k_2 - k_2')y_2} \int_{-b}^{b} dy_3 \, e^{i(k_3 - k_3')y_3} \Big|^2$$

$$= 16a^2b^2 \left[\frac{\sin(k_2 - k_2')a}{(k_2 - k_2')a} \right]^2 \left[\frac{\sin(k_3 - k_3')b}{(k_3 - k_3')b} \right]^2. \quad (5.4.26)$$

The function

$$f(\xi) = \frac{\sin^2 \xi}{\xi^2} \quad (5.4.27)$$

appearing here shows the typical behavior of diffraction patterns: It has a main maximum at $\xi = 0$ and minima (in fact zeros) at

$$\xi = n\pi, \quad n = \pm 1, \pm 2, \ldots \tag{5.4.28}$$

(see Fig.21). Between the zeros there are small secondary maxima. The angular distance δ between the minima is given by

$$k \cdot \delta \cdot a = \pi, \quad \text{i.e.} \quad \delta = \frac{\lambda}{2a}, \tag{5.4.29}$$

because $k = 2\pi/\lambda$. This is small for $a \gg \lambda$.

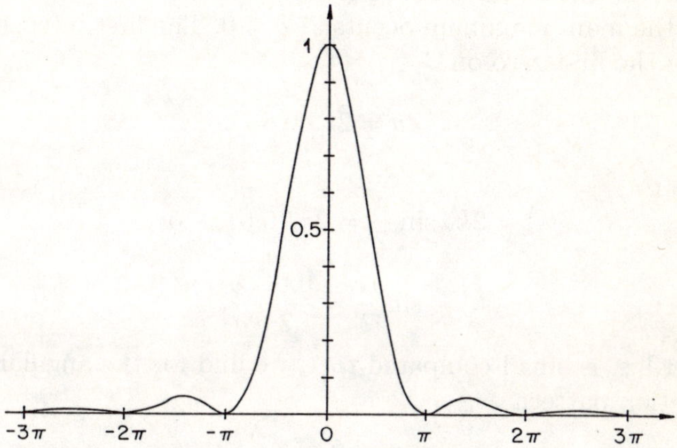

Fig. 21. The diffraction function $f(\xi) = \sin^2 \xi / \xi^2$

A **spherical aperture** is important for many applications. If a is the radius of the hole, we obtain for the diffracted intensity (5.4.23)

$$I \sim \left| \int_0^a dr\, r \int_0^{2\pi} d\varphi \exp[i(k_2 - k_2')r\cos\varphi + i(k_3 - k_3')r\sin\varphi] \right|^2 . \tag{5.4.30}$$

Here φ is the azimuthal angle of the planar polar coordinates. The quantities in the exponent are expressed as follows

$$k_2 - k_2' = q\cos\vartheta, \quad k_3 - k_3' = q\sin\vartheta. \tag{5.4.31}$$

Then q is the absolute value

$$q = |\boldsymbol{k}' - \boldsymbol{k}| = 2k \sin\frac{\alpha}{2}, \tag{5.4.32}$$

where α is the angle between $\boldsymbol{k}$ and $\boldsymbol{k}'$, the so-called diffraction angle. Then (5.4.30) becomes

$$I \sim \left| \int_0^a dr\, r \int_0^{2\pi} d\varphi \exp[iqr\cos(\varphi - \vartheta)] \right|^2 . \tag{5.4.33}$$

The φ-integral gives the Bessel function $J_0(qr)$ (3.3.31)

$$I \sim \left| \int_0^a dr\, r J_0(qr) \right|^2 = \left| \frac{1}{q^2} \int_0^{aq} J_0(s) s\, ds \right|^2$$

$$= \left| \frac{a}{q} J_1(aq) \right|^2, \tag{5.4.34}$$

where J_1 is the Bessel function of order 1.

Again the main maximum occurs at $q = 0$. The first intensity minimum is given by the first zero on J_1

$$aq = 2\pi \cdot 0.61... \tag{5.4.35}$$

This leads to

$$2ka \sin \frac{\alpha}{2} = 2\pi \cdot 0.61, \quad \text{or}$$

$$\sin \frac{\alpha}{2} = \frac{0.61}{2} \frac{\lambda}{a}. \tag{5.4.36}$$

Since the r.h.s. is small compared to 1, we find for the angular aperture of the diffraction pattern

$$\alpha = 0.61... \frac{\lambda}{a}. \tag{5.4.37}$$

This result gives the limit of resolution in a microscope, a telescope and in the eye. The sensitive cells in the retina of the eye are just a little bit bigger than the diffraction limit caused by the iris. This is a nice example for the economy of nature. The quantity

$$\frac{\alpha}{\lambda} = \frac{\triangle k}{2\pi} \tag{5.4.38}$$

can be interpreted as an uncertainty in momentum $\triangle k$ of the radiation, and, by momentum conservation, as a momentum uncertainty of the particle looked at through the microscope. Taking $a = \triangle x$ as an uncertainty in position, we obtain the uncertainty relation

$$\triangle x \cdot \triangle k = 2\pi \cdot 0.61... \tag{5.4.39}$$

from (5.4.37). With de Broglie's basic relation of wave mechanics

$$p = \hbar k, \quad \hbar = \frac{h}{2\pi}, \tag{5.4.40}$$

where p is the momentum and h is Planck's constant, we arrive at Heisenberg's uncertainty relation of wave mechanics

$$\triangle x \cdot \triangle p \sim h. \tag{5.4.41}$$

This relation limits the accuracies in any position and momentum measurement, not only with a microscope, as a consequence of quantum mechanics.

We now turn to **diffraction at a grating**. A grating is a periodic arrangement of lines on glass, so that a stripe region where light can path alternates with an impenetrable stripe. We assume the stripes in the z-direction and the wave vector $\boldsymbol{k}$ of the incident wave in the xy-plane (Fig.22). The diffracted wave from the n-th slit F_n is given by

$$u_n \sim \int_{F_n} \exp i\Big[(k_2 - k_2')y_2 + (k_3 - k_3')y_3\Big]\, dy_2 dy_3, \tag{5.4.42}$$

according to (5.4.22, 23). The last term in the exponent vanishes because there is no diffraction in the z-direction. Introducing

$$k_2 = k \sin\alpha_0, \quad k_2' = k\sin\alpha, \quad k = \frac{2\pi}{\lambda}, \tag{5.4.43}$$

we get

$$u_n \sim \int_{-a/2}^{a/2} dy \exp[ik(\sin\alpha_0 - \sin\alpha)(nd + y)], \tag{5.4.44}$$

where a is the width of a slit and d is the total period of the grating, the so-called lattice constant. Then the amplitude from the n-th slit can be written as follows

$$u_n = f(\alpha_0, \alpha) \exp[ik(\sin\alpha_0 - \sin\alpha)nd]. \tag{5.4.45}$$

Let

$$\triangle = kd(\sin\alpha_0 - \sin\alpha) \tag{5.4.46}$$

be the phase difference between adjacent slits, then the total amplitude of the diffracted wave is equal to

$$\begin{aligned} u &= f(\alpha_0,\alpha)\sum_{n=0}^{N-1} e^{in\triangle} = f(\alpha_0,\alpha)\frac{1-\exp(iN\triangle)}{1-\exp(i\triangle)} \\ &= f(\alpha_0,\alpha)\,\frac{\exp(iN\triangle/2)}{\exp(i\triangle/2)}\,\frac{\sin(N\triangle/2)}{\sin(\triangle/2)}. \end{aligned} \tag{5.4.47}$$

The intensity is then given by the absolute square

$$I \sim |f(\alpha_0,\alpha)|^2 \,\frac{\sin^2(N\triangle/2)}{\sin^2(\triangle/2)}. \tag{5.4.48}$$

The factor $|f|^2$ is called structure factor because it depends on the details of one period of the grating. This factor is slowly varying with the angle α,

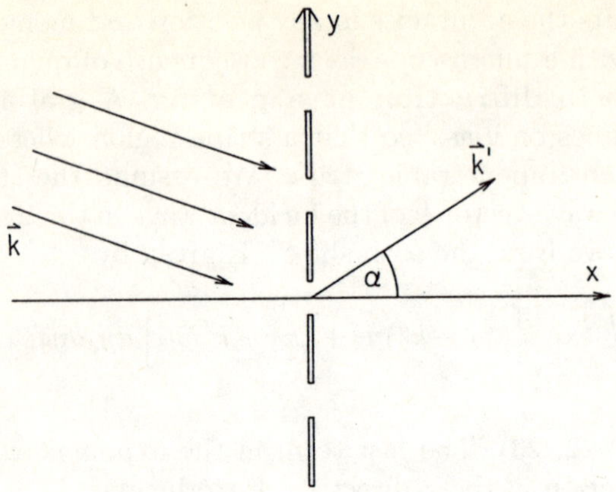

Fig. 22. Diffraction at a grating

in contrast to the other so-called lattice factor. The latter has big maxima $= N^2$ if

$$\triangle = 2\pi m, \quad m = 0, 1, 2, \ldots \tag{5.4.49}$$

these are the main diffraction maxima. There is a first secondary maximum at $\triangle = \pi/N$ of magnitude

$$\frac{\sin^2 \pi/2}{\sin^2(\pi/2N)} \approx \frac{1}{(\pi/2N)^2} \approx \frac{N^2}{2.5}. \tag{5.4.50}$$

Normally this cannot be resolved and appears as a shoulder of the main maximum. The second secondary maximum at $\triangle = 3\pi/N$ is already much smaller

$$\frac{\sin^2(3\pi/2)}{\sin^2(3\pi/2N)} \approx \frac{N^2}{9\pi^2/4} \approx \frac{N^2}{22}. \tag{5.4.51}$$

Thus, the intensity quickly decreases to zero and remains practically zero until the next main maximum (5.4.49) is reached. A grating with large N shows a very sharp diffraction pattern.

If the light source at $L = \boldsymbol{y}_0$ and the point of observation $P = \boldsymbol{x}$ have a finite distance from the screen, one speaks of **Fresnel's diffraction**. Then the quadratic terms in (5.4.22) cannot be neglected. To simplify the computation, we choose the origin on the line LP in the aperture (Fig.23). Then we have $\alpha_2' = \alpha$, $\alpha_3' = \alpha_3$ and the linear terms in (5.4.22) vanish. Hence,

$$u(\boldsymbol{x}) = -\frac{i}{\lambda} \frac{u_0}{|\boldsymbol{x}||\boldsymbol{y}_0|} \cos(-\boldsymbol{n}, \boldsymbol{x}) \exp[ik(|\boldsymbol{x}| + |\boldsymbol{y}_0|)]$$

$$\times \int_F \exp ik \Big\{ \frac{1}{2}\Big(\frac{1}{|\boldsymbol{x}|} + \frac{1}{|\boldsymbol{y}_0|}\Big) \Big[y_2^2 + y_3^2 - (\cos\alpha_2 y_2 + \cos\alpha_3 y_3)^2 \Big] \Big\} \, dy_1 dy_3. \tag{5.4.52}$$

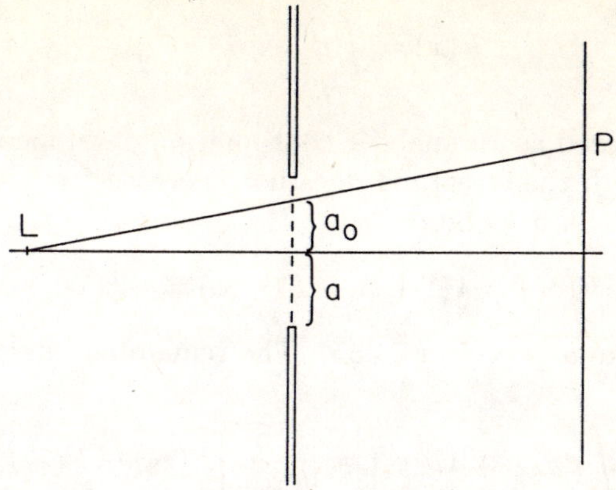

Fig. 23. Light source L and point P of observation in Fresnel diffraction

We will use this result to calculate the diffraction pattern for a slit of width $2a$, while its height $2b$ is assumed to be very big. The line LP is chosen in the plane perpendicular to the screen, so that $\cos\alpha_3 = 0$. We set

$$\cos^2\alpha_2 \overset{\text{def}}{=} 1-\alpha^2, \quad \alpha = \cos(-\boldsymbol{n},\boldsymbol{x}). \tag{5.4.53}$$

Then (5.4.52) simplifies to

$$u \sim \int_F \exp\Big[ik\frac{1}{2}\Big(\frac{1}{|\boldsymbol{x}|}+\frac{1}{|\boldsymbol{y}_0|}\Big)(y_2^2+y_3^2-y_2^2+\alpha^2 y_2^2)\Big]\, dy_2 dy_3. \tag{5.4.54}$$

Let the light source L be in the symmetrical position in front of the slit (Fig.23), then the origin $y_2 = 0$ on the line LP has a certain distance a_0 from the symmetry axis. The integral (5.4.53) over the slit can therefore be written as follows

$$u \sim \int_{-a-a_0}^{a-a_0} dy_2 \exp(i\varphi_2 y_2^2) \int_{-b}^{b} dy_3 \exp(i\varphi_3 y_3^2), \tag{5.4.55}$$

where

$$\varphi_2 = \frac{k}{2f}\alpha^2, \quad \varphi_3 = \frac{k}{2f}, \tag{5.4.56}$$

and

$$\frac{1}{f} = \frac{1}{|\boldsymbol{x}|} + \frac{1}{|\boldsymbol{y}_0|} \tag{5.4.57}$$

is the so-called inverse focal distance.

The integrals appearing in (5.4.55) are so-called **Fresnel integrals**. One defines the function

$$F(x) = \int_0^x dt \, \exp\left(i\frac{\pi}{2}t^2\right). \tag{5.4.58}$$

It can be considered as an analytic continuation of an incomplete Gaussian integral. Although the integrand does not decrease for $t \to \infty$, the integral has a finite limit for $x \to \infty$

$$F(\infty) = \tfrac{1}{2}(1+i), \quad F(-\infty) = -\tfrac{1}{2}(1+i). \tag{5.4.59}$$

Then we can take $b \to \infty$ in (5.4.55). The remaining integral becomes

$$\int_{a_1}^{a_2} dy_2 \, \exp[i\varphi_2 y_2^2] = \sqrt{\frac{\pi}{2\varphi_2}}\Big[F(z_2) - F(z_1)\Big], \tag{5.4.60}$$

where

$$z_2 = a_2\sqrt{\frac{2}{\pi}\varphi_2} = \alpha(a - a_0)\sqrt{\frac{2}{\pi}\frac{k}{2f}} = \alpha(a-a_0)\sqrt{\frac{2}{\lambda f}}, \tag{5.4.61}$$

$$z_1 = -\alpha(a + a_0)\sqrt{\frac{2}{\lambda f}} \tag{5.4.62}$$

depend on the direction (5.4.53). We note that

$$z_2 - z_1 = 2\alpha a\sqrt{\frac{2}{\lambda f}} = \text{const.} \tag{5.4.63}$$

independent of a_0. The diffracted intensity behind the slit is proportional to

$$|u|^2 \sim |F(z_2) - F(z_1)|^2. \tag{5.4.64}$$

It is shown in Fig.24. An elegant graphical method for its construction is discussed in Problem 10.

Finally we want to discuss **diffraction by a** special **three-dimensional lattice**. Such lattices are given to us by nature in the form of crystalline solids. The lattice constant has values $d \sim 10^{-8}$cm. Consequently, one cannot use optical light to get diffraction. One has to use X-rays which have wavelengths of about the same order of magnitude. To find the diffracted wave under most general assumptions, we do not rely on Kirchhoff's theory. Supposing for simplicity a scalar wave theory, the scattered wave from the atom n at the place $\boldsymbol{R}_n$ is given by (5.2.33)

$$u_n(\boldsymbol{x}) = f_n\frac{e^{ikR}}{R}\, e^{i\boldsymbol{k}\cdot\boldsymbol{R}_n}. \tag{5.4.65}$$

Here $\boldsymbol{R}$ is the vector from the atom to the point of observation $\boldsymbol{x} = \boldsymbol{R}_n + \boldsymbol{R}$.

$$f_n = f_n(\boldsymbol{k}, \boldsymbol{k}') \tag{5.4.66}$$

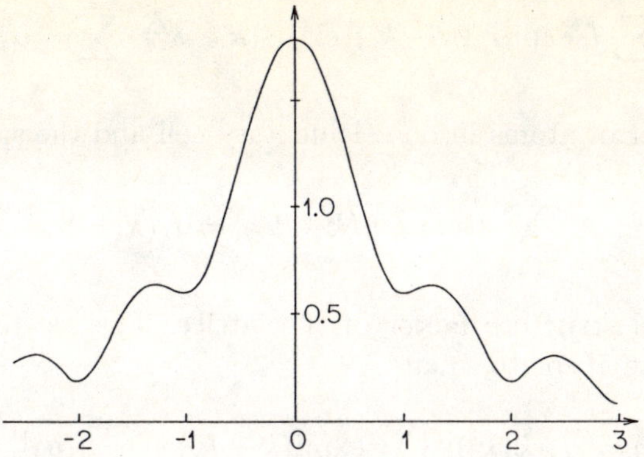

Fig. 24. Fresnel diffraction behind a slit

is the so-called scattering amplitude of one atom with

$$\boldsymbol{k} = \frac{2\pi}{\lambda}\boldsymbol{e}, \quad \boldsymbol{k}' = \frac{2\pi}{\lambda}\boldsymbol{e}', \tag{5.4.67}$$

where $\boldsymbol{e}$ is the direction of the incident wave and $\boldsymbol{e}'$ the direction of the scattered wave. This scattering amplitude modulates the spherical wave in (5.4.65), and the last factor $\exp(i\boldsymbol{k}\cdot\boldsymbol{R}_n)$ represents the phase shift of the incoming plane wave at the position $\boldsymbol{R}_n$ of the scattering atom.

We consider again the scattering (Frauenhofer) limit $R, x \gg R_n$. Then we may neglect the last term in

$$R^2 = x^2 - 2xR_n\cos(\boldsymbol{x}, \boldsymbol{R}_n) + R_n^2, \tag{5.4.68}$$

which leads to

$$kR = kx - (\boldsymbol{k}'\cdot\boldsymbol{R}_n) \tag{5.4.69}$$

and

$$u_n(\boldsymbol{x}) = f_n e^{i\boldsymbol{R}_n\cdot(\boldsymbol{k}-\boldsymbol{k}')}\frac{e^{ikx}}{x}. \tag{5.4.70}$$

Every atom l belongs to a certain elementary cell of the lattice, and its position within this cell is specified by a vector $\boldsymbol{r}_l$. Then the positions of the atoms are given by

$$\boldsymbol{R}_{n,l} = \boldsymbol{r}_l + \sum_{j=1}^{3} n_j\boldsymbol{a}_j, \tag{5.4.71}$$

where n_j are integers and $\boldsymbol{a}_j$ are the three primitive lattice translations. The last sum in (5.4.71) is a lattice vector that specifies the position of the elementary cell.

Substituting (5.4.71) into (5.4.70) and summing over all atoms, we find the following expression for the **total scattering wave**

$$u(\boldsymbol{x}) = \sum_n \sum_{l=1}^{s} f_l \exp[i\boldsymbol{r}_l(\boldsymbol{k}-\boldsymbol{k}')] \exp\Big[i(\boldsymbol{k}-\boldsymbol{k}')\cdot \sum n_j \boldsymbol{a}_j\Big] \frac{e^{ikx}}{x}. \quad (5.4.72)$$

s is the number of atoms in one elementary cell and the sum

$$\sum_{l=1}^{s} f_l \exp[i\boldsymbol{r}_l(\boldsymbol{k}-\boldsymbol{k}')] = f_s(\boldsymbol{k},\boldsymbol{k}') \quad (5.4.73)$$

is the so-called structure factor of the lattice. The scattering intensity is obtained by the absolute square

$$I(\boldsymbol{k},\boldsymbol{k}') = |f_s(\boldsymbol{k},\boldsymbol{k}')|^2 \Big|\sum_n \exp[i(\boldsymbol{k}-\boldsymbol{k}')\cdot \sum n_j \boldsymbol{a}_j]\Big|^2. \quad (5.4.74)$$

The sum herein has the same form as the lattice factor (5.4.47) of the plane lattice, hence

$$I(\boldsymbol{k},\boldsymbol{k}') = |f_s(\boldsymbol{k},\boldsymbol{k}')|^2 \prod_{j=1}^{3} \left(\frac{\sin \frac{1}{2} N_j \triangle \cdot \boldsymbol{a}_j}{\sin \frac{1}{2} \triangle \cdot \boldsymbol{a}_j}\right)^2, \quad (5.4.75)$$

where

$$\triangle = \boldsymbol{k}' - \boldsymbol{k} = \frac{2\pi}{\lambda}(\boldsymbol{e}' - \boldsymbol{e}) \quad (5.4.76)$$

is the change of the wave vector in the scattering process. As in case of the plane lattice (5.4.48), the intensity is the product of a structure factor and the lattice factor.

We now get three conditions for a principal maximum instead of one

$$\begin{aligned} \triangle \cdot \boldsymbol{a}_1 &= 2\pi m_1 \\ \triangle \cdot \boldsymbol{a}_2 &= 2\pi m_2 \\ \triangle \cdot \boldsymbol{a}_3 &= 2\pi m_3, \end{aligned} \quad (5.4.77)$$

where m_j must be integer. In the form

$$\frac{1}{\lambda}(\boldsymbol{e}' - \boldsymbol{e}) \cdot \boldsymbol{a}_j = m_j, \quad j = 1, 2, 3, \quad (5.4.78)$$

these equations are called **Laue's equations**. These are three linear equations, however, the unit vector $\boldsymbol{e}'$ describing the direction of the scattered wave has only two free parameters. Consequently, in (5.4.78) one additional parameter must be varied, in order to fulfill all three conditions. There are three different methods in use to do this: (i) one can vary λ by using "white" X-rays, i.e. a continuous X-ray spectrum, this is the original method of von Laue. (ii) One can vary $\boldsymbol{e}$ by rotating the crystal, this is the method of Bragg. (iii) One can simply vary $\boldsymbol{e}$ by using a poly-crystalline powder instead of a single crystal, this is the method of Debye and Scherrer. In any case the crystalline structure follows from the positions of the principal maxima. From the intensity distribution one gets the absolute square of the structure factor (5.4.73) which gives the more important information about the structure of the elementary cell. But for the complete reconstruction of the elementary cell the phase of $f_s(\boldsymbol{k}, \boldsymbol{k}')$ is also necessary. This is the missing phase problem.

5.5 The Laser – An Optical Trumpet

In this last section we leave electrodynamics and field theory, but nevertheless, we discuss a highly important subject of optics. In our treatment of electrodynamics and optics we have repeatedly considered matter in interaction with radiation. In all these discussions we have assumed that the atoms in the material emit their radiation independently, so that only one atom is interacting with the radiation field at a time. As is well-known, the situation in acoustics may be quite different. If a trumpet is playing a note, all parts of the instrument oscillate coherently. As a result, almost all oscillatory energy in the walls of the trumpet is converted into the oscillation of the air in the instrument which leads to the strong sound emission. Besides, the sound is so strong because the air oscillates in a single mode of oscillation with the characteristic frequency of the note that is played. Alternatively one could try to excite the trumpet to oscillate by striking it with a hammer. Then the energy is distributed over many modes and much is lost by damping, so that the sound output is poor and ugly, and the trumpet is ruined.

The laser is a well played trumpet in optics. To achieve this, one has first to use an optical cavity, so that the radiation field can only oscillate in a few modes. The losses of the cavity must be small enough. Furthermore, the characteristic frequency of the cavity mode must nearly agree with the transition frequency of the active atoms in the material that is contained in the cavity. To play the trumpet, a pump mechanism is necessary which excites the atoms to an excited state. When they fall back into the ground state, the radiation is emitted. If there are many excited atoms present at a certain moment, they can all radiate coherently by the process of stimulated emission.

It is clear that the full description of this system requires quantum theory, in particular for the treatment of the atoms. But most basic features are already contained in the classical limit of this theory, which is again a mechanical theory. Its basic equations are ordinary differential equations in time, the so-called semi-classical laser equations (see *H.Haken, Laser Theory, Springer Verlag 1984*, for example). A rigorous derivation of those equations and a careful discussion of the classical limit in this context has been given by *K.Hepp and E.Lieb, Helv.Phys.Acta 46, 573 (1973).* The only mathematical difference to classical mechanics is that the basic quantities are complex functions due to their quantum mechanical origin. The radiation field is described by an amplitude $a(t)$, assuming a single oscillating mode in the cavity. The field intensity is proportional to $|a(t)|^2$. The atoms are represented by three polarization functions s_x, s_y, s_z, if, for simplicity, we assume atoms with only two states, a ground state and one excited state. These quantities come from a spin 1/2 description of the two-level atoms, and this "spin" corresponds classically to a polarization. (One considers an electric dipole transition and the polarization is proportional to the dipole moment.) The **semi-classical laser equations** then read

$$a'(t) = -(i\omega + \kappa)a - i\lambda s_- - i\mu s_+ \tag{5.5.1}$$

$$s_z'(t) = -2\gamma(s_z - \eta) + (i\lambda s_- a^* + i\mu s_- a + c.c) \tag{5.5.2}$$

$$s_+'(t) = (i\Omega - \gamma)s_+ - 2i\lambda s_z a^* - 2i\mu s_z a \tag{5.5.3}$$

$$s_-'(t) = -(i\Omega + \gamma)s_- + 2i\lambda s_z a + 2i\mu s_z a^*. \tag{5.5.4}$$

The prime denotes the derivative with respect to time t, and

$$s_\pm = s_x \pm i s_y. \tag{5.5.5}$$

ω is the cavity frequency, while Ω is the transition frequency of the active atoms which may differ from ω. κ and γ are damping constants, κ accounts for the cavity losses and γ for the losses in the atomic system. λ and μ are coupling constants. $s_z(t)$ is proportional to the number of atoms in the excited state, sometimes called population inversion,

$$-\frac{1}{2} \leq \eta \leq \frac{1}{2} \tag{5.5.6}$$

is the pump power, as we will see in the sequel. The range of variation in (5.5.6) comes again from the spin 1/2 description. The field amplitude $a(t)$ is not directly coupled to $s_z(t)$, but first to the two other polarization components (5.5.5). Those then couple to $s_z(t)$ according to (5.5.2-4). This coupling is proportional to the field amplitude $a(t)$, $a^*(t)$, which is typical for stimulated emission, *c.c.* means complex conjugate.

One **stationary solution of the semi-classical laser equations** (5.5.1-4) can immediately be written down

$$a = 0, \quad s_z = \eta, \quad s_- = 0 = s_+. \tag{5.5.7}$$

This is a solution without radiation which describes the laser below the threshold of laser action. The trumpet does not play if one blows too weakly. The question is whether this solution becomes unstable, if the pump rate η is increased. We want to investigate this in the case $\mu = 0$ which is the so-called rotating wave approximation,

$$a' = -(i\omega + \kappa)a - i\lambda s_-. \tag{5.5.8}$$

To study linear stability of (5.5.7), we write

$$s_z = \eta + \tilde{s}_z \tag{5.5.9}$$

and linearize the equations, assuming $\tilde{s}_z$, a and s_- to be small:

$$\tilde{s}'_z = -2\gamma\, \tilde{s}_z \tag{5.5.10}$$

$$s'_- = -(i\Omega + \gamma)s_- + 2i\lambda\eta a. \tag{5.5.11}$$

The forth equation (5.5.3) is just the complex conjugate of (5.5.11). The linear system (5.5.8), (5.5.10, 11) has exponential solutions where all three functions are proportional to $\sim \exp(\nu t)$. A non-trivial solution is only possible if the characteristic determinant vanishes:

$$\begin{vmatrix} -i\omega - \kappa - \nu & -i\lambda & 0 \\ 2i\lambda\eta & -i\Omega - \gamma - \nu & 0 \\ 0 & 0 & -2\gamma - \nu \end{vmatrix} =$$

$$= (-2\gamma - \nu)\Big[(i\omega + \kappa + \nu)(i\Omega + \gamma + \nu) - 2\lambda^2\eta\Big] = 0. \tag{5.5.12}$$

The rows and columns have been written down in the following order: a, s_-, s_z. This gives a quadratic equation for the characteristic exponent ν:

$$\nu^2 + [\kappa + \gamma + i(\omega + \Omega)]\nu - 2\lambda^2\eta + (i\omega + \kappa)(i\Omega + \gamma) = 0, \tag{5.5.13}$$

the third solution $\eta = -2\gamma$ is always stable. The solutions of (5.5.13) are

$$\nu_{1,2} = -\frac{1}{2}(\kappa + \gamma) - \frac{i}{2}(\omega + \Omega) \pm \sqrt{\tfrac{1}{4}[\kappa - \gamma - i(\Omega - \omega)]^2 + 2\lambda^2\eta}. \tag{5.5.14}$$

A solution becomes unstable, if the real part of ν gets positive. This is only possible with the plus sign of the square root. Abbreviating the expression under the square root by $\alpha + i\beta$ and using

$$\mathrm{Re}\sqrt{\alpha + i\beta} = \sqrt{\tfrac{1}{2}\left(\sqrt{\alpha^2 + \beta^2} + \alpha\right)}, \tag{5.5.15}$$

we get the following condition for instability

$$\frac{1}{2}\sqrt{\alpha^2 + \beta^2} + \frac{1}{2}\alpha \geq \frac{1}{4}(\kappa + \gamma)^2, \tag{5.5.16}$$

which leads to

$$\eta \geq \frac{\kappa\gamma}{2\lambda^2}\left(1 + \frac{(\Omega - \omega)^2}{(\kappa + \gamma)^2}\right) \stackrel{\text{def}}{=} \eta_c. \tag{5.5.17}$$

This is the so-called threshold condition. **For $\eta > \eta_c$, the above solution** (5.5.7) **of the laser equations becomes unstable, so that there must**

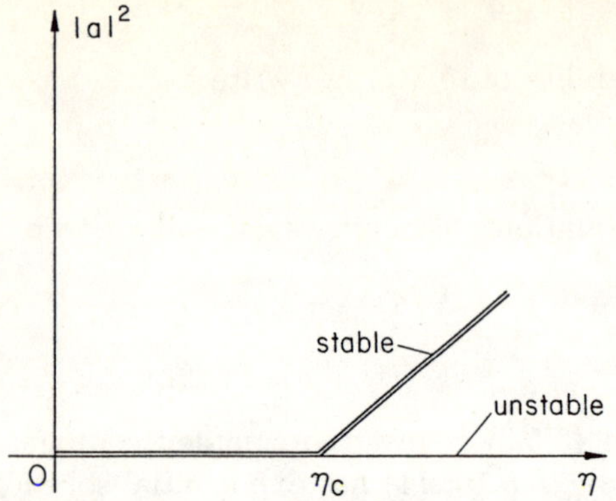

Fig. 25. Hopf bifurcation at laser threshold

exist a second stable solution, for physical reasons. This is an example of a so-called Hopf bifurcation (Fig.25).

To find the new solution, we make an exponential ansatz for the full non-linear equations of the following form:

$$a(t) = ae^{-i\varepsilon t} \tag{5.5.18}$$

$$s_-(t) = s_- e^{-i\varepsilon t}, \quad s_+(t) = s_+ e^{i\varepsilon t}. \tag{5.5.19}$$

The quantities a, s_-, s_+ on the r.h.s. are assumed to be time-independent. Substituting this into (5.5.8), we get a relation between the time-independent amplitudes

$$s_- = \frac{1}{\lambda}(\varepsilon - \omega + i\kappa)a. \tag{5.5.20}$$

From eq.(5.5.4) (with $\mu = 0$) we shall obtain

$$-i\varepsilon s_- e^{-i\varepsilon t} = -(i\Omega + \gamma)s_- e^{-i\varepsilon t} + 2i\lambda s_z a e^{-i\varepsilon t}. \tag{5.5.21}$$

The exponential factors cancel out, so that s_z must also be independent of time:

$$s_z = \frac{1}{2\lambda}(\Omega - \varepsilon - i\gamma)\frac{s_-}{a} = \frac{1}{2\lambda^2}(\Omega - \varepsilon - i\gamma)(\varepsilon - \omega + i\kappa). \tag{5.5.22}$$

Finally eq.(5.5.2) must be fulfilled

$$\begin{aligned} 0 &= -2\gamma(s_z - \eta) + i\lambda s_- a^* - i\lambda s_+ a \\ &= -\frac{\gamma}{\lambda^2}(\Omega - \varepsilon - i\gamma)(\varepsilon - \omega + i\kappa) + 2\gamma\eta \\ &\quad + i(\varepsilon - \omega + i\kappa)|a|^2 - i(\varepsilon - \omega + i\kappa)|a|^2. \end{aligned} \tag{5.5.23}$$

The imaginary part of this equation

$$\frac{\gamma^2}{\lambda^2}(\varepsilon-\omega)-\frac{\gamma\kappa}{\lambda^2}(\Omega-\varepsilon)=0 \tag{5.5.24}$$

determines the frequency ε:

$$\varepsilon=\frac{\gamma\omega+\kappa\Omega}{\gamma+\kappa}. \tag{5.5.25}$$

Hence, **above threshold, the laser oscillates** with a frequency ε that is the weighted mean between the cavity frequency ω and the atomic frequency Ω.

The result (5.5.25) for ε implies

$$(\Omega-\varepsilon)(\varepsilon-\omega)=\frac{\kappa\gamma(\Omega-\omega)^2}{(\gamma+\kappa)^2}. \tag{5.5.26}$$

Substituting this into (5.5.22), we find the population inversion of the atomic system

$$\begin{aligned} s_z &= \frac{\kappa\gamma}{2\lambda^2}\left[1+\frac{(\Omega-\varepsilon)(\varepsilon-\omega)}{\kappa\gamma}\right] \\ &= \frac{\kappa\gamma}{2\lambda^2}\left[1+\frac{(\Omega-\omega)^2}{(\gamma+\kappa)^2}\right]=\eta_c. \end{aligned} \tag{5.5.27}$$

This is just the critical pump power (5.5.17). The interesting feature of this result is that above threshold $\eta>\eta_c$, s_z is independent of the pump power η. The atomic system does not absorb more energy than the critical pump power η_c. Consequently, the additional energy, pumped into the laser, must go into the radiation field, the laser starts to oscillate. In fact, from the real part of (5.5.23) we shall obtain

$$\frac{\gamma}{\lambda^2}(\varepsilon-\Omega)(\varepsilon-\omega)-\frac{\gamma^2\kappa}{\lambda^2}+2\gamma\eta-2\kappa|a|^2=0. \tag{5.5.28}$$

This determines the intensity of the radiation

$$\begin{aligned} |a|^2 &= \frac{\gamma}{\kappa}\left[\eta+\frac{1}{2\lambda^2}(\varepsilon-\Omega)(\varepsilon-\omega)-\frac{\kappa\gamma}{2\lambda^2}\right] \\ &= \frac{\gamma}{\kappa}(\eta-\eta_c), \end{aligned} \tag{5.5.29}$$

where (5.5.26) has been used. **The radiation intensity grows linearly with the pump power** $\eta-\eta_c$ (Fig.25). The trumpet is playing.

It can be shown that this stationary solution of the semi-classical laser equations for $\eta>\eta_c$ is stable for not to big pump strength η. For larger η there exist further bifurcations and finally the system shows chaotic behavior. These very high pump powers are not interesting for practical applications. – Although the semi-classical laser equations give a simple explanation of the basic facts of laser action, they are not the whole story. We want to illustrate the limitation of the semi-classical equations by investigating some **non-stationary solutions**.

Let us consider the equations of motion (5.5.1-4) in the rotating wave approximation $\mu = 0$, but without losses $\kappa = 0 = \gamma$ and pump $\eta = 0$. For simplicity, we also assume resonance $\Omega = \omega$ of the cavity with the atomic transition. Then we have the following simple non-linear equations

$$a'(t) = -i\omega a - i\lambda s_- \tag{5.5.30}$$

$$s'_-(t) = -i\omega s_- + 2i\lambda s_z a \tag{5.5.31}$$

$$s'_z(t) = i\lambda s_- a^* - i\lambda s_+ a. \tag{5.5.32}$$

For completeness we add the two complex conjugated equations

$$a'^*(t) = i\omega a + i\lambda s_+ \tag{5.5.33}$$

$$s'_+(t) = i\omega s_+ - 2i\lambda s_z a^*. \tag{5.5.34}$$

We are able to integrate this system, because we can **find** enough **conserved quantities**. A first one is the square of the "spin"

$$\boldsymbol{s}^2 \overset{\text{def}}{=} s_x^2 + s_y^2 + s_z^2 = s_+ s_- + s_z^2. \tag{5.5.35}$$

In fact,

$$\begin{aligned} \frac{d}{dt}\left(s_+ s_- + s_z^2\right) &= s'_+ s_- + s_+ s'_- + 2 s_z s'_z \\ = i\omega s_+ s_- - 2i\lambda s_z s_- a^* &- i\omega s_+ s_- + 2i\lambda s_+ s_z a \\ &+ 2i\lambda s_z s_- a^* - 2i\lambda s_z s_+ a = 0, \end{aligned}$$

thus

$$\boldsymbol{s}^2 = \text{const.} \overset{\text{def}}{=} s^2. \tag{5.5.36}$$

A second conserved quantity follows from

$$\begin{aligned} \frac{d}{dt}\left(|a|^2 + s_z\right) &= a' a^* + a a^{*\prime} + s'_z \\ = -i\omega|a|^2 - i\lambda s_- a^* &+ i\omega|a|^2 + i\lambda s_+ a \\ &+ i\lambda s_- a^* - i\lambda s_+ a = 0. \end{aligned}$$

This conserved quantity e

$$e \overset{\text{def}}{=} |a|^2 + s_z = \text{const.} \tag{5.5.37}$$

is connected with the total energy in the system. If all atoms are in the ground state, $s_z = -s$, the energy in the radiation field, proportional to

$$|a|^2 = e - s_z = e + s \overset{\text{def}}{=} r, \tag{5.5.38}$$

must agree with the total energy. The latter is denoted by r.

To integrate the above equations (5.5.30-32), we first remove the diagonal terms by means of the following ansatz:

$$a(t) = \alpha(t)e^{-i\omega t} \tag{5.5.39}$$

$$s_-(t) = \sigma_-(t)e^{-i\omega t}. \tag{5.5.40}$$

Then the equations of motion assume the following form

$$\alpha' = -i\lambda\sigma_-, \quad \alpha^{*\prime} = i\lambda\sigma_+ \tag{5.5.41}$$

$$\sigma_-' = 2i\lambda s_z\alpha, \quad \sigma_+' = -2i\lambda s_z\alpha^* \tag{5.4.42}$$

$$s_z' = i\lambda\sigma_-\alpha^* - i\lambda\sigma_+\alpha. \tag{5.5.43}$$

All dynamical quantities herein are still time-dependent. We want to eliminate

$$s_z = e - |\alpha|^2. \tag{5.5.44}$$

Substituting this into (5.5.42), we get

$$\sigma_-' = 2i\lambda\alpha(e - |\alpha|^2) \tag{5.5.45}$$

$$\sigma_+' = -2i\lambda\alpha^*(e - |\alpha|^2), \tag{5.5.46}$$

and, in addition, we note that

$$\sigma_+\sigma_- = s^2 - s_z^2 = s^2 - (e - |\alpha|^2)^2. \tag{5.5.47}$$

Differentiating (5.5.41) with respect to t again and using (5.5.45, 46), we shall obtain

$$\alpha'' = 2\lambda^2\alpha(e - |\alpha|^2), \tag{5.5.48}$$

and

$$\alpha^{*\prime\prime} = 2\lambda^2\alpha^*(e - |\alpha|^2). \tag{5.5.49}$$

Multiplying the first equation (5.5.48) by α^* and the second by α and taking

$$\alpha'\alpha^{*\prime} = \lambda^2\sigma_-\sigma_+ = \lambda^2\Big[\sigma^2 - (e - |\alpha|^2)^2\Big] \tag{5.5.50}$$

into account, we arrive at

$$\frac{d^2}{dt^2}|\alpha(t)|^2 = 4\lambda^2|\alpha|^2(e - |\alpha|^2) + 2\lambda^2[s^2 - (e - |\alpha|^2)^2]. \tag{5.5.51}$$

We introduce the quantity

$$n(t) \stackrel{\text{def}}{=} |\alpha(t)|^2 = |a(t)|^2, \tag{5.5.52}$$

which is proportional to the radiation intensity or to the photon number. This is the reason for using the symbol $n(t)$. Now eq. (5.5.51) assumes the following form

$$\frac{d^2n}{dt^2} = -6\lambda^2 n^2 + 8\lambda^2 \varepsilon n + 2\lambda^2(s^2 - e^2). \tag{5.5.53}$$

One integration of this second order equation can immediately be carried out by multiplication with the first derivative dn/dt:

$$\frac{1}{2}\frac{d}{dt}\left(\frac{dn}{dt}\right)^2 = \frac{d}{dt}\Big[-2\lambda^2 n^3 + 4\lambda^2 en^2 + 2\lambda^2(s - e^2)n\Big]. \tag{5.5.54}$$

This leads to

$$\frac{dn}{dt} = \sqrt{-4\lambda^2 n^3 + 8\lambda^2 en^2 + 4\lambda^2(s - e^2)n}, \tag{5.5.55}$$

where we have put the constant of integration equal to 0, to get the solution we are interested in. We separate the variables

$$\frac{dn}{\sqrt{R(n)}} = \lambda\, dt, \tag{5.5.56}$$

here

$$\begin{aligned} R(t) &= -4n^3 + 8en^2 + 4(s - e^2)n = \\ &= 4n(n - e + s)(e + s - n) \end{aligned} \tag{5.5.57}$$

is a cubic polynomial. It follows from this equation that $R(n)$ is positive for

$$e - s < n < e + s = r. \tag{5.5.58}$$

If $n = e - s$, the photon number is minimal and all atoms are excited, whereas for $n = r$ the total energy is stored in the radiation field and all atoms are in the ground state. Consequently, the condition (5.5.58) covers the physical range of the photon number $n(t)$.

Let us suppose that at time $t = 0$ all atoms are excited, $n = e - s$. Then we find by integration of (5.5.56)

$$\lambda t = \int\limits_{e-s}^{n} \frac{dn}{2\sqrt{n(n - e + s)(e + s - n)}}. \tag{5.5.59}$$

This is an elliptic integral of the first kind. It is usually denoted by $F(\varphi, k)$, where φ and k are given in terms of the zeros of the cubic polynomial under the square root. In our case (5.5.59) we shall obtain (see *M.Abramowitz, I.A.Stegun, Handbook of Mathematical Functions, Dover 1965*, for example)

$$\sin\varphi = \sqrt{\frac{(e + s)(n - e + s)}{2sn}} \tag{5.5.60}$$

$$k = \sqrt{\frac{2s}{e + s}}. \tag{5.5.61}$$

Then (5.5.59) is equal to

$$\lambda t = \frac{1}{\sqrt{e+s}}\, F(\varphi, k). \tag{5.5.62}$$

The inversion of this elliptic integral leads to the Jacobian elliptic function sn :

$$\operatorname{sn} u = \sin\varphi, \quad \text{if} \quad u = F(\varphi, k). \tag{5.5.63}$$

Then we get from (5.5.62)

$$\operatorname{sn}(\sqrt{e+s}\,\lambda t) = \sqrt{\frac{(e+s)(n-e+s)}{2sn}}, \tag{5.5.64}$$

where the parameter of the elliptic function is given by

$$k^2 = \frac{2s}{e+s} = \frac{2s}{r}. \tag{5.5.65}$$

This leads to the following final result for the radiation output

$$n(t) = \frac{e^2 - s^2}{e + s - 2s\operatorname{sn}^2(\sqrt{e+s}\,\lambda t)}. \tag{5.5.66}$$

Since the elliptic function sn (μt) is periodic in t, the radiation intensity $n(t)$ oscillates periodically between its minimal and maximal value. The minimal value is equal to

$$n(0) = \frac{e^2 - s^2}{e+s} = e - s. \tag{5.5.67}$$

The maximum is attained at the quarter period $t = T$ where sn(...) = 1, thus

$$n(T) = \frac{e^2 - s^2}{e-s} = e + s. \tag{5.5.68}$$

Then all atoms are in the ground state so that the radiation is maximal. This oscillatory behavior of the photon number $n(t)$ is shown in Fig.26.

According to the elementary theory of elliptic functions (see *M.Abramowitz, I.A.Stegun, cited above*, for example), the quarter period T is given by the complete elliptic integral of the first kind

$$K(k^2) = \sqrt{e+s}\,\lambda T = \int_0^1 \frac{dx}{\sqrt{(1-x^2)(1-k^2x^2)}}. \tag{5.5.69}$$

Let us now consider the limit $e \downarrow s$, that means the curve in Fig.26 moves downwards towards $n = 0$. It follows from (5.5.65) that $k^2 \uparrow 1$. In this limit, the elliptic integral (5.5.69) behaves as follows

$$K(k^2) = \frac{1}{2}\log\frac{16}{1-k^2} + o(1). \tag{5.5.70}$$

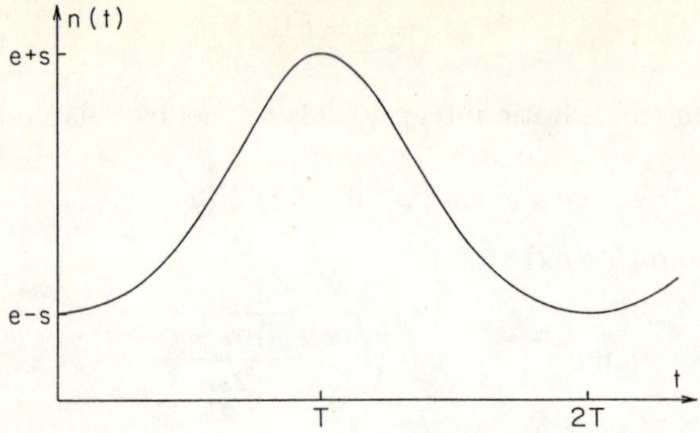

Fig. 26. Nonlinear oscillations of the photon number in the lossless laser

This goes to infinity, if $k^2 \uparrow 1$. Consequently, the period of oscillations gets longer and longer, when the initial number of photons $n(0)$ is made smaller. For $n(0) = 0$, T is infinite which means that the fully excited atomic system would not radiate at all, it would remain in the unstable excited state for ever. This is of course absurd. In reality the system radiates by spontaneous emission. But the latter is neglected in the semi-classical theory, which takes only induced emission into account. Hence, for small photon numbers where spontaneous emission becomes important, the semi-classical theory fails and quantum theory must be used. The quantum mechanical calculation corresponding to the above classical one can be found in the paper *Time Evolution of a Quantum Mechanical Maser Model, by G.Scharf, Annals of Phys. 83 (1974) 71.*

5.6 Problems

1. Show that the boundary conditions that have not been used in the derivation of Fresnel's formulae are automatically satisfied. Choose the polarization a) parallel, b) perpendicular to the incident plane.
2. Show that the reflection coefficient r^2 is equal to the ratio of the reflected to the incident energy flow. Discuss energy conservation at the interface with help of Fresnel's formulae in both cases of polarization.
3. Consider a monochromatic plane wave in vacuum which is perpendicularly incident on a metallic surface $z = 0$. The metal has a real conductivity $\sigma < \infty$. Calculate the current density in the metal and the energy flow of the incident and reflected waves. Show that the energy loss is converted into Joule's heat.
4. Construct over a fixed straight line AB all triangles ABC with angles α, β at B and A such that

$$\frac{\sin\alpha}{\sin\beta} = n = \text{const.} \tag{5.6.1}$$

is constant according to the refraction law. Show that the points C lie on a circle (Apollonius). How are A and B situated with respect to this circle?

5. A thin convex lens with focal length $f_1 = 5$ cm is placed 3 cm in front of a thin concave lens with $f_2 = -10$ cm. How big is the total focal length and where are the principal planes of the whole system, if the principal planes of the individual lenses are assumed to agree with the (thin) lenses. This is a tele-objective.
6. In Young's interference experiment one uses a screen with two slits of distance 1 mm.
   a) How big is the distance of the diffraction maxima on the image screen at a distance of 1 m?
   b) The light source is 50 cm in front of the first screen. How small must the light source be, in order that the displacement of the diffraction pattern remains smaller than 1/4 of the distance in a)?
7. Consider a grating where the transparent parts have exactly the same length as the non-transparent parts. Calculate the ratio of the intensities of the diffracted light of different order.
8. How does a Frauenhofer diffraction pattern change if the aperture is expanded by a factor α in one direction?
9. Sirius has a diameter of $1.5 \cdot 10^6$ km and a distance of 8.8 light years from the earth. How big should the aperture of a telescope be in order to see Sirius as a finite disk, and not only as a diffraction image of a light point?
10. Consider the conformal mapping $F = F(z)$, $z \in \mathbb{C}$, where $F(z)$ is the Fresnel integral (5.4.58). Show that the real z-axis is mapped on a spiral, whereby the lengths are unchanged (Cornu's spiral). How can one construct Fresnel's diffraction pattern (5.4.64) using the arc length and the length of the straight line between two points z_1, z_2 on the spiral?
11. Simplification of the wave equation in optics: Consider wave propagation in the x-direction. For an optical frequency ω the complex amplitude can be written as

$$u(t,x) = v(t,x)e^{-i(\omega t - kx)}, \quad k = \frac{\omega}{c}, \tag{5.6.2}$$

where u is a solution of the wave equation. In non-stationary optical problems the relevant time T is usually large $T \gg 1/\omega$. Show that

$$v(t,x) = v_0(x - ct) + O\Big(\frac{1}{\omega T}\Big), \tag{5.6.3}$$

where $v_0(x)$ is given by the initial wave packet $u(0,x)$.
Hint: Use the stationary phase method to derive the estimate (5.6.3).

6. Epilogue: Quantum Electrodynamics

The two most important fundamental constants in nature are c, the velocity of light and Planck's constant h. All theories in physics can be classified according to whether one or the other, or both constants appear. A theory without c, which means physically that the light speed is assumed to be infinite, is called non-relativistic, a theory without h, or rather $h = 0$, is called classical in contrast to quantum. Newtonian mechanics and usual hydrodynamics are classical and non-relativistic. They are restricted to small velocities and macroscopic bodies. For high velocities $v \sim c$, one needs relativistic mechanics or hydrodynamics, but one has still a macroscopic description ($h = 0$). Classical electrodynamics belongs to the same category because c is present from the very beginning and h is not. On the other hand, microscopic bodies at low velocities are described by non-relativistic quantum mechanics ($h \neq 0$, $c = \infty$). The most universal and fundamental theories are the relativistic quantum theories with $h \neq 0$ and $c < \infty$, where quantum electrodynamics (Q.E.D.) is the prototype. It contains the non-relativistic and classical theories as limiting cases. One needs Q.E.D. if the energy of radiation becomes comparable with the relativistic energy mc^2 of an electron, that means $h\nu \approx mc^2$, where ν is the radiation frequency. This is indeed a condition which contains both constants h and c.

How can Planck's constant h enter into electrodynamics? A first occasion is through the material sources of the electromagnetic field. In our treatment of classical electrodynamics we have almost always assumed that the sources are given external charge and current densities. An exception was the discussion of electro-hydrodynamics at the end of Sect.4.3 (4.3.26), where charge and current densities were treated by their own field equations. In Q.E.D. the situation is now always like this, however, since the sources are usually microscopic (for example electrons), they obey quantum equations containing h. Here h enters for the first time.

For electrons the dynamical equation is simple, because they are elementary and not composite particles (like protons). Their quantum mechanical and Lorentz covariant equation is a system of first order linear partial differential equations (like Maxwell's equations !), the famous Dirac equation, which can be written as follows:

$$i\frac{\hbar}{c}\frac{\partial\psi(t,\boldsymbol{x})}{\partial t} = \Big[mc\beta - \boldsymbol{\alpha}\Big(i\hbar\frac{\partial}{\partial \boldsymbol{x}} + \frac{e}{c}\boldsymbol{A}\Big) + \frac{e}{c}V\Big]\psi(t,\boldsymbol{x}). \tag{6.1}$$

Here m is the electron mass, e the elementary charge, $\hbar = h/2\pi$, $\psi = (\psi_a), a = 1, 2, 3, 4$ is a 4-component complex-valued function (Dirac spinor) and $\beta, \boldsymbol{\alpha} = (\alpha^1, \alpha^2, \alpha^3)$ are four Hermitian 4×4 matrices which satisfy the following anticommutation relation

$$\alpha^j \alpha^k + \alpha^k \alpha^j = 2\delta_{jk}, \quad \beta^2 = 1$$

$$\alpha^j \beta + \beta \alpha^j = 0, \quad j, k = 1, 2, 3. \tag{6.2}$$

$V(t, \boldsymbol{x})$ and $\boldsymbol{A}(t, \boldsymbol{x})$ are old friends, namely the electromagnetic potentials. The Dirac equation is the quantum substitute for the relativistic Euler equations (4.3.28). In fact, if one substitutes the semi-classical ansatz (5.3.13)

$$\psi(x) = a(x) \exp \frac{-i}{\hbar} \int^x (mu^\mu + eA^\mu) dx_\mu \tag{6.3}$$

into (6.1), where $A^\mu = (\frac{1}{c}V, \boldsymbol{A})$ is the four-vector potential and u^μ the four-velocity, one recovers (4.3.28) to leading order in $\hbar$.

Let us first discuss the Dirac equation in the simple manner, assuming $V(\boldsymbol{x})$ and $\boldsymbol{A}(\boldsymbol{x})$ to be given time-independent external potentials. The most prominent example of this situation is an attractive Coulomb potential $V(\boldsymbol{x}) \sim -e/|\boldsymbol{x}|$ and $\boldsymbol{A} = \boldsymbol{0}$. This describes the motion of the electron around the proton in the hydrogen atom. The relevant solutions of the Dirac equation (6.1) then give very accurate results for the spectrum of hydrogen, including the fine-structure, and various other quantities. The reason for this success is the fact that the Dirac equation correctly predicts the spin of the electron and its corresponding magnetic moment (3.2.7)

$$\mu_B = \frac{e\hbar}{2mc}, \tag{6.4}$$

we are using Gauss' system in this section. This was the first triumph of Q.E.D. by Dirac in 1928.

But strictly speaking, this was not yet Q.E.D. because only the electron was treated dynamically, not the electromagnetic field. One might think that the full theory is given by the coupled Dirac and Maxwell's equations, but nature is more complicated. If the potentials $A^\mu(x)$ in (6.1) become strongly time-dependent, this time-dependent electromagnetic field can produce particles (electron–positron pairs), similarly as time-dependent currents produce radiation (Sect.3.4). This particle production forces us to interpret the Dirac field $\psi(x)$ in a radically new way. It is no longer a complex function describing one electron, but becomes an operator on an appropriate Hilbert space, which describes the creation and annihilation of particles. Here we leave classical field theory and enter quantum field theory. Let us first sketch the procedure for the free Dirac field without interaction ($A^\mu = 0$). A classical solution of the free Dirac equation can be represented by a Fourier integral

$$\psi(x) = (2\pi)^{-3/2} \int d^3p \left[b_s(\boldsymbol{p}) u_s(\boldsymbol{p}) e^{-ipx} + d_s(\boldsymbol{p})^+ v_s(\boldsymbol{p}) e^{ipx} \right], \tag{6.5}$$

where $p^0 = \sqrt{c^2\boldsymbol{p}^2 + m^2c^4}$ and the sum $s = \pm 1$ goes over the two spin states. The 4-component functions $u_s(\boldsymbol{p})$, $v_s(\boldsymbol{p})$ are the positive and negative frequency eigen-solutions of the Fourier transformed Dirac operator

$$(mc^2\beta + c\boldsymbol{\alpha} \cdot \boldsymbol{p}) u_s(\boldsymbol{p}) = p^0 u_s(\boldsymbol{p}) \tag{6.6}$$

$$(mc^2\beta + c\boldsymbol{\alpha} \cdot \boldsymbol{p}) v_s(-\boldsymbol{p}) = -p^0 v_s(-\boldsymbol{p}), \tag{6.7}$$

and $b_s(\boldsymbol{p})$ and $d_s(\boldsymbol{p})^+$ are arbitrary amplitudes. Now comes the crucial step: The classical solution (6.5) becomes a quantum field if these amplitudes are considered as operators satisfying the following anticommutation relations:

$$b_s(\boldsymbol{p}) b_{s'}(\boldsymbol{q})^+ + b_{s'}(\boldsymbol{q})^+ b_s(\boldsymbol{p}) = \hbar \delta_{ss'} \delta(\boldsymbol{p} - \boldsymbol{q}) \tag{6.8}$$

$$d_s(\boldsymbol{p})^+ d_{s'}(\boldsymbol{q}) + d_{s'}(\boldsymbol{q}) d_s(\boldsymbol{p})^+ = \hbar \delta_{ss'} \delta(\boldsymbol{p} - \boldsymbol{q}), \tag{6.9}$$

and all other anticommutators vanish. The crosses in (6.5,8,9) denote the Hermitian adjoint operators, $\delta(\boldsymbol{p})$ is the 3-dimensional δ-distribution. These anticommutation relations look very strange. As a motivation we mention that there is a certain correspondence to Poisson brackets in classical mechanics, or field theory. But the true understanding of these relations comes from their implications: It follows that the operator $b_s(\boldsymbol{p})$ in (6.5) annihilates an electron with momentum $\boldsymbol{p}$ in spin state s, $d_s(\boldsymbol{p})^+$ creates a positron; $d_s(\boldsymbol{p})$ annihilates a positron and $b_s(\boldsymbol{p})^+$ creates an electron. These operators operate in the Hilbert space of free electrons and positrons, the so-called Fock space $\mathcal{F}$, which also contains the vacuum state Ω. It is conceptually simpler to first construct this space $\mathcal{F}$ and to define the creation and annihilation operators, and then to verify that they indeed satisfy the anticommutation relations (6.8,9). Since $b_s(\boldsymbol{p})^+ b_s(\boldsymbol{p})^+ = 0$, one cannot generate two electrons in the same state. This is Pauli's exclusion principle, electrons and positrons obey Fermi statistics.

The whole procedure is called second quantization because there is the second place (6.8, 9) where Planck's constant $\hbar$ and commutators or anticommutators appear. (The first quantized Dirac equation (6.1) is related to the canonical commutation relations between momentum and position operators.) The great importance of second quantization lies in the fact that it introduces particles into field theory in a natural way, whereby the distinction between particles and fields disappears. Both are different aspects of the same object $\psi(x)$. In classical electrodynamics we introduced particles by hand as sources in Maxwell's equations. But a full consistent dynamical theory of classical interacting particles and fields does not seem to exist. Einstein, who believed in classical field theory and did much work on these problems, once characterized the situation as follows: "The electron is a foreigner in electrodynamics." Quantum field theory, on the other hand, is very successful and the electron is not at all a foreigner here. A further example is the natural explanation of the existence of antiparticles. The prediction of the positron by Dirac was the second triumph of Q.E.D.

Now we turn to the electromagnetic field. It must become a quantized field too, otherwise there would be a strange asymmetry in nature (and one would even get inconsistencies). We shall work exclusively with the four-potential $A^\mu(x)$ in the Lorentz gauge, so that the free field satisfies the wave equation

$$\frac{1}{c^2}\frac{\partial^2 A^\mu}{\partial t^2} - \left(\frac{\partial}{\partial \boldsymbol{x}}\right)^2 A^\mu = 0. \tag{6.10}$$

One might hesitate to use this classical equation for a quantum field because Planck's constant $\hbar$ is lacking. The reason for this is simple because $\hbar$ drops out of any free equation if the mass is zero. As before we write the solution of (6.10) as a Fourier integral

$$A^\mu(x) = (2\pi)^{-3/2} \int \frac{d^3k}{\sqrt{2\omega}} \left(a^\mu(\boldsymbol{k})e^{-ikx} + a^\mu(\boldsymbol{k})^+ e^{ikx}\right), \tag{6.11}$$

where $\omega = c|\boldsymbol{k}|$. This becomes a quantum field, if $a^\mu(\boldsymbol{k})$ are operators satisfying the commutation relations

$$a^\mu(\boldsymbol{k})a^\nu(\boldsymbol{k}')^+ - a^\nu(\boldsymbol{k}')^+ a^\mu(\boldsymbol{k}) = \hbar\delta(\boldsymbol{k}-\boldsymbol{k}'), \tag{6.12}$$

and all other commutators vanish. Since the solutions of (6.10) describe radiation (Sect.3.3), we have quantized the radiation field. We have disregarded a subtlety with the zeroth component $A^0(x)$. Again (6.12) implies that $a^\mu(\boldsymbol{k})^+$ generates a particle (photon) with momentum $\boldsymbol{k}$ and $a^\mu(\boldsymbol{k})$ annihilates it. The photon is its own antiparticle. Since commutators appear in (6.12) instead of anticommutators, there is no exclusion principle for photons. But the states must be symmetric under permutation of the particles (Bose statistics). The fact that the electron field (spin 1/2) must be quantized with anticommutators and obey Fermi statistics, whereas photons (spin 1) are quantized with commutators and obey Bose statistics is the theorem of spin and statistics.

So far we have only considered electrons and photons without interaction and the reader is certainly keen to see how interactions are treated. In analogy to classical field theory, most authors start by introducing interacting quantum fields. But this is entering a jungle: Lots of formal computations hang around, wild animals called divergences are lurking in the bush which one hopes to slay "by cutoff", and until now, nobody has come out alive (as far as a physical theory like Q.E.D. is concerned). Although one has learnt to live with the divergences, we ought to state that the very reason for them must be an incorrect treatment, namely, either a mistake or the use of ill-defined quantities. This is not an attractive alternative. Therefore, we will work with well-defined quantities, namely free fields, only; then divergences cannot appear (if we make no mistake). The free fields and the corresponding Fock space $\mathcal{F}$ can certainly be used to represent the particles in the asymptotic region, which means long before and after collisions. The results of scattering experiments are described by the so-called scattering matrix (S-matrix) S, which relates the state of the particles $\Phi_{\rm in} \in \mathcal{F}$ before the collision to the state $\Phi_{\rm out} \in \mathcal{F}$ after the collision

$$\Phi_{\rm out} = S\Phi_{\rm in}. \tag{6.13}$$

It is our aim to find S. One might not be satisfied with S alone and ask for more details of the temporal evolution of the relativistic microscopic system. Well, such details are hardly observable. However, it is possible to derive interacting fields from the S-matrix in a rigorous way, which answers the question. The S-matrix, in any case, is the primary object.

It is a remarkable feature that the S-matrix can be expressed in terms of the free fields constructed above. However, this is not so simple because, strictly speaking, the free fields are not operators, but operator-valued distributions. This is obvious from the appearance of δ-distributions in (6.8,9,12). Consequently, one must smear out with test functions $g(x)$, say in Schwartz space $S(\mathbb{R}^4)$, so that the S-matrix is a functional $S(g)$. The physical meaning of the test function $g(x)$ is the following: Since $g(x) \to 0$ for $x^0 \to \pm\infty$, this test function switches the interaction on and off. The measurable quantities (scattering cross sections) are obtained in the so-called adiabatic limit $g(x) \to 1$. Let us give an example: In the case of electron scattering, $\Phi_{\rm in}$ and an observed outgoing state $\Psi_{\rm out}$ contain two incoming and outgoing electrons, respectively, with four-momenta p_i, q_i and p_f, q_f. Then the S-matrix element is of the following form

$$(\Psi_{\rm out}, S\Phi_{\rm in}) = \delta(p_f + q_f - p_i - q_i)\, M, \tag{6.14}$$

due to energy–momentum conservation. The differential cross section $d\sigma/d\Omega$ in the center-of-mass system is the ratio of the number of scattered electrons per unit time and solid angle $d\Omega$ divided by the number of incoming electrons per unit time and unit area, which has indeed the dimension of an area. Its relation to the S-matrix is now simply given by

$$\frac{d\sigma}{d\Omega} = (2\pi)^2 \frac{E^2}{4}\, |M|^2, \tag{6.15}$$

where $2E$ is the center-of-mass energy. Strictly speaking the relation is more complicated: The outgoing electrons are always accompanied by soft photons (Bremsstrahlung (3.5.75)), irrespectively whether they are measured or not. They must be included in the definition of the electron cross section (inclusive cross section). It is this inclusive cross section that exists in the adiabatic limit $g \to 1$.

Now we turn to the main problem, how one gets the S-matrix $S(g)$. In the fifties Stückelberg and others made the important observation that $S(g)$ is strongly restricted by causality: If g_1, g_2 are two test functions with disjoint supports, such that all points in $\operatorname{supp} g_1$ are earlier than all points of $\operatorname{supp} g_2$ in some Lorentz frame, then the S-matrix factorizes

$$S(g_1 + g_2) = S(g_2)\, S(g_1). \tag{6.16}$$

This expresses causality in the sense that what happens earlier $(S(g_1))$ is not influenced by what happens later $(S(g_2))$. To work out the consequences of (6.16) we follow the idea of Feynman and expand $S(g)$ in a perturbation series

$$S(g) = \mathbb{1} + \sum_{n=1}^{\infty} \frac{1}{n!} \int d^4x_1 \dots d^4x_n \; T_n(x_1, \dots, x_n) g(x_1) \dots g(x_n). \tag{6.17}$$

Here T_n must be symmetric under permutation of the arguments $x_1, \dots, x_n$. The causality condition (6.16) now implies

$$T_n(x_1, \dots, x_n) = T_m(x_1, \dots x_m) T_{n-m}(x_{m+1}, \dots x_n), \tag{6.18}$$

if all $\{x_1, \dots x_m\}$ are later than all $\{x_{m+1}, \dots x_n\}$. We claim that this condition together with translation invariance determines all T_n's supposing the first order $T_1(x)$ is given. Let us illustrate this striking fact by looking at the step from T_1 to T_2. We consider the distribution

$$A_2(x_1, x_2) = T_2(x_1, x_2) - T_1(x_1) T_1(x_2), \tag{6.19}$$

which vanishes if $x_1^0 > x_2^0$ by (6.18). One therefore calls this an advanced distribution. Similarly

$$R_2(x_1, x_2) = T_2(x_2, x_1) - T_1(x_2) T_1(x_1) \tag{6.20}$$

vanishes for $x_1^0 < x_2^0$ and defines a retarded distribution. In the difference

$$D_2(x_1, x_2) = R_2 - A_2 = T_1(x_1) T_1(x_2) - T_1(x_2) T_1(x_1) \tag{6.21}$$

the unknown $T_2(x_1, x_2) = T_2(x_2, x_1)$ drops out, hence, this distribution can be calculated from T_1. Since R_2 and A_2 have disjoint supports, they are obtained by splitting D_2 (6.21) into a retarded and advanced part. Naively, this distribution splitting is carried out by multiplication with a step function

$$R_2(x_1, x_2) \approx \Theta(x_1^0 - x_2^0) D_2(x_1, x_2). \tag{6.22}$$

Then T_2 is obtained from (6.20)

$$\begin{aligned} T_2(x_1, x_2) &= R_2 + T_1(x_2) T_1(x_1) \\ &\approx \Theta(x_1^0 - x_2^0) T_1(x_1) T_1(x_2) + \Theta(x_2^0 - x_1^0) T_1(x_2) T_1(x_1) \\ &\stackrel{\text{def}}{=} \mathrm{T}\{T_1(x_1) T_1(x_2)\}. \end{aligned} \tag{6.23}$$

The result is the so-called time-ordered product of the two factors T_1, where the factor with the larger time coordinate stands to the left. In n-th order one would get a T-product of n factors T_1.

As the wavy equality signs in (6.22, 23) indicate, there is a serious problem in this widely used procedure. It shows up in the well-known ultraviolet divergences of quantum field theory, which are contained in the naive T-products like (6.23). Since all quantities we have begun with are well-defined, the divergences must be due to an incorrect manipulation. In fact, the product of a distribution with a discontinuous step function (6.22) is not always defined. For example, $\Theta(x^0)\delta(x)$ makes no sense. However, the splitting of a causal distribution $D_2(x_1, x_2)$ into retarded and advanced parts is always a well-defined operation. If it is done correctly, one gets well-defined time-ordered products without any divergency. The correct theory is due to H.Epstein and V.Glaser (*Annales de l'Institut Poincaré A 19, 211 (1973)*). We have worked out their method for gauge theories, in particular for Q.E.D. (*G.Scharf, Finite Quantum Electrodynamics, Texts and Monographs in Physics, Springer Verlag 1989*). This book contains the second part of the story of electrodynamics.

After the calculation of the n-point distributions T_n, we are able to predict the results of scattering experiments. For illustration we give a list of the most important processes: Electron scattering was already mentioned, one can also scatter photons on electrons (Compton scattering). A particular case is the scattering of electrons on a nucleus with the emission of bremsstrahlung. These processes have counterparts in classical electrodynamics. The examples which are typical for quantum field theory are annihilation of an electron–positron pair, giving two photons, and the inverse process of pair production. An electron–positron pair can also go over into a muon pair or a pair of tau-leptons. All these processes have been measured and are in perfect agreement with the theory. Higher orders of perturbation theory give rise to small corrections in the hydrogen spectrum and to the magnetic moment of the electron (6.4), where one must calculate at least up to $n = 6$. These are the most precise tests of Q.E.D. The agreement between experiment and theory is always excellent, so that the story of electrodynamics, finally, has a happy end.

Subject Index

The numbers in this index are related to the equations next to the subject. The first number refers to the chapter, the second one to the section and the third one is the number of the equation. The numbers in italics are page numbers.

Springer-Verlag and the Environment

We at Springer-Verlag firmly believe that an international science publisher has a special obligation to the environment, and our corporate policies consistently reflect this conviction.

We also expect our business partners – paper mills, printers, packaging manufacturers, etc. – to commit themselves to using environmentally friendly materials and production processes.

The paper in this book is made from low- or no-chlorine pulp and is acid free, in conformance with international standards for paper permanency.